AF851212

VITAMIN K2 - VITAL FOR HEALTH AND WELLBEING

Edited by **Jan Oxholm Gordeladze**

Vitamin K2 - Vital for Health and Wellbeing
http://dx.doi.org/10.5772/61430
Edited by Jan Oxholm Gordeladze

Contributors

Masood Sadiq Butt, Aurang Zeb, Muhammad Yasin, Michio Kurosu, Munechika Enjoji, Yoshiharu Karube, Jiro Takata, Kazuhisa Matsunaga, Brian Meehan, Jon Beckwith, Hsin-Jung Ho, Michio Komai, Hitoshi Shirakawa, Michael Donaldson, Shinya Fujii, Hiroyuki Kagechika, Yohei Miyamoto, Jan Oxholm Gordeladze, Mona Møller, Kotaro Azuma, Satoshi Inoue, Robin Soper, Andrew Pitsillides, Stephen Hodges, Bethany Scarrott, Jonathan Elliott, Ragnhild Svalheim, Jayde O'Neil, Magali Chollet, Barbara Walther, Niels Erik Frandsen

Published by InTech

Janeza Trdine 9, 51000 Rijeka, Croatia

© The Editor(s) and the Author(s) 2017

The moral rights of the editor(s) and the author(s) have been asserted.
All rights to the book as a whole are reserved by InTech. The book as a whole (compilation) cannot be reproduced, distributed or used for commercial or non-commercial purposes without InTech's written permission. Enquiries concerning the use of the book should be directed to InTech's rights and permissions department (permissions@intechopen.com).
Violations are liable to prosecution under the governing Copyright Law.

Individual chapters of this publication are distributed under the terms of the Creative Commons Attribution 3.0 Unported License which permits commercial use, distribution and reproduction of the individual chapters, provided the original author(s) and source publication are appropriately acknowledged. More details and guidelines concerning content reuse and adaptation can be found at http://www.intechopen.com/copyright-policy.html.

Notice

Statements and opinions expressed in the chapters are these of the individual contributors and not necessarily those of the editors or publisher. No responsibility is accepted for the accuracy of information contained in the published chapters. The publisher assumes no responsibility for any damage or injury to persons or property arising out of the use of any materials, instructions, methods or ideas contained in the book.

Publishing Process Manager Edi Lipovic
Technical Editor SPi Global
Cover InTech Design team

First published March, 2017
Printed in Croatia
Legal deposit, Croatia: National and University Library in Zagreb

Additional hard copies can be obtained from orders@intechopen.com

Vitamin K2 - Vital for Health and Wellbeing, Edited by Jan Oxholm Gordeladze
p. cm.
Print ISBN 978-953-51-3019-2
Online ISBN 978-953-51-3020-8

We are IntechOpen, the world's largest scientific publisher of Open Access books.

3,250+
Open access books available

106,000+
International authors and editors

112M+
Downloads

151
Countries delivered to

Our authors are among the
Top 1%
most cited scientists

12.2%
Contributors from top 500 universities

WEB OF SCIENCE™

Selection of our books indexed in the Book Citation Index
in Web of Science™ Core Collection (BKCI)

Interested in publishing with us?
Contact book.department@intechopen.com

Numbers displayed above are based on latest data collected.
For more information visit www.intechopen.com

For EU product safety concerns:
IN TECH d.o.o., Prolaz Marije Krucifikse Kozulić 3, 51000 Rijeka, Croatia,
info@intechopen.com or visit our website at intechopen.com.

Meet the editor

Jan Oxholm Gordeladze, PhD (born on the 25th of April, 1950), holds a triple professor competence (medical biochemistry, physiology, and pharmacology) and is presently working as a professor at the Inst. of Basic Medical Science, Dept. for Molecular Medicine, Section for Biochemistry, University of Oslo, Norway. He has previously been employed as the medical director of Merck Sharp & Dohme (MSD), Norway, serving two years as a Fulbright scholar at the NIH, Bethesda, Maryland, USA, and from 2006 to 2009, he has been employed as associate professor at the University of Montpellier, France. He has published more than 130 scientific articles and reviews/book chapters and presented more than 260 abstracts/posters/talks at conferences worldwide. At present, he and his research team are focused on research related to stem cell differentiation, cell phenotype stabilization by manipulating regulatory looping systems consisting of key genes and microRNAs, as well as the impact of vitamin K2 on cellular and organ functions, including osteoblasts (bone), chondrocytes (cartilage), striated muscle cells (skeletal muscle), and adipocytes (white and brown fat tissues).

Contents

Preface

Vitamin K2 is a vitamin with multiple biological functions, ranging from calcification of the bone; decalcification of soft tissues (like blood vessels and heart valves); positive effects (i.e., functional normalization) of fat tissue, the pancreas, skeletal muscles, liver, and brain; and many others. Furthermore, it dampens or blocks untoward inflammatory processes induced by immune cells (like Th-1 and Th-17 cells), thus aiding our body in combating the detrimental effects of interleukins and TNF-α on cell phenotypes, i.e., alteration and/or loss of specific cellular or tissue functions.

Hence, vitamin K2 may be construed as a factor or a hormone enabling our body in stabilizing the so-called interorgan cross talk, in order to "normalize" a plethora of organ functions, such as bone and tooth differentiation/mineralization and soft tissue decalcification, a reduction in the activity of inflammatory cells in general (which otherwise may destabilize cellular phenotypes and therefore organ functioning), stimulation of adipocytes to develop into a beige type of fat cells, thus reducing the chances for developing adiposity and/or the degree to which fat and carbohydrate metabolism brings about unhealthy weight gain, as well as a remedy to prevent and or (perhaps also) cure conditions, like liver and prostate cancer, as well as dementia.

In this respect, vitamin K2 ingestion may be embraced as one factor, which supports longevity, as it so elegantly is described in the article by Dr. Lara Pizzorno: http://www.lmreview.com/articles/view/Vitamin-K2-Essential-for-Prevention-of-Age-Associated-Chronic-Disease/. It is therefore important to explore the multitude of effects of vitamin K2 further, in order to gain more information about the biological effect of this little molecule.

And finally, the medical community is indebted to the research performed by the group of Dr. Satoshi Inoue, who founded the modern concept of vitamin K2, by creating experimental animals/cells with genetic manipulations of the receptor SXR/PXR, which binds and brings about the effect of vitamin K2 in cells and whole organisms. We are honored to present some of his work in the present compilation of articles on vitamin K2.

I would also like to thank all the authors who contributed to this book, as well as Kappa Bioscience of Norway, which helped make this book possible, while still ensuring full academic freedom for each and every team of scientists who shared their special knowledge of vitamin K2.

Jan Oxholm Gordeladze, professor PhD
Institute of Basic Medical Science,
Dept. of Molecular Medicine,
Section for Biochemistry, Norway

Section 1

Introduction

Chapter 1

Introductory Chapter: Vitamin K2

Jan Oxholm Gordeladze

Additional information is available at the end of the chapter

http://dx.doi.org/10.5772/66384

Before going into detail the biological effects of vitamin K2, it must be emphasized that vitamin K2, also known as menaquinones with varying side chain lengths, may in fact associate with other genes and modulate their effects substantially.

In fact, vitamin K2 may associate with a protein nuclear factor or intra-nuclear hormone receptor, just like what is known for vitamin A and vitamin D. This protein has many names, such as NR1I2, PXR or SXR. Vitamin K2 with its isoprenoid side chain of varying length will, for reasons of simplification, now be referred to as MK-7, even though there are moieties which are both shorter and longer. MK-7 is chosen, due to the fact that this molecule might be the more abundant, as well as the more active moiety.

First, it might be of interest to see which the genes are, and thus, what are the cellular or biological functions being 'determined' or 'modulated' by MK-7 and its 'relatives'. A scrutiny of the interactions of NR1I2 taken from 'Gene Cards': http://string-db.org/cgi/network.pl?taskId=I2KNgtQWbVsT looks like this (**Figure 1**).

It has been well established that NR1I2-PXR-SXR is the receptor, which binds vitamin K2 analogues of different chain lengths, e.g. MK-4 and MK-7, and NR1I2 may thus communicate the effects of vitamin K2 via associating with RXRA, forming a heterodimer. Furthermore, one may postulate that vitamin K2 may influence several other genes indirectly, by 'impinging' on elements or members of a 'gene lattice' like the one shown in **Figure 1**. Some of the genes being putatively strongly affected by vitamin K2 are (text from the same web-page as referred to **Figure 1**):

RPS6KB1 ribosomal protein S6 kinase, 70 kDa, polypeptide 1; Serine/threonine-protein kinase that functions downstream of signalling by mTOR responding to growth factors, as well as nutrients in order to sustain cell proliferation, growth and progression of the cell cycle. It modulates protein synthesis via phosphorylation of EIF4B, RPS6 and EEF2K, and ensures cell survival via repression of pro-apoptotic functions of BAD. When nutrient depletion occurs, inactive forms associate with EIF3 to build a translation initiation complex.

INTECH
open science | open minds

© 2017 The Author(s). Licensee InTech. This chapter is distributed under the terms of the Creative Commons Attribution License (http://creativecommons.org/licenses/by/3.0), which permits unrestricted use, distribution, and reproduction in any medium, provided the original work is properly cited.

PPARGC1A peroxisome proliferator-activated receptor γ, co-activator 1α; Serves as a transcriptional co-activator of steroid hormone receptors, as well as nuclear receptors. It greatly enhances transcription of the PPARG and thyroid hormone receptors, and can regulate mitochondrial genes, contributing to adaptive thermogenesis. It plays a pivotal role in metabolic adaptation in response to the availability of various diets via coordinating the expression of a large spectrum of genes mandatory for glucose and fatty acid metabolism.

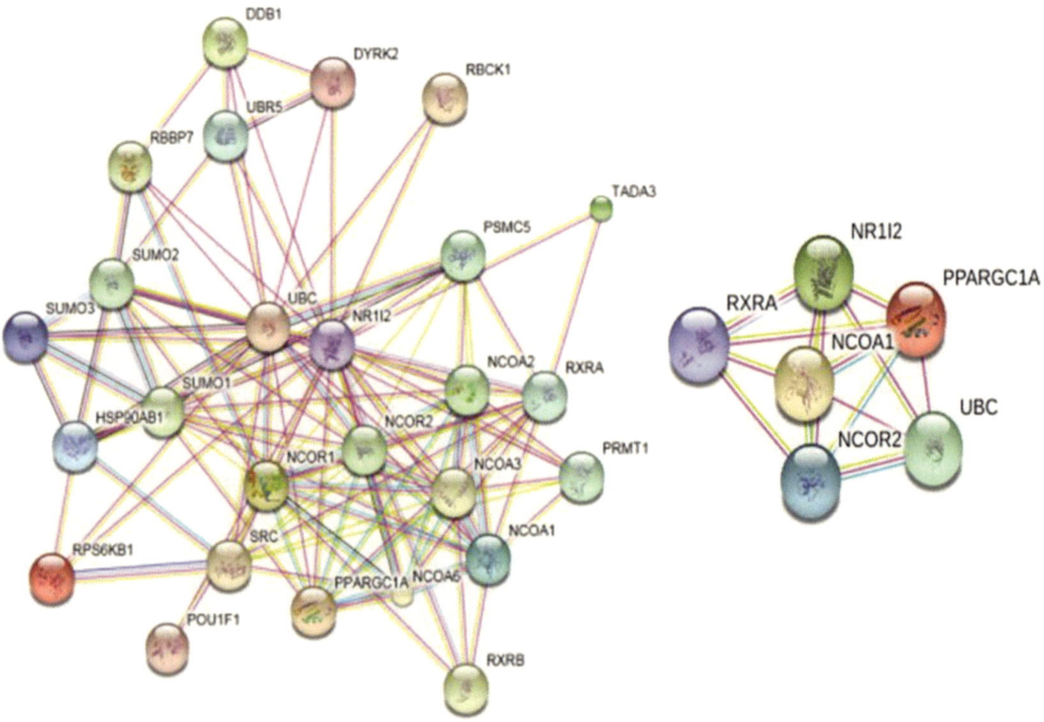

Figure 1. Genes interacting with the vitamin K2 (MK-7 or MK-4), which bind to the nuclear (transcription factor) NR1I2 being connected to a network of associated genes. Right panel: high stringency emulation and left panel: intermediate stringency.

NR1I2 nuclear receptor subfamily 1, group I, member 2; this is the nuclear receptor actually binding vitamin K2, and also goes by the names of SXR and PXR.

DYRK2 dual-specificity tyrosine-(Y)-phosphorylation regulated kinase 2; this serine/threonine-PK is involved in modulatory control of mitosis, proliferation of cellular entities, apoptosis, as well as cytoskeleton organization encompassing neurite outgrowth. It functions partly via its role in ubiquitin-dependent degradation of proteasomes. It works downstream of ATM to phosphorylate p53/TP53 at position 'Ser-46', thus contributing to induce apoptosis as a response to DNA damage. It phosphorylates NFATC1, thus inhibiting its accumulation in the nucleus, as well as its activity as a transcription factor.

RBCK1 RanBP- and C3HC4-type zinc finger containing 1; E3 ubiquitin-protein ligase, accepting ubiquitin delivered by E2 ubiquitin-conjugating enzymes, one of which is UBE2L3/

UBCM4, thereafter transferred to its substrates. It serves as E3 ligase for oxidized IREB2, while heme and oxygen are both necessary for the ubiquitination of IREB2. It promotes ubiquitin-coupling to TAB2 and IRF3, as well as their proteasome-mediated degradation. It is a component of the LUBAC complex, which conjugates ('M-1'-linked) polyubiquitin to substrate molecules, and plays a key role in NF- kappa-B activation and modulation of inflammation.

SRC v-src sarcoma viral oncogene homolog; non-receptor protein tyrosine kinase being activated subsequent to engaging of several various classes of cellular receptors, encompassing immune response receptors and integrins as well as adhesion receptors, TyrK-receptor protein kinases, G protein-sensitive receptors, and cytokine receptors. It participates in signalling, controlling a plethora of biological phenomena encompassing transcription of genes, immune responses, adhesion of cells, progression of the cell cycle, apoptosis, migration as well as transformation.

NCOA6 nuclear receptor co-activator 6; co-activator of nuclear receptor, directly binding nucleoceptors, stimulating transcriptional characteristics in a hormone-like manner. It co-activates expression in an agonist- and AF2-mediated fashion. It is implicated as co-activator of various nuclear receptors, like those binding steroids (GR & ERs), retinoids (RARs & RXRs), thyroid hormone (TRs), cholecalciferol (D3, VDR), as well as prostanoids (PPARs). It probably serves as a ubiquitous co-activator, instead of just another nucleoceptor co-activator. It may furthermore serve as a co-activator within the NF-kappa-B signalling pathway.

RXRA & RXRB retinoid X receptor, alpha & beta; Retinoic acid receptor. These receptors associate as heterodimeric entities to target response elements, elicited by their ligands, which are all-trans or 9-cis retinoic acid, thus regulating the expression of genes responsible for various biological processes. The heterodimers of RAR/RXR associate with response elements specific for retinoic acids (RAREs) comprising tandem 5′-AGGTCA-3′ binding sites, DR1-DR5 (by similarity). They preferentially associate with 9-cis retinoic acid.

RBBP7 retinoblastoma binding protein 7; nucleocore histone-associating subunit targeting the chromatin-bound remodelling factors, histone acetyltransferases, as well as and histone deacetylases to histone substrates in a way compatible to 'its' nucleosomal DNA. It is a constituent of a plethora of complexes, regulating chromatin turnover, which include the type B histone acetyltransferase (HAT) complex, required for chromatin assembly subsequent to DNA replication; core histone deacetylase (HDAC) complexes, promoting deacetylation of histones and consequently transcriptional repression.

SUMO1&2 SMT3 'suppressor of mif two 3, homolog 1'; ubiquitin-like protein which may be covalently bound to proteins as a monomeric entity or as a lysine-linked polymer. It is covalently associated via an iso-peptide bond to its substrates–necessitates prior activation by an E1 complex (SAE1-SAE2) and firm linkage to the E2 enzyme UBE2I. It can be promoted/facilitated by E3 ligases, such as PIAS1-4, RANBP2 or CBX4. The present post-translational alteration of lysine-containing protein residues serves a decisive role in a series of cellular processes, such as intra-nuclear transport, replication and repair of DNA, mitosis, as well as signal transduction.

NCOA2 nuclear receptor co-activator 2; transcriptional co-activator for both steroid receptors in general, as well as any type of nuclear receptors. It is the co-activator of the AF-2 steroid binding domain, but not of the AF-1 modulating N-terminal domain. It is required along with NCOA1 to master the cellular energy balance between white and brown adipose tissues.

UBR5 ubiquitin protein ligase E3 component n-recognin 5; E3 ubiquitin-protein ligase, which serves as one member of the N-end rule pathway. It recognizes and associates with specific N-terminal protein residues, destabilizing in accordance with the 'N-end rule', leading to a final ubiquitination with subsequent degradation. It is involved in the maturation and/or transcriptional modulation of mRNAs by activating CDK9 (via multiple ubiquitinations). It may serve a role in the progression and control of the cell cycle. It may also exert tumor suppressor activity. It modulates the DNA topoisomerase II binding protein (TopBP1) activity within in the DNA.

A short and non-exhaustive summary (for details, see above) of the potential biological impact exerted by MK-7 is quite impressive, and encompasses phenomena like the regulation of: protein synthesis, the cell cycle, cell survival, cell adhesion, nuclear transport, DNA replication and repair, cell phenotype regulation and stabilization via impact on histone deacetylases (HDACs), as well as inflammatory processes [1]. As seen from the few genes described above, it may be asserted that vitamin K2 is also able to influence bodily phenomena like energy (lipid and glucose) metabolism, due to a balance between the amount of bodily distribution of white and brown/beige adipocyte phenotypes. It is well known that white adipocytes store fat, while brown and beige adipocytes are either directly burning fatty acids only or performing both tasks simultaneously, in order to balance the characteristics of the other two. A few articles describe the versatile biological effects of vitamin K2 analogues in detail, and are referred to in the following paragraphs.

The nuclear receptors (NRs), the pregnane X receptor (PXR) and the related constitutive androstane receptor (CAR), play important roles as part of the xenobiotic detoxification reactions by modulating the expression of drug-metabolizing enzymes and transport molecules, to aid in the degradation and excretion of foreign chemical substances, as well as endogenously produced metabolites. The present survey is seeking to expand on the perceived biomedical relevance of both the PXR and CAR beyond their established role as master xenosensors, rather focusing on disease-oriented subjects, and with emphasis on their ability to be modulated by small molecules. These nuclear receptors are apparently involved in both the development and the treatment of non-alcoholic fatty liver disease (NAFLD). These receptors are, in fact, transcription factors (TFs), that are able to sense altering environmental and/or hormone-like signals, thus effectuating transcriptional changes in order to balance vital functions like cell/organ growth, but also development and reproduction. To be able to sustain this function, the following ligand-induced activation by xenobiotics (but also by liganding vitamin K2 (e.g. MK-7), 'members of subfamily 1 nuclear receptors' (NR1s) will heterodimerize with retinoid X receptor (RXR) and thus regulate gene transcription modulated processes being engaged in energy metabolism, but also inflammation. Many of these receptors, including PPARs (peroxisome proliferator-activated receptors), PXR, CAR, LXR and FXR (pregnane and xenobiotic, constitutive androstane, liver and X) receptors serve as key regulatory elements of the gut-liver-adipose biological axis, but also coordinate metabolic responses within organs, when oscillating between the fed and fasting states [2].

Non-alcoholic fatty liver disease (NAFLD) happens to be the most common liver ailment, which may eventually progress to cirrhosis and thereafter develop into hepatocellular carcinoma. NAFLD is characterized by insufficient nuclear receptor activity, leading to disturbed signalling through the gut-liver-adipose axis, which encompasses obesity, increased 'leakage' or permeability of the bowel system, with ensuing systemic inflammation, 'derangements' of the hepatic lipid metabolism, as well as insulin resistance. Unfortunately, environmental chemicals may complicate the issue by directly interfering with these nuclear receptors, conferring 'metabolic confusion' and thus the inability to distinguish feeding from starvation hours. Clinical investigations including conducted in the past (cfr. the PIVENS and FLINT trials) have shown that treatments aimed at these nuclear receptors may cause paradoxical reactions characterized by a dissociation of phenomena like inflammation, fibrosis, insulin resistance, dyslipidemia, steatosis and obesity. However, novel strategies (e.g. tissue-specific ligand molecules and/or dual receptor agonists) may be mandatory to be able to separate beneficial effects of nuclear receptor activation from untoward metabolic adverse effects [3].

In another investigation, one was looking into the effect of vitamin K2 on aortic calcification induced by warfarin via the Gas6/Axl survival pathway. A calcification rat model was established where warfarin was given to rats, which were divided into the following groups: controls and calcification groups, where some of the animals received vitamin K2. And the effect measurements/analyses were as follows: aortic calcium depositions (with Alizarin red); alkaline phosphatase activity in serum; apoptosis was evaluated by the TUNEL assay; and protein expression levels of Gas6, Axl, phosphorylated Akt (p-Akt) and Bcl-2 were analysed using western blotting. To summarize the results: the calcium content, calcium depositions, ALP activity and apoptosis were significantly higher in the calcification groups than control group. Furthermore, Gas6, Axl, p-Akt and Bcl-2 expression was lower in the calcification group than in the control group. Interestingly, 100 μg/g vitamin K2 administration reduced calcium depositions, ALP activity, as well as apoptosis, while Gas6, Axl, p-Akt and Bcl-2 expression were enhanced. Furthermore, vitamin K2 reversed almost half of the calcification. Finally, there was a positive correlation between formation calcification and apoptosis with $P < 0.0001$. This data therefore provides a sound theoretical basis for future treatments of aortic calcification [4].

The impact of vitamin K2 on apoptosis in different types of cancer cells have been shown in previous studies. But, the apoptotic effect of K2 on bladder cancer cells has not yet been evaluated. Hence, apoptotic activity and the underlying mechanism of K2 in bladder cancer cells were investigated. It was shown that K2 induces apoptosis in these cells via the 'mitochondria pathway' (i.e. membrane potential, cytochrome C release and the caspase-3 cascade). Also, phosphorylation of c-Jun N-terminal kinase (JNK) and p38 MAPK was detected in the vitamin K2-treated cells. Generation of ROS (reactive O_2 species) was observed in the cancer cells, however, treatment with K2 and the anti-oxidant N-acetyl cysteine virtually blocked the K2-triggered apoptosis, loss of mitochondrial membrane potential, as well as the JNK and p38 MAPK phosphorylation. These findings show that vitamin K2 clearly induces apoptosis in bladder cancer cells via the ROS-mediated JNK/p38 MAPK and mitochondrial pathways [5].

The pregnane X receptor (PXR) was until recently considered to serve as a nuclear receptor deemed to be specialized for detecting exposure from xenobiotics. In concurrence with this characteristic, PXR was identified to modulate drug-metabolizing enzymes. During the previous decades, PXR shown to harbour a broader spectrum of features. It is now evident that ligand-activated PXR modulates hepatic glucose turnover and lipid metabolism, while also affecting metabolic homeostasis throughout the whole body. At present, the consequences of PXR-elicited modulation on overall metabolic health, are not fully investigated, however, it has been shown that Rifampicin as well as St. John's wort, both serving as prototypical human PXR agonists, impair glucose tolerance in healthy individuals. Therefore, chronic exposure to naturally occurring PXR-agonists could be construed as a risk factor for diabetes and the metabolic syndrome [6].

As one associate of heterodimeric couples (ligand-receptor complexes), the retinoid X receptor (RXR) plays a leading role within the superfamily of nuclear receptors (NRs). Some heterodimers, e.g. PPAR&RXR, LXR&RXR, as well as and FXR&RXR are perceived as 'permissive', i.e. they turn into active transcriptional moieties, when an RXR-selective ligand ('rexinoid') or an NR partner ligand is present. In contrast, the so-called 'non-permissive' heterodimers (such as RAR&RXR, VDR&RXR or TR&RXR) are non-responding (inert) to the 'rexinoids' in question, when alone. Nonetheless, the agonists turn into 'transcriptional activators', when appearing together with a synergizing partner. However, despite their constellations assumed, when serving within the heterodimer formation/activation of multiple pathways, RXR appears as a target for drug discoveries. Interestingly, a rexinoid is applied in the clinic for the treatment of cutaneous T-cell lymphoma. More importantly, a plethora of RXR modulators also beholds a therapeutical ability for the treatment of metabolic diseases. The modulatory 'skill' of the rexinoids lies in the ligand-receptor complex conformation, as well as the wide spectrum and extent of their association with co-regulators, thus sustaining the specificity of the physiological response elicited. Interestingly, collected genetic and pharmacological data, emanating from investigations of insulin sensitivity, diabetes and obesity, definitely conclude that RXR agonists and antagonists show great promise as anti-obesity agents [7].

Unfortunately, the 'therapy' with rexinoids enhances plasma triglycerides, suppresses the hypothalamic-pituitary-thyroid hormone axis, as well as induces hepatomegaly, which complicates further search for compounds to treat insulin resistance and type 2 diabetes mellitus (T2DM). The recently developed PPARγ/RXR and LXR/RXR heterodimer-selective rexinoids, acting differently from PPARγ or LXR agonists alone, could possibly circumvent these limitations. Suffice to say, therapy with vitamin K2 alone (either as MK-7 or MK-4) in combination with a suitable rexinoid, might overcome the treatment 'paradox' described above [7].

The steroid and xenobiotic receptor (SXR) and its murine orthologue, the pregnane X receptor (PXR), are, as mentioned many times over, expressed in the liver and intestine where, they function as xenobiotic sensors involved in detoxification and drug excretion. However, it has recently been shown that both SXR and PXR are present in osseous tissue, where they facilitate bone metabolism [8]. It was shown that a deletion of PXR provokes in an ageing-dependent wearing of the articular cartilage in knee joints. A histomorphometrical analysis demonstrated a marked diminution of the width of, as well as an enlarged gap between, articular cartilage of the femur and tibia in PXR-knockout mice. It was therefore speculated that the up-regulated SXR in the chondrocytes play a protective role in articular cartilage. Interestingly, the Fam20a (family-with-sequence-similarity-20a) gene proved to be an SXR-sensitive gene, known to be induced by SXR ligands, such as rifampicin and vitamin K2 (MK-7) [8].

Furthermore, it was proven that the expression of Fam20a was primarily seen in articular chondrocytes. Consistent with the epidemiological data given, the present results strengthen the notion that SXR/PXR may protect against ageing-dependent detrimental wearing of articular cartilage, and, therefore, one may assert that ligands for SXR/PXR (such as vitamin K2) may prevent the induction of osteoarthritis due to 'old age' [9].

The fat-soluble vitamin K1 is involved in blood coagulation mediated by maintaining the activity of coagulation factors in the liver. However, the vitamin K2 variety exerts extrahepatic 'abilities' known to prevent bone fractures. In addition, epidemiological studies suggest that a lack of vitamin K (mainly K2) is associated with several geriatric diseases, including osteoporosis (bone brittleness), osteoarthritis (inflamed joints), dementia (mental retardation) and arteriosclerosis (blood vessel stiffness/narrowing). Furthermore, it was demonstrated that vitamin K may serve as an important factor in the prevention and/or treatment various types of cancers [10].

Recently, a novel role was discovered for vitamin K, serving as a ligand of the nuclear, steroid and xenobiotic receptor (SXR), as well as its murine, pregnane X receptor (PXR) orthologue. As a supplement to its published function as a co-factor for the γ-glutamyl carboxylase (GGCX), which effectuates a plethora of post-transcriptional changes, vitamin K also displays an additional mode of action, which is conveyed via transcriptionally modulations of the SXR/PXR target genes. Investigations of bone chips from PXR-deficient animals (mice) demonstrated that the osseous protection impact of vitamin K is partially conveyed via SXR/PXR-dependent signals. The discovery of new ways, by which vitamin K acts, has unravelled novel hope that this little molecule might come in handy in the prevention and/or treatment of a series of diseases affecting the elderly [10].

The steroid and xenobiotic/pregnane X receptor (SXR/PXR) is a nuclear receptor, which is located mainly in hepar and intestine, where they serve as xenobiotic sensors. However, many groups have identified SXR/PXR as a ubiquitous mediator of bone homeostasis. It was shown that systemic deletion of PXR brings about a marked loss of bone tissue (osteopenia) characterized by mechanical fragility in experimental animals (mice) down to 4 months of age. Values for BMD (bone mineral density) of PXR knockout (PXRKO) animals decreased substantially, when compared with BMD-values of wild-type animals. Micro-computed tomography measurements of trabecular bone (femur) from PXRCO animals unravelled a 3D-bone volume fraction markedly lower than what was found for WT mice [9].

Furthermore, histomorphometrical measurements of the trabecular bone material from the proximal tibia demonstrates a marked diminution bone mass from the PXRKO mice. Furthermore, bone formation was reduced, whereas bone resorption was augmented in the PXRKO mice. Histomorphometrical measurements of femoral cortical bones unravelled larger cortical areas in control animals than in PXRKO animals. The WT mice displayed an augmented cortical width, compared with PXRKO controls, and the so-called 'three-point bending test' showed that the 'new' morphological phenotypes in fact led to significant mechanical fragility. And, not to forget, serum levels of phosphate, as well as calcium and alkaline phosphatase remained unaltered in the PXRKO mice, as compared with their WT littermates. Consequently, it was concluded that SXR/PXR-activation augments the bone formation/resorption ratio, thus carving out a role for SXR/PXR as the key 'provider' of bone health [9].

The steroid and xenobiotic receptor (SXR = PXR = NR1I2) is activated by a plethora of endogenous hormones, pharmaceutics (drugs), as well as xenobiotics. SXR exhibits an enlarged, flexible and hydrophobic domain (LBD) for the ligand binding, and this is markedly divergent across different species of mammals. SXR also displays a pronounced difference in its pharmaco-dynamics and -kinetics amongst different mammalian species. The versatile response profile of SXR has paved the way for launching a 'steroid and xenobiotic binding hypothesis'. SXR is well-established as a xenobiotic sensor, which in a coordinated fashion rules the clearance of xenobiotics from the liver and intestine, through the up-regulation of genes being instrumental in both drug and xenobiotic turnover and excretion. Recently, it was revealed that SXR (most unexpectedly so) in fact makes a major contribution to (1) subduing inflammation, (2) bone homeostasis, (3) vitamin D turnover, (4) lipid metabolism, (5) energy homeostasis, as well as (6) the body's defence against cancer. Hence, the discovery of SXR as more than a xenobiotic sensor enables the use of a powerful tool for scrutinizing new mechanisms via which dietary factors, chemicals and the environment ultimately impact health, as well as disease development, prevention and treatment [10–13].

It has since long been shown that the xenobiotic receptors (XRs) have functionally evolved into cellular sensors for both endogenous and exogenous stimuli. Over the past decade, it has been demonstrated that regulating of the transcription of genes encoding drug-metabolizing enzymes and transporters, as well as the genes involving energy homeostasis, cell proliferation and/or immune responses are sensitive to the XRs. Unlike prototypical steroid hormone receptors, XRs are activated through both direct ligand-binding and ligand-independent (indirect) mechanisms by a plethora of structurally unrelated chemicals. In particular, scrutiny has been on the signalling control of the constitutive androstane receptor (CAR), the pregnane X receptor (PXR), as well as the aryl hydrocarbon receptor (AhR), which, in this context, function together in a synergistic manner. The latter (AhR) normally activates natural killer (NK) cells residing within the liver [10–13].

Ever since the identification of the retinoid X receptor (RXR) being a member of the nuclear receptor (NR) superfamily, the interest it has created has unravelled new understanding of its physiological regulation by nuclear receptors. Biologically, RXR serves an important role via its potent ability to associate with a variety of nuclear receptors. RXR possesses the ability to modulate nutrient metabolism via forming so-called 'permissive' heterodimers with PPAR (peroxisome proliferator-activated receptor), the liver-X-receptor (LXR), the farnesoid X receptor (FXR), the pregnane X receptor (PXR) and the constitutive androstane receptor (CAR), of which all of them start being functional, when ligands are associated to one or both of these heterodimers [10–13].

On the other hand, RXR will form 'non-permissive' heterodimers with the VDR (vitamin D receptor), TR (thyroid receptor), as well as RAR (retinoic acid receptor), which function properly only when vitamin D, T3 and retinoic acid, respectively, are present. Furthermore, RXR can form homodimers in the presence of a selective agonist, rexinoid, to regulate gene expression and to either inhibit proliferation or induce apoptosis in human cancers. Thus, over the last 25 years there have been several reports on the design and synthesis of small molecule RXR selective agonists, rexinoids [10–13].

A summary of some of the end point effects of agonists of the make: PXR/LXR (**Figure 2**, left panel), which appear to be working through the PI3K-AKT signalling pathway (**Figure 2**, right panel).

http://www.biochemj.org/content/453/1/71 http://www.genome.jp/kegg-bin/show_pathway?hsa04151

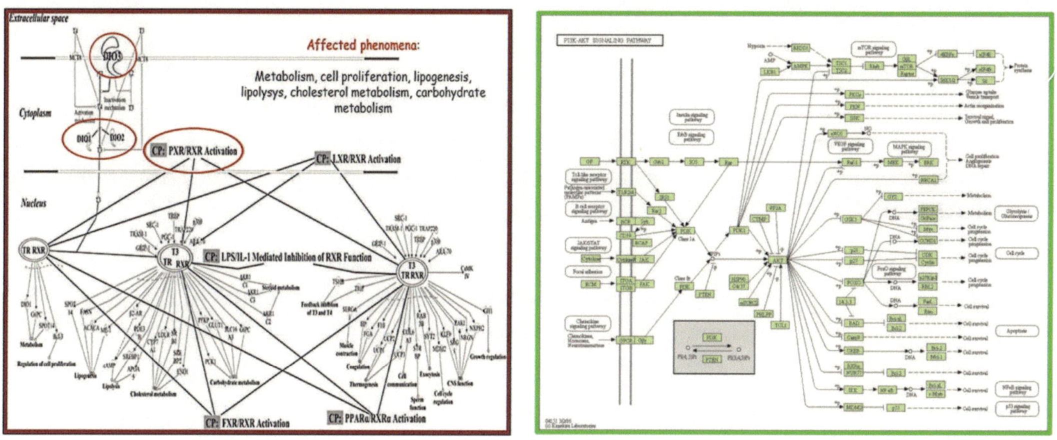

Figure 2. Summaries of genes impinging on metabolic pathways in general, cell proliferation, lipid (including cholesterol) metabolism with emphasis on the impact of DIO3 (deiodinase 3), and 'KEGGs Pathways' showing how the PI3K-AKT regulatory system affects a plethora of metabolic functions. *Source:* http://www.biochemj.org/content/453/1/71; http://www.genome.jp/kegg-bin/show_pathway?hsa04151.

On the left-hand figure, the pathways of T4 (via the deiodinase DIO3) and T3 are indicated. The keen reader will definitely study the chart for her- or himself, however, suffice to emphasize that there are the biological pathways or cellular functions on which receptors PXR/RXR and LXR/RXR exert powerful biological impacts: metabolism in general, cell proliferation, lipogenesis, lipolysis, cholesterol metabolism, carbohydrate metabolism, steroid metabolism, feedback inhibition of T3 and T4, muscle contraction, coagulation, thermogenesis, cell communication, sperm function, exocytosis, cell cycle regulation, and CNS function [14]. On the right-hand figure, the reader can observe that the PI3K-AKT signalling pathway, used by both osteoblasts or fibroblasts/HUVEC/VIC cell phenotypes to ensure or block the development of mineralizing properties [1], or described as modulators of GSK3, Bcl-2, RXRα and NFκB, and the FOXO-family of transcription factors, which are involved in biological phenomena summarized as glycolysis/gluconeogenesis, cell cycle, apoptosis, and NFκB/p53- signalling, of which all of them can and should be construed as adaptors of cell 'lifetime' = survival and 'phenotype' = functional stabilization [14].

As can be seen from **Figure 2**, it should be quite clear (or rather obvious?) that vitamin K2 (e.g. MK-4 or MK-7) exerts a spectrum of effects on various organs, and not only serves as a vitamin/hormone with an impact on coagulation mineralization of bone and demineralization of soft tissues like blood vessel walls (ref). And there are numerous reports to be found asserting a much broader function of vitamin K2, here are but a few:

(1) 'The steroid and Xenobiotic receptor (SXR), beyond xenobiotic metabolism' [15], which describes the impact of vitamin K2 on cholesterol and lipid homeostasis, bile acid homeostasis, inflammatory bowel disease (featuring SXR and NFκB), interplay between SXR and other receptors like CAR, FXR, VDR, the implication of SXR in cancer development and treatment, the implication (synergy) of SXR, and FoxO1, and FoxA2 in energy homeostasis.

(2) 'Role of Pregnane X Receptor in Obesity and Glukose Homeostasis in Male mice' [16], where PXR (the receptor binding vitamin K2) knock-out mice display weight increase, with a concomitant increase in liver weight (hepatomegalia), hyperinsulinemia and hypoadiponectinemia, as well as a loss of R2 receptors for adiponectin. These trends are all associated with the development of type II diabetes (T2DM).

(3) 'Targeting xenobiotic receptors PXR and CAR for metabolic diseases' [17], where the pregnane X receptor (PXR) and the constitutive androstane receptor (CAR) was found to be involved in normalizing metabolic disorders like the metabolic syndrome, T2DM, dyslipidemia, as well as atherosclerosis.

(4) 'Novel Functions of PXR in cardiometabolic disease' [18], where the authors in the abstract asserts that recent studies have revealed novel and unexpected roles of PXR in modulating obesity, insulin sensitivity, lipid homeostasis, atherogenesis, and vascular functions. These studies suggest that PXR signalling may contribute significantly to the pathophysiological effects of many known xenobiotics on cardiometabolic disease in humans. Genes mentioned in this article are many, and classified under functions like: 'cholesterol and lipid metabolism', 'lipoprotein metabolism', glucose metabolism', cholesterol and lipid metabolism', 'inflammation and atherosclerosis', detoxification and oxidative stress protection'.

(5) 'Regulation of PXR and CAR by protein-protein interaction and signalling crosstalk' [19], where the authors assert that 'it is clear that PXR and CAR perform a much broader range of cellular functions through protein-protein interactions and signalling crosstalk, which typically mutually affect function of all the partners involved. Future research on PXR and CAR should, therefore, look beyond their xenobiotic functions'.

(6) 'Mechanism of xenobiotic receptor activation: Direct vs indirect' [20], where the authors focus on the nuclear receptor (NR; remember that SXR = PXR also goes by the name NR1/2 = Nuclear Receptor 1/2) and its mode of activation, via pairing with a different nuclear receptor partner (activators/inhibitors, such as RXR, CCRP, NCOR, SMRT and several others). PXR may also be activated indirectly, which contributes to a meticulous and intricate regulatory system. Suffice to say that vitamin K2 (e.g. MK-7 or MK-4) is, but one factor contributing to the activation of NR1I2 = SXR = PXR, and is probably dependent on a certain intra-cellular/intra-nuclear level (as a 'cut-point' for proper activation) to exert a significant/maximal effect.

These six reports (of several more) signify that vitamin K2 may and will be of significance as to optimize and/or maximize certain toward effects, beneficial to organ systems in the body, as well as optimization of inter-organ cross-talk for the benefit of organ health and homeostasis. Finally, it should be mentioned that microRNAs play an important part in the regulatory network determining the impact of nuclear proteins responsible for proper gene expression—metabolic function—cell phenotype 'development' and stabilization.

1. Role of microRNAs in the determination of cellular phenotype

MicroRNAs (miRNAs or miRs) are conserved, small non-coding RNAs (18–25 nucleotides long) instrumental in the regulation of gene expression, and serve as a part of a network of factors, including transcription factors determining the phenotype of a certain cell in the body (ref). The transcription factors, such as the SXR = PXR = NR1I2 serve as receptors, much the same way as the receptors for vitamin A (RXR) and D (VDR), and may modulate gene transcription to determine the cell phenotype with its defined phenotypic characteristics (e.g. mineralizing osteoblast or non-mineralizing vessel-lining epithelial cells or heart valve fibroblast). It is well known that vitamin K2 (MK-7) stabilizes the two phenotypes (also in the presence of an inflammatory environment), thus preserving 'correct' inter-organ cross-talk, as would be expected in a healthy organism.

First, it should be emphasized that vitamin K2 binds to a nuclear receptor, which is part of a regulatory network consisting of microRNAs and other transcription factors. Second, this network represents a minimal lattice of regulatory factors, which may be manipulated in order to breach the stability of a certain cell phenotype (e.g. a cancer cell), or doing the opposite: reinforcing the phenotype in question (e.g. mineralizing osteoblast and non-mineralizing fibroblast during renal failure, uremia, for instance). Without going into detail, it is asserted that the regulatory networks presented here bargain for: (1) how vitamin K2 is involved in the stabilization of the osteoblastic phenotype, and (2) which are the major players (microRNAs and genes) determining whether a cell will adapt mineralizing properties or not (osteoblast or fibroblast).

Suffice to say (with reference to **Figure 3**), the master transcription factor JUN, impinges on a set of microRNAs (let-7 species) in a hierarchical structure of traditional genes and other microRNAs, which are well known in the literature, as being part of the WNT-Notch, the TGFβ and the BMP pathways (determining the osteoblast phenotype) (see KEGGs pathways), where specific markers like WNT6, DKK1, and CTNNB1 (β-catenin, activator of Runx2, the most referred marker of osteoblastic cells), are represented, and where some of the major microRNA-species, like miR-125, miR-21, miR-221, miR-27 and miR-23, known to be important in the differentiation and stabilization of the osteoblast phenotype is ensured.

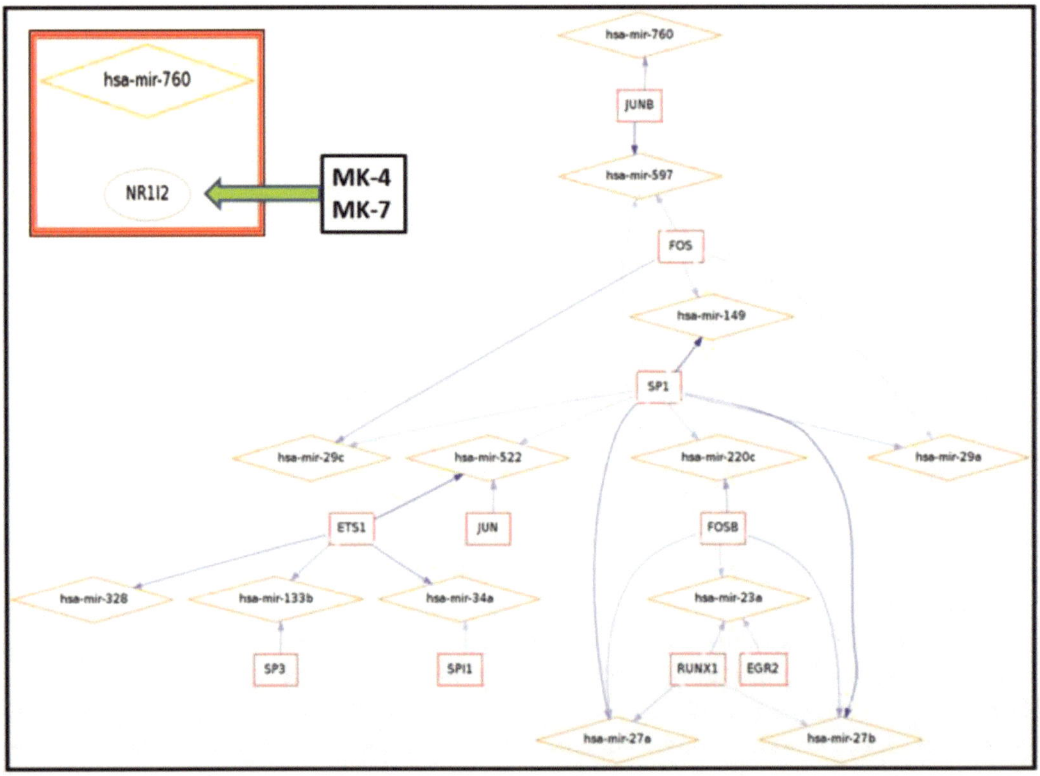

Figure 3. The involvement of vitamin K2 (MK-4 and MK-7), binding to NR1I2 = SXR = PXR, exerting its effect via hsa-miri-760 on regulatory loops in osteoblasts (according to the **Mir@nt@n** algorithm). This chart shows how vitamin K2 may affect the regulatory system determining the phenotype of osteoblast as a mineralizing cell in the body, involving microRNAs, transcription factors (e.g. FOS, JUN, SP1 & SP3) and 'functional' or 'marker' genes, like RUNX1.

In very much the same way, one may analyse the gene-transcription factor—microRNA axis in conditions like non-alcoholic fatty liver disease (NAFLD) or non-alcoholic steatohepatitis (NASH), and hopefully arrive at blood-born microRNAs, representing the diseases in question and/or their severity. MicroRNA species and putative target genes related to liver disease (NAFLD & NASH), as well as cardio (vascular) affection were pooled from three different articles (**Figure 5**).

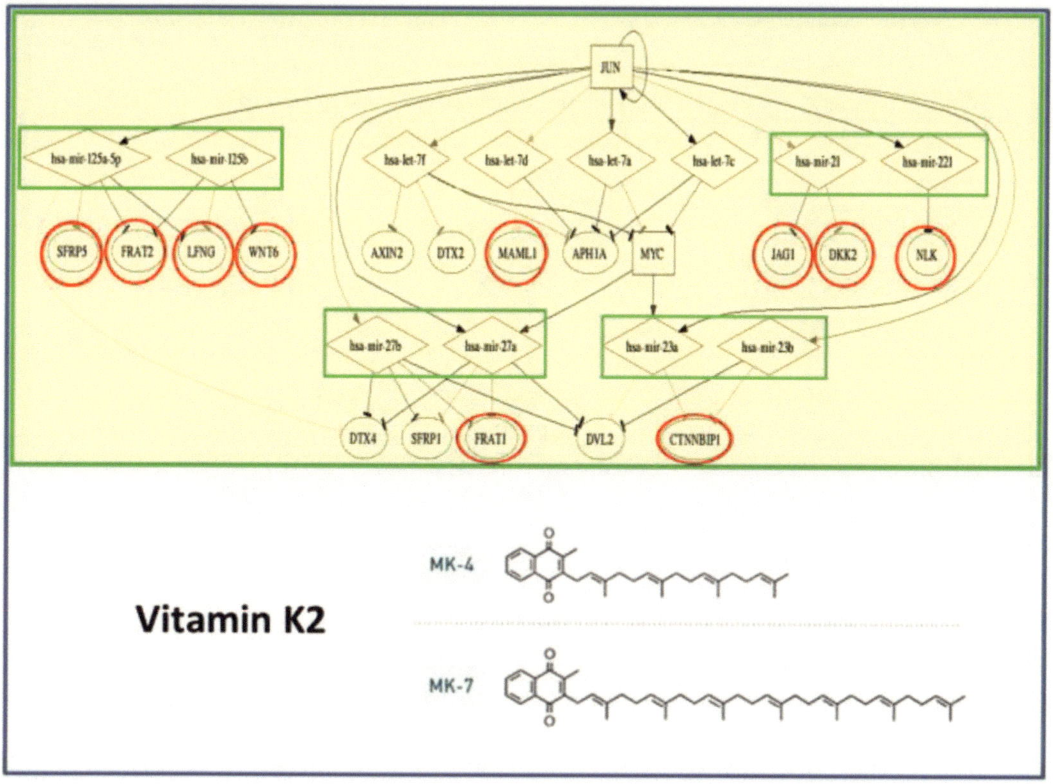

Figure 4. Hierarchical regulatory system consisting of microRNAs, transcription factors and genes involved in the mineralization phenotype of osteoblastic cells, but also present in fibroblasts having attained untoward/unwanted mineralizing properties during exposure to inflammatory cytokines and/or co-cultured with Th-1 and Th-17 cells.

At highest stringency applied, there are but a few genes and microRNAs emerging as 'connected': hsa-mir-122 with FBXO32 (F-Box protein 32; involved in FOXO-mediated signalling) and STAT4 (transcription factor); hsa-mir-144 with ABCA1 (ATP binding cassette subfamily A member 1, involved in cholesterol and sphingolipids transport from golgi and ER to the apical membrane and regulation of lipid metabolism by PPARα), hsa-mir-33b with ABCA1 and SLC25A25 (solute carrier family 25 member 25), and finally hsa-mir-145 with TFAM (transcription factor A, mitochondrial). Without going into details, all three articles [21–23], emphasize microRNA species 122, 144, and 33b, as instrumental in regulating hepatic lipid metabolism, with emphasis on hsa-mir-122. This microRNA-species is instrumental in the optimization of fatty acid oxidation vs synthesis, cholesterol production, as well as VLDL secretion to the circulation, and thus determining the health status of any individual in terms of risk of incurring atherosclerosis. The fact that this regulatory system, shown in **Figure 4**, lacks reciprocal regulatory loops makes it more vulnerable and unstable, when threatened by 'disease states', like NAFLD/NASH, than systems found in the osteoblast, which apparently appears more resilient to change, when exposed to conditions where inflammation prevails.

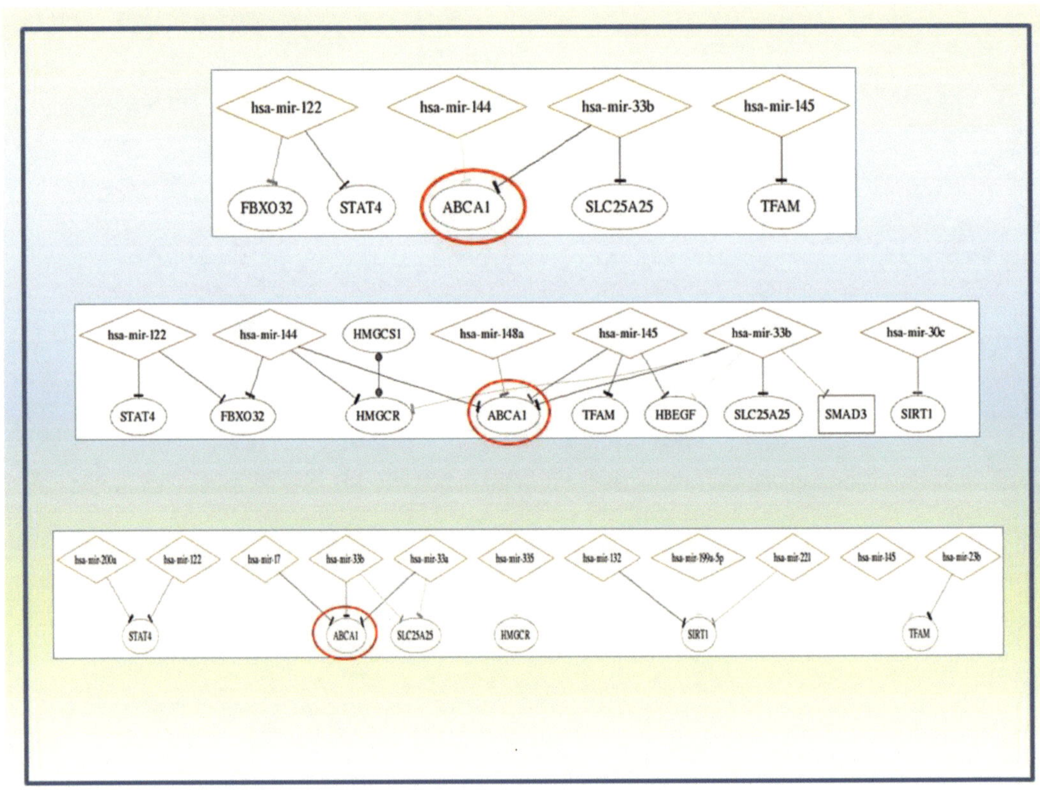

Figure 5. MicroRNAs, transcription factors and 'functional' genes related to liver function in patients with NAFLD/NASH with or without metabolic cardiovascular disease (see Refs. [21–23]). The three charts represent decreasing stringency/from top to bottom), and it should be emphasized that the regulatory system does not contain any reciprocal regulatory feedback systems, as was shown for the 'stabilization' of the osteoblast (or mineralizing phenotype).

Finally, it should be emphasized that bioinformatics analyses of the 'NRI2-relative' NR1I3, the constitutive androstane receptor (CAR), which interacts with NR1I2 = SXR = PXR, is also biologically interfering with many of the same factors (e.g. PPARα, CEPBα, STATs and T3) [24], thus linking them together in a very tight regulatory network, affecting lipid metabolism. Understanding the impact of vitamin K2 on these regulatory systems seems to be mandatory to grasp and acknowledge the idea that this fat-soluble molecule exerts such a tremendous effect on biological processes compatible with organ health, disease free old age, and thus 'longevity'.

http://dx.doi.org/10.5772/66384

Addendum 1

List of microRNAs and genes used as 'input' into the Mir@nt@n-algorithm, asking the program to 'retrieve' regulatory networks

hsa-mir-223, hsa-mir-320, hsa-mir-122, hsa-mir-27b, hsa-mir-30c, hsa-mir-27a, hsa-mir-378, hsa-mir-335, hsa-mir-148a, hsa-mir-370, hsa-mir- 33a, hsa-mir-33b, hsa-mir-144, hsa-mir-10b, hsa-mir-17, hsa-mir-19b, hsa-mir-26, hsa-mir-33a, hsa-mir-33b, hsa-mir-93, hsa-mir-101, hsa-mir-106b, hsa-mir-128, hsa-mir-144, hsa-mir- 145, hsa-mir-148a, hsa-mir-302a, hsa-mir-758, hsa-mir-223, hsa-mir 320, hsa-mir-122, hsa-mir- 613, hsa-mir-155, hsa-mir-143, hsa-mir-132, hsa-mir-145, hsa-mir-161, hsa-mir-211, hsa-mir-241, hsa-mir-146a, hsa-mir-30c, hsa-mir-99a, hsa-mir-122, hsa- mir-29c, hsa-mir- 320a, hsa-mir- 133a, hsa-mir-7, hsa-mir- 93, hsa-mir-98, hsa-mir-161, hsa-mir-191, hsa-mir- 195, hsa-mir-203, hsa-mir- 425, has-mir-422, hsa-mir-200a, hsa mir- 519d, hsa-mir-23b, hsa-mir-199a-5p, hsa-mir-199a-3p, hsa-mir-107, hsa-mir-103, hsa-mir-203, hsa-mir-21, hsa-mir-23, hsa-mir-24, hsa-mir-221, hsa-mir-223, hsa-mir-224, and hsa-mir-320b,

with or without:

hsa-mir-122, hsa-mir- 33a, hsa-mir-33b, hsa-mir- 132, hsa-mir-145, hsa-mir- 211, hsa-mir-144, hsa-mir-148a, hsa-mir-161, hsa-mir- 241, hsa-mir-146a, hsa-mir-30c, hsa-mir-223, hsa-mir-24, and hsa-mir-29,

combined with:

FOXO3, p300, fbxo32, ABCA1, SREBP1, SREBP2, PGC-1α, SLC25A25, NRF1, TFAM, HMGCS1, SMO1 SMO2 , LDLR, HMGCS2, hfe, hjv, LPL, HMGCR, LDLR, MTP, ASO, LPL, smad3, c/EBP, VIM, ADRP, DGAT, CPTA1, FABP4, LOX, ago2, STAT4, HBEGF, and Sirt1,

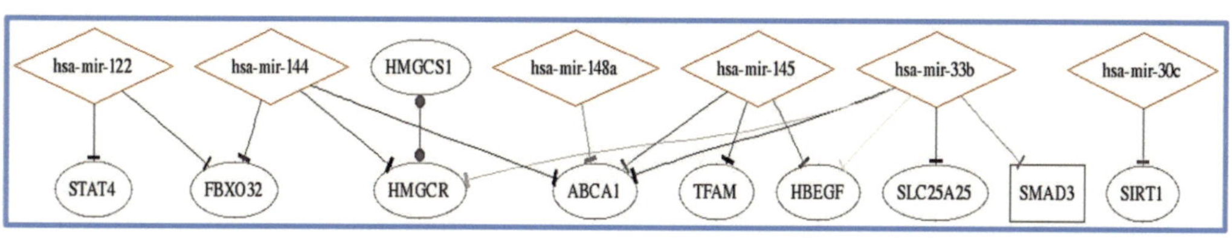

Author details

Jan Oxholm Gordeladze

Address all correspondence to: j.o.gordeladze@medisin.uio.no

Institute of Medical Biochemistry, Medical Faculty, University of Oslo, Oslo, Norway

References

[1] Gordeladze JO, Stakkestad Ø, Haugen S, Reseland JE, Syversen U, Johnsen GF, Haugen HJ, Lyngstadaas SP, MØller M, and Valtieri M. Regulatory Loops Consisting of Transcription Factors and microRNA Species Determine the Mineralizing Characteristics of Cell Phenotypes – Implications for Bone Engineering and Prevention of Soft Tissue Mineralization. InTech Publishing Company, http://dx.doi.org/10.5772/59149.

[2] Chai SC, Cherian MT, Wang YM, Chen T. Small-molecule modulators of PXR and CAR. Biochim Biophys Acta. 2016 Sep;1859(9):1141–1154. doi: 10.1016/j.bbagrm.2016.02.013. Epub 2016 Feb 24.

[3] Cave MC, Clair HB, Hardesty JE, Falkner KC, Feng W, Clark BJ, Sidey J, Shi H, Aqel BA, McClain CJ, Prough RA. Nuclear receptors and nonalcoholic fatty liver disease. Biochim Phys Acta. (2016);1859:1083–1099.

[4] Jiang X, Tao H, Qiu C, Ma X, Li S, Guo X, Lv A, Li H. Vitamin K2 regression aortic calcification induced by warfarin via Gas6/Axl survival pathway in rats. Eur J Pharmacol. (2016);786:10–18.

[5] Duan F, Yu Y, Guan R, Xu Z, Liang H, Hong L. Vitamin K2 induces mitochondria-related apoptosis in human bladder cancer cells via ROS and JNK/p38 MAPK signal pathways.11. PLoS One. 2016 Aug 29;11(8):e0161886. doi: 10.1371/journal.pone.0161886. eCollection 2016.

[6] Hakkola J, Rysä J, Hukkanen J. Regulation of hepatic energy metabolism by the nuclear receptor PXR. Biochim Biophys Acta. 2016 Sep;1859(9):1072–1082. doi: 10.1016/j.bbagrm.2016.03.012. Epub 2016 Apr 8.

[7] Pérez E, Bourguet W, Gronemeyer H, de Lera AR. Modulation of RXR function through ligand design. Biochim Biophys Acta. 2012 Jan;1821(1):57–69. doi: 10.1016/j.bbalip.2011.04.003. Epub 2011 Apr 16. Review.

[8] Azuma K, Casey SC, Ito M, Urano T, Horie-Inoue K, Ouchi Y, Kirchner S, Blumberg B, Inoue S. Pregnane X receptor knockout mice display osteopenia with reduced bone formation and enhanced bone resorption. J Endocrinol. 2010 Dec;207(3):257–263. doi: 10.1677/JOE-10-0208. Epub 2010 Sep 27.

[9] Azuma K, Casey SC, Urano T, Horie-Inoue K, Ouchi Y, Blumberg B, Inoue S. Pregnane X eceptor Knockout mce diplay agig-depedent wearig of articuar cartilae. PLoS One. 2015 Mar 6;10(3):e0119177. doi: 10.1371/journal.pone.0119177. eCollection 2015.

[10] Vitamin K2: Optimal Levels Essential for the Prevention of Age-Associated Chronic Disease by Lara Pizzorno - http://www.lmreview.com/

[11] Milnes MR, Garcia A, Grossman E, Grün F, Shiotsugu J, Tabb MM, Kawashima Y, Katsu Y, Watanabe H, Iguchi T, Blumberg B.20. Activation of steroid and xenobiotic receptor (SXR, NR1I2) and its orthologs in laboratory, toxicologic, and genome model species. Environ Health Perspect. 2008 Jul;116(7):880–885. doi: 10.1289/ehp.10853.

[12] Xu C, Huang M, Bi H. PXR- and CAR-mediated herbal effect on human diseases. Biochim Biophys Acta. 2016 Sep;1859(9):1121–1129. doi: 10.1016/j.bbagrm.2016.02.009. Epub 2016 Feb 22.

[13] Banerjee M, Robbins D, Chen T. Targeting xenobiotic receptors PXR and CAR in human diseases. Trends Pharmacol Sci. 2012 Oct;33(10):552–558. doi: 10.1016/j.tips.2012.07.003. Epub 2012 Aug 10.

[14] Dudek KM, Suter L, Darras VM, Marczylo EL, Gant TW. Decreased translation of Dio3 mRNA is associated with drug-induced hepatotoxicity. Biochem. J. (2013);453:71–82.doi: 10.1042/BJ20130049.

[15] Zhou C, Verma, S. The steroid and xenobiotic receptor (SXR), beyond xenobiotic metabolism. Nucl Recept Signal 2009;7:1–21.DOI: 10.1621/nrs.07001.

[16] Spruiell K1, Richardson RM, Cullen JM, Awumey EM, Gonzalez FJ, Gyamfi MA. Role of pregnane X receptor in obesity and glucose homeostasis in male mice. J Biol Chem. 2014 Feb 7;289(6):3244–3261. doi: 10.1074/jbc.M113.494575. Epub 2013 Dec 20.

[17] Gao J, Xie W. Targeting xenobiotic receptors PXR and CAR for metabolic diseases. Trends Pharmacol Sci. 2012 Oct;33(10):552–558. doi: 10.1016/j.tips.2012.07.003. Epub 2012 Aug 10.

[18] Zhou C. Novel functions of PXR in cardiometabolic disease. Novel functions of PXR in cardiometabolic disease. Biochim Biophys Acta. 2016 Sep;1859(9):1112–1120. doi: 10.1016/j.bbagrm. 2016.02.015. Epub 2016 Feb 26.

[19] Oladimeji P, Cui H, Zhang C, Chen T. Regulation of PXR and CAR by protein-protein interaction and signaling crosstalk. Expert Opin Drug Metab Toxicol. 2016 Sep;12(9): 997–1010. doi: 10.1080/17425255.2016.1201069. Epub 2016 Jun 23.

[20] Mackowiak B, Wang H. Mechanism of xenobiotic receptor activation: direct vs. indirect. Biochim Biophys Acta. 2016 Sep;1859(9):1130–1140. doi: 10.1016/j.bbagrm.2016.02.006. Epub 2016 Feb 10. PMID: 26877237.

[21] Szabo G, Csak T. Role of microRNAs in NAFLD/NASH. Dig Dis Sci. 2016 May;61(5): 1314–1324. doi: 10.1007/s10620-015-4002-4. Epub 2016 Jan 14. Review.

[22] Mehta R, Otgonsuren M, Younoszai Z, Allawi H, Raybuck B, Younossi Z. Circulating miRNA in patients with non-alcoholic fatty liver disease and coronary artery disease. BMJ Open Gastroenterol. 2016 Jul 26;3(1):e000096. doi: 10.1136/bmjgast-2016-000096. eCollection 2016.

[23] Willeit P, Skroblin P, Kiechl S, Fernández-Hernando C, Mayr M. Liver microRNAs: potential mediators and biomarkers for metabolic and cardiovascular disease? Eur Heart J. 2016 Apr 20. pii: ehw146. [Epub ahead of print].

[24] Hao R, Su S, Wan Y, Shen F, Niu B, Coslo DM, Albert I, Han X, Omiecinski CJ. Bioinformatic analysis of microRNA networks following the activation of the constitutive androstane receptor (CAR) in mouse liver. Biochim Biophys Acta. 2016 Sep;1859(9):1228–1237. doi: 10.1016/j.bbagrm.2016.04.002. Epub 2016 Apr 11.

Chapter 2

Vitamin K, SXR, and GGCX

Kotaro Azuma and Satoshi Inoue

Additional information is available at the end of the chapter

http://dx.doi.org/10.5772/63983

Abstract

Vitamin K was discovered in 1929 as a substance essential for blood coagulation and had been clinically utilized before the precise mechanism of action became aware in 1970s. The function as a cofactor of γ-glutamyl carboxylase (GGCX) was the mechanism firstly discovered with the identification of several substrate proteins including blood coagulation factors and osteocalcin. Recently, we and others have shown that vitamin K has other modes of function, such as ligand of nuclear receptor SXR (steroid and xenobiotic receptor) and its murine ortholog PXR (pregnane X receptor) and modulator of protein kinase A (PKA) activity. Besides its importance in blood coagulation, involvement of vitamin K has been shown in two major aging-related diseases, osteoporosis and osteoarthritis. Based on clinical and epidemiological studies, vitamin K is shown to have protective roles for both of them. Interestingly, clinical studies concerning single nucleotide polymorphisms (SNPs) of GGCX and γ-carboxylated status of osteocalcin suggested relationship between GGCX activity and bone-protective effect, while recent findings from basic research indicated that vitamin K functions mediated by SXR/PXR as well as GGCX are important in the bone metabolism. We also suggested that cartilage-protective effect is mediated by SXR/PXR signaling by animal experiments using *Pxr* knockout mice.

Keywords: γ-glutamyl carboxylase (GGCX), steroid and xenobiotic receptor (SXR), pregnane X receptor (PXR), protein kinase A (PKA), osteocalcin, osteoporosis, osteoarthritis

1. Introduction

In 1929, a Danish biochemist, Dr. Henrik Dam predicted a fat-soluble diet substance which is essential for blood coagulation. The substance was referred as "Koagulationsvitamin" in

INTECH
open science | open minds

© 2017 The Author(s). Licensee InTech. This chapter is distributed under the terms of the Creative Commons Attribution License (http://creativecommons.org/licenses/by/3.0), which permits unrestricted use, distribution, and reproduction in any medium, provided the original work is properly cited. (cc) BY

German; thus it is called vitamin K in English named after the initial letter of its German word. He shared the Nobel Prize in Physiology or Medicine in 1943 with an American biochemist Dr. Edward A. Doisy who later identified the structure of vitamin K. During the 1970s, the mechanism of vitamin K began to be revealed with the discovery, namely, vitamin K was necessary for γ-carboxylation of some coagulation factors which is catalyzed by an enzyme called γ-glutamyl carboxylase (GGCX) [1, 2]. Interestingly, warfarin, which inhibits vitamin K function, was in medical use since 1954, and vitamin K administration to newborn babies for preventing intracranial hemorrhage started in many countries in the 1960s before the enzymatic mechanisms of vitamin K function had been clarified.

Recently, epidemiological and clinical studies suggested that vitamin K is related to various physiological and pathological processes besides coagulation. Based on these studies, vitamin K was approved to be used as a drug preventing osteoporotic fracture in several Asian countries. Moreover, for these two decades, another mode of vitamin K action has been elucidated. We discovered vitamin K functions as a ligand for a nuclear receptor, SXR (steroid and xenobiotic receptor), and its murine ortholog, PXR (pregnane X receptor) [3], which have physiological or pathological significance. Summing up, vitamin K plays important roles in wide variety of biological process in various modes of actions.

In this chapter we are going to introduce novel mechanism of vitamin K action mediated by SXR/PXR as well as recent findings concerning classical vitamin K action mediated by GGCX. Then we would like to discuss the functions of vitamin K in some aging-related diseases based on recent discoveries.

2. Multiple mechanisms of vitamin K function

The classical function of vitamin K is a cofactor of GGCX which was clarified in the 1970s [1, 2]. GGCX catalyzes the addition of a carboxyl group to glutamate residues in the substrate proteins, which is coupled by oxidization of vitamin K hydroquinone to vitamin K epoxide. Vitamin K-dependent coagulation factors (II, VII, IX, and X) are well known substrates for GGCX. They become active when several glutamate residues are γ-carboxylated. So far, 18 human proteins are reported to be γ-carboxylated and their functions are regulated by γ-carboxylation status in most of them. It is known that cyclic use of vitamin K is necessary for its function as a cofactor for GGCX [4]. To be recycled, vitamin K epoxide should be reduced by an enzyme called vitamin K epoxide reductase (VKOR). Warfarin, which has an anticoagulant activity, inhibits VKOR, causing a subsequent decrease in GGCX activity (**Figure 1**).

http://dx.doi.org/10.5772/63983

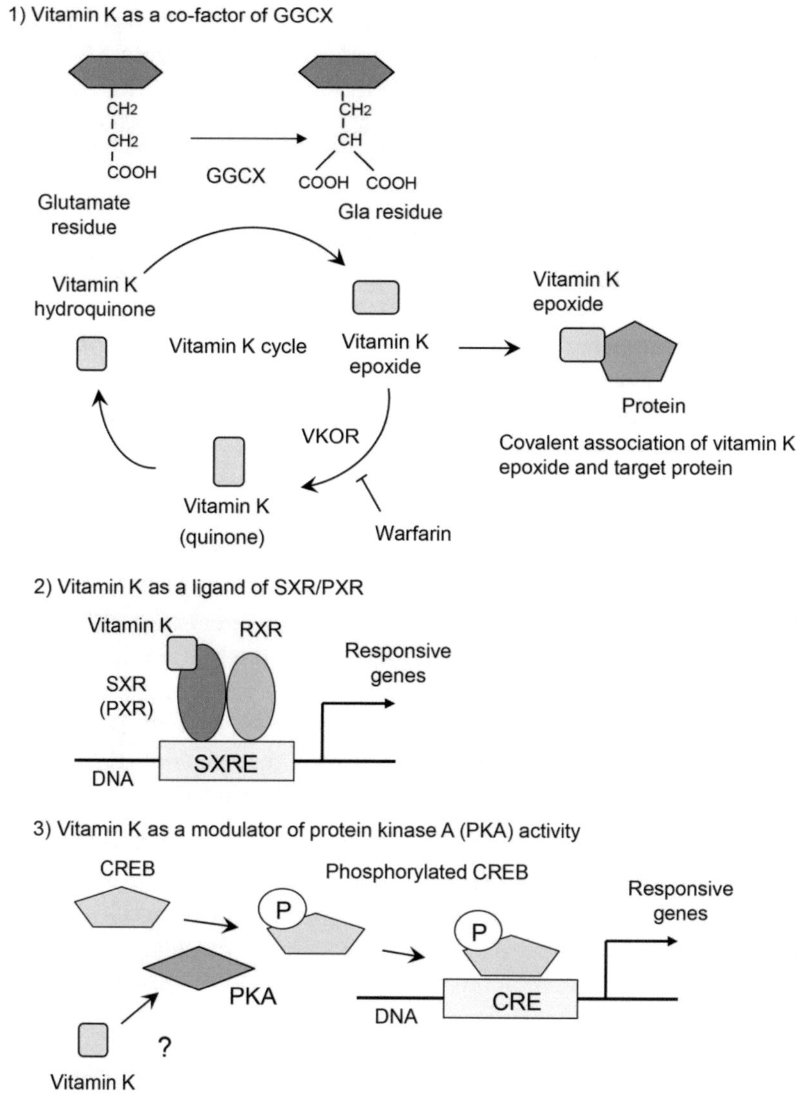

Figure 1. Multiple mechanisms of vitamin K actions. (1) GGCX catalyzes conversion of glutamate residues into Gla residues by incorporating an additional carboxyl group to glutamate. This reaction requires cyclic use of vitamin K. Vitamin K epoxide reductase (VKOR) is required for recycling vitamin K which is oxidized during γ-glutamyl carboxylation. Warfarin inhibits VKOR and vitamin K recycling, thereby suppressing GGCX activity. Covalent binding of vitamin K epoxide and a target protein is also proposed as a novel mode of vitamin K action which is dependent on GGCX activity. (2) Vitamin K also functions as a ligand of steroid and xenobiotic receptor (SXR) and its murine homolog, pregnane X receptor (PXR). On vitamin K binding, SXR/PXR forms heterodimers with 9-cis-retinoid acid receptor (RXR), and this complex binds to SXR-responsive elements (SXRE) within the promoter or enhancer regions of target genes. (3) Vitamin K also activates protein kinase A (PKA) with unknown mechanism. This action was suppressed by PKA inhibitor but not affected by stimulation with SXR agonist or knocking down of GGCX. Typical substrate of PKA is CREB (cyclic AMP-responsive element binding protein) and it binds to CRE (cyclic AMP-responsive element) within the promoter or enhancer regions of target genes when CREB is phosphorylated.

Recently, another mode of GGCX-dependent vitamin K function was reported in the study of proapoptotic effect of vitamin K. Handa et al. found proapoptotic protein Bak was covalently modified by vitamin K epoxide and regulated by its modification [5]. This function is dependent on GGCX-mediated vitamin K function since GGCX activity is required to generate vitamin K epoxide (**Figure 1**).

On the other hand, we discovered GGCX-independent mode of vitamin K function mediated by transcriptional regulation [3] as compared to posttranscriptional modifications explained above. Vitamin K was found to be one of the ligands of the nuclear receptor, SXR, and its murine ortholog, PXR. This receptor is also called NR1I2 according to standardized nomenclature designated by the nuclear receptor committee. In 1998, SXR/PXR was cloned as a novel nuclear receptor that is mainly expressed in the liver and intestine [6]. At first, its functions were characterized as a ligand-dependent transcription factor which is activated by various pharmaceutical agents and xenobiotic compounds [7]. It was originally classified as an orphan receptor since the endogenous ligand was not known when it was cloned. It was later shown that some kinds of secondary bile acids (such as lithocholic acid) could be endogenous ligands for this receptor [8, 9]. It forms a heterodimer with 9-cis-retinoid acid receptor (RXR) on ligand stimulation. This complex then binds to SXR-responsive elements (SXRE) in the promoter or enhancer regions of target genes (**Figure 1**). Some of its target genes are the drug-metabolizing enzyme, such as *CYP3A4*, and the ABC (ATP-binding cassette) family transporter, *MDR1*. Because of that, a function of SXR/PXR is considered as a xenobiotic sensor-inducing genes involved in detoxification and drug excretion [10] and named as such. The discovery of novel vitamin K function as a ligand for SXR/PXR indicated that physiological and pathological processes mediated by PXR/SXR would be affected by vitamin K.

There is another mode of vitamin K function which modulates activation of signal transduction pathway. This is inferred by existence of some genes induced by vitamin K, not by SXR agonist, rifampicin [11]. This induction was not affected by knocking down of GGCX suggesting that this is γ-carboxylation-independent pathway. Expression of those genes was suppressed by protein kinase A (PKA) inhibitor, showing the novel vitamin K function as a modulator of PKA activity (**Figure 1**).

Inhibition of another protein kinase, protein kinase C (PKC) α and ε, by vitamin K was also reported [12]. Inhibition of IKK (inhibitor of nuclear factor kappa B kinase) and subsequent inhibition of NFkB (nuclear factor kappa B) were observed. Whether this function of vitamin K is independent of mechanisms described above remains to be elucidated.

3. Epidemiological and clinical studies on vitamin K and aging-related skeletal diseases

A traditional Japanese food, "natto" (fermented soybeans) contains high concentrations of MK-7, a form of vitamin K2 (menaquinone), synthesized by microorganisms. Epidemiological study conducted in Japan revealed negative correlation of Natto intake and incidence of hip fracture [13], which drew attention toward possible link between vitamin K and osteoporosis.

Later, among several nutrients including vitamin D and calcium, vitamin K was shown to be the only nutrient that is significantly correlated with hip fracture incidence in Japanese population [14]. Furthermore, the fracture-preventing effect of vitamin K was observed in several clinical studies in Japan, which was confirmed by meta-analysis [15]. Based on these results, vitamin K2 is used for treatment of osteoporosis in several Asian countries. We previously reported a functional single nucleotide polymorphism (SNP) in *GGCX* that causes higher enzymatic activity correlated with higher bone mineral density in elderly Japanese women [16], suggesting bone-protective function of vitamin K is related to GGCX activity. Osteocalcin, one of the substrates of GGCX, is specifically expressed in osteoblastic lineage. The concentration of undercarboxylated form of osteocalcin (ucOC) in serum was reported to be positively correlated with fracture risk [17]. Measurement of ucOC has been clinically used to decide the indication of vitamin K for treatment of osteoporosis in Japan. These support the contribution of GGCX activity to bone-protective effect.

Vitamin K also has some epidemiological evidences in relationship with another skeletal disease, osteoarthritis. Low vitamin K intake was correlated to the prevalence of osteoarthritis both in North America and in Japan [18–20]. Unfortunately, therapeutic effect of vitamin K for established osteoarthritis was not proven by a trial [21], suggesting that the study period was too short or vitamin K has only preventive effect.

4. Paradoxical GGCX-mediated vitamin K functions on bone metabolism

It is difficult to evaluate vitamin K function on bone tissue mediated by GGCX *in vivo* due to its dominant effect on coagulation activity. For example, it is impossible to measure bone mineral density of adult mice systemically lacking GGCX because *Ggcx*-knockout mice die before birth or on the day of birth with massive bleeding [22]. To overcome this obstacle, we utilized Cre/loxP system which enables tissue-/organ-specific knockout of GGCX dependent on promoter activity [23] and generated osteoblast-specific GGCX-deficient mice by crossing with *Col1a1*-Cre mice [24]. Since osteoblasts express several substrates of GGCX including osteocalcin, we assumed bone-protective effect of vitamin K is mediated by GGCX activity in osteoblasts. Surprisingly, the bone mineral density was increased in osteoblast-specific GGCX-deficient mice and aberrant mineralization was observed in these mice by ultrastructural analysis. This result indicates that GGCX in osteoblast may not contribute to bone-protective effect of vitamin K. Moreover, it is contradicting to the clinical studies on *GGCX* SNPs or ucOC described above. We speculate that GGCX activity in other tissue is responsible for bone-protective effect of vitamin K and/or vitamin K function mediated by SXR/PXR that would be more important in the bone tissue. Further studies are necessary to clarify this enigma. It is noteworthy that osteocalcin-deficient mice have been shown to have mechanically stronger bone than wild-type mice [25], suggesting that the decrease of carboxylated osteocalcin, rather than increase of ucOC, has "bone strengthening effect."

5. SXR-mediated vitamin K functions on bone and cartilage

As described above, we proposed another mode of vitamin K function as a ligand of a nuclear receptor, SXR, and its murine ortholog, PXR. We showed that SXR is also expressed in osteoblastic cell lines and is activated by vitamin K2 [3]. We further identified SXR-dependent vitamin K-responsive genes by microarray analysis using human osteoblastic cell line, MG63 cells stably overexpressing SXR [26]. The identified genes included *Tsukushi* which encodes a protein that has a collagen-accumulating effect [27], *Matrilin2* which encodes a protein comprising extracellular matrix like collagen [28], and *Cd14* which regulates osteoblastogenesis [29] and osteoclastogenesis by inducing differentiation of B cells [30, 31]. These genes are induced even in the presence of warfarin, indicating their induction is independent of GGCX activity.

The involvement of SXR/PXR signaling in bone metabolism *in vivo* was suggested by the bone phenotype of systemic *Pxr* knockout mice [32]. We showed that 4-month-old female *Pxr* knockout mice had lower bone mineral density in femoral bone. Micro-CT analyses revealed fragile structure in the femoral trabecular bones of *Pxr* knockout mice. By histomorphometrical analyses, enhanced bone resorption and suppressed bone formation were observed in *Pxr* knockout mice. The mechanical strength of bone from *Pxr* knockout mice was weaker than that of wild-type mice. Negishi et al. reported the phenotype of systemic *Pxr* knockout mice from different origins [33, 34]. They also observed lower bone mineral density in *Pxr* knockout mice. They proposed a mechanism involving SLC34A2, a transporter of inorganic phosphate, expressed in the intestine. They showed *Slc34a2* is a PXR-responsive gene in the intestine and this was supported by the observation that serum levels of inorganic phosphate were significantly decreased in *Pxr* knockout mice. In con-

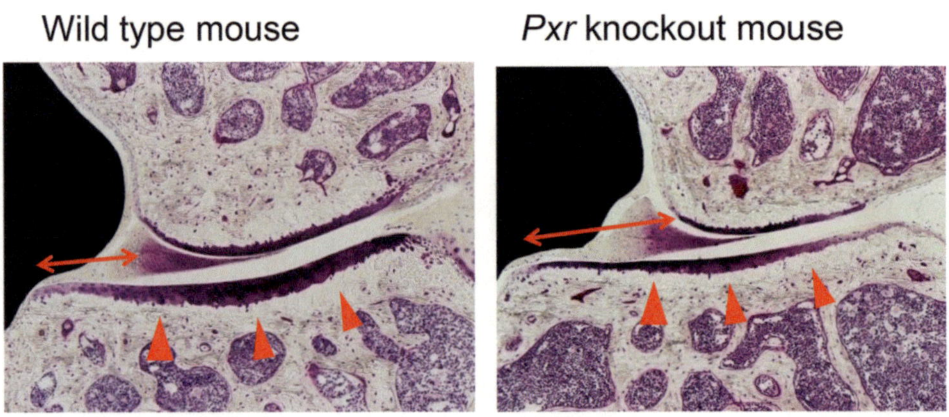

Figure 2. Aging-dependent wearing of articular cartilage of the knee joint in *Pxr* knockout mice. Representative microscopic images of articular cartilage of 13-month-old wild-type and *Pxr* knockout mice are shown. Arrowheads indicate lateral articular cartilage of the tibia. This difference was not significant in 4-month-old mice, suggesting this is aging-dependent process. Cited from Azuma et al. [35].

trast, we did not observe difference in the serum levels of inorganic phosphate between *Pxr* knockout mice and wild-type mice [32], suggesting the existence of different mechanisms according to the mouse strains and/or environment.

We also proposed SXR/PXR-dependent mechanism concerning vitamin K effect on articular cartilage [35]. We found that systemic *Pxr* knockout mice displayed aging-dependent wearing of articular cartilage of knee joints (**Figure 2**). Remarkable reduction of width and an enlarged gap between femoral and tibial articular cartilage were observed in *Pxr* knockout mice in 8-month-old and 13-month-old mice, but not in 4-month-old mice, indicating this is an aging-dependent process. With microarray analyses using ATDC5 chondrocytic cells overexpressing human SXR, we identified *Fam20a* (family with sequence similarity 20a) as an SXR-dependent gene induced by SXR ligands. We showed FAM20A related to the higher expression of COL2A1, a main component of extracellular matrix of the articular cartilages, suggesting the cartilage-protective effect of FAM20A. These results are consistent with epidemiological studies showing relationship between vitamin K intake and osteoarthritis and supporting the potential roles of vitamin K in preventing osteoarthritis caused by aging.

6. Conclusion

In this chapter, we described multiple mechanisms of vitamin K functions clarified so far and their involvement in aging-related skeletal diseases as examples for their biological significance. Besides blood coagulation, osteoporosis, and osteoarthritis, it became gradually aware that many physiological and pathological phenomena, such as fertility [36], atherosclerosis [37–39], brain development [40], dementia [41], and glucose metabolism [42–44], are related to the status of vitamin K sufficiency. We sincerely hope that vitamin K study leads to discoveries of new biological mechanisms and targets for disease prevention and treatment and eventually contributes to human culture and welfare.

Author details

Kotaro Azuma[1] and Satoshi Inoue[1,2*]

*Address all correspondence to: sinoue@tmig.or.jp

1 Genomics for Longevity and Health, Tokyo Metropolitan Institute of Gerontology, Tokyo, Japan

2 Division of Gene Regulation and Signal Transduction, Research Center for Genomic Medicine, Saitama Medical School, Saitama, Japan

References

[1] Nelsestuen GL, Zytkovicz TH, Howard JB: The mode of action of vitamin K. Identification of gamma-carboxyglutamic acid as a component of prothrombin. J Biol Chem. 1974;249:6347–6350.

[2] Stenflo J, Fernlund P, Egan W, Roepstorff P: Vitamin K dependent modifications of glutamic acid residues in prothrombin. Proc Natl Acad Sci U S A. 1974;71:2730–2733.

[3] Tabb MM, Sun A, Zhou C, Grün F, Errandi J, Romero K Pham H, Inoue S, Mallick S, Lin M, Forman BM, Blumberg B: Vitamin K2 regulation of bone homeostasis is mediated by the steroid and xenobiotic receptor SXR. J Biol Chem. 2003;278:43919–43927. DOI: 10.1074/jbc.M303136200.

[4] Stafford DW: The vitamin K cycle. J Thromb Haemost. 2005;3:1873–1878. DOI: 10.1111/j.1538-7836.2005.01419.x.

[5] Karasawa S, Azuma M, Kasama T, Sakamoto S, Kabe Y, Imai T, Yamaguchi Y, Miyazawa K, Handa H: Vitamin K2 covalently binds to Bak and induces Bak-mediated apoptosis. Mol Pharmacol. 2013;83:613–620. DOI: 10.1124/mol.112.082602.

[6] Blumberg B, Sabbagh W Jr, Juguilon H, Bolado J Jr, van Meter CM, Ong ES, Evans RM: SXR, a novel steroid and xenobiotic-sensing nuclear receptor. Genes Dev. 1998;12:3195–3205. DOI: 10.1101/gad.12.20.3195.

[7] Zhou C, Verma S, Blumberg B: The steroid and xenobiotic receptor (SXR), beyond xenobiotic metabolism. Nucl Recept Signal. 2009;7:e001. DOI: 10.1621/nrs.07001.

[8] Staudinger JL, Goodwin B, Jones SA, Hawkins-Brown D, MacKenzie KI, LaTour A, Liu Y, Klaassen CD, Brown KK, Reinhard J, Willson TM, Koller BH, Kliewer SA: The nuclear receptor PXR is a lithocholic acid sensor that protects against liver toxicity. Proc Natl Acad Sci U S A. 2001;98:3369–3374. DOI: 10.1073/pnas.051551698.

[9] Xie W, Radominska-Pandya A, Shi Y, Simon CM, Nelson MC, Ong ES, Waxman DJ, Evans RM: An essential role for nuclear receptors SXR/PXR in detoxification of cholestatic bile acids. Proc Natl Acad Sci U S A. 2001;98:3375–3380. DOI: 10.1073/pnas.051014398.

[10] Synold TW, Dussault I, Forman BM: The orphan nuclear receptor SXR coordinately regulates drug metabolism and efflux. Nat Med. 2001;7:584–590. DOI: 10.1038/87912.

[11] Ichikawa T, Horie-Inoue K, Ikeda K, Blumberg B, Inoue S: Vitamin K2 induces phosphorylation of protein kinase A and expression of novel target genes in osteoblastic cells. J Mol Endocrinol. 2007;39:239–247. DOI: 10.1677/JME-07-0048.

[12] Xia J, Matsuhashi S, Hamajima H, Iwane S, Takahashi H, Eguchi Y, Mizuta T, Fujimoto K, Kuroda S, Ozaki I: The role of PKC isoforms in the inhibition of NF-kB activation by

vitamin K2 in human hepatocellular carcinoma cells. J Nutr Biochem. 2012;23:1668–1675. DOI: 10.1016/j.jnutbio.2011.11.010.

[13] Kaneki M, Hodges SJ, Hosoi T, Fujiwara S, Lyons A, Crean SJ, Ishida N, Nakagawa M, Takechi M, Sano Y, Mizuno Y, Hoshino S, Miyao M, Inoue S, Horiki K, Shiraki M, Ouchi Y, Orimo H: Japanese fermented soybean food as the major determinant of the large geographic difference in circulating levels of vitamin K2: possible implications for hip-fracture risk. Nutrition. 2001;17:315–321. DOI: 10.1016/S0899-9007(00)00554-2.

[14] Yaegashi Y, Onoda T, Tanno K, Kuribayashi T, Sakata K, Orimo H: Association of hip fracture incidence and intake of calcium, magnesium, vitamin D, and vitamin K. Eur J Epidemiol. 2008;23:219–225. DOI: 10.1007/s10654-008-9225-7.

[15] Cockayne S, Adamson J, Lanham-New S, Shearer MJ, Gilbody S, Torgerson DJ: Vitamin K and the prevention of fractures: systematic review and meta-analysis of randomized controlled trials. Arch Intern Med. 2006;166:1256–1261. DOI: 10.1001/archinte.166.12.1256.

[16] Kinoshita H, Nakagawa K, Narusawa K, Goseki-Sone M, Fukushi-Irie M, Mizoi L, Yoshida H, Okano T, Nakamura T, Suzuki T, Inoue S, Orimo H, Ouchi Y, Hosoi T: A functional single nucleotide polymorphism in the vitamin-K-dependent gamma-glutamyl carboxylase gene (Arg325Gln) is associated with bone mineral density in elderly Japanese women. Bone. 2007;40:451–456. DOI: 10.1016/j.bone.2006.08.007.

[17] Szulc P, Chapuy MC, Meunier PJ, Delmas PD: Serum undercarboxylated osteocalcin is a marker of the risk of hip fracture in elderly women. J Clin Invest. 1993;91:1769–1774. DOI: 10.1172/JCI116387.

[18] Neogi T, Booth SL, Zhang YQ, Jacques PF, Terkeltaub R, Aliabadi P, Felson DT: Low vitamin K status is associated with osteoarthritis in the hand and knee. Arthritis Rheum. 2006;54:1255–1261. DOI: 10.1002/art.21735.

[19] Oka H, Akune T, Muraki S, En-yo Y, Yoshida M, Saika A, Sasaki S, Nakamura K, Kawaguchi H, Yoshimura N: Association of low dietary vitamin K intake with radiographic knee osteoarthritis in the Japanese elderly population: dietary survey in a population-based cohort of the ROAD study. J Orthop Sci. 2009;14:687–692. DOI: 10.1007/s00776-009-1395-y.

[20] Misra D, Booth SL, Tolstykh I, Felson DT, Nevitt MC, Lewis CE, Torner J, Neogi T: Vitamin K deficiency is associated with incident knee osteoarthritis. Am J Med. 2013;126:243–248. DOI: 10.1016/j.amjmed.2012.10.011.

[21] Neogi T, Felson DT, Sarno R, Booth SL: Vitamin K in hand osteoarthritis: results from a randomized clinical trial. Ann Rheum Dis. 2008;67:1570–1573. DOI: 10.1136/ard.2008.094771.

[22] Zhu A, Sun H, Raymond RM Jr, Furie BC, Furie B, Bronstein M, Kaufman RJ, Westrick R, Ginsburg D: Fatal hemorrhage in mice lacking gamma-glutamyl carboxylase. Blood 2007;109:5270–5275. DOI: 10.1182/blood-2006-12-064188.

[23] Azuma K, Tsukui T, Ikeda K, Shiba S, Nakagawa K, Okano T, Urano T, Horie-Inoue K, Ouchi Y, Ikawa M, Inoue S: Liver-specific γ-glutamyl carboxylase-deficient mice display bleeding diathesis and short life span. PLoS One. 2014;9:e88643. DOI: 10.1371/journal.pone.0088643.

[24] Azuma K, Shiba S, Hasegawa T, Ikeda K, Urano T, Horie-Inoue K, Ouchi Y, Amizuka N, Inoue S: Osteoblast-specific γ-glutamyl carboxylase-deficient mice display enhanced bone formation with aberrant mineralization. J Bone Miner Res. 2015;30:1245–1254. DOI: 10.1002/jbmr.2463.

[25] Ducy P, Desbois C, Boyce B, Pinero G, Story B, Dunstan C, Smith E, Bonadio J, Goldstein S, Gundberg C, Bradley A, Karsenty G: Increased bone formation in osteocalcin-deficient mice. Nature. 1996;382:448–452. DOI: 10.1038/382448a0.

[26] Ichikawa T, Horie-Inoue K, Ikeda K, Blumberg B, Inoue S: Steroid and xenobiotic receptor SXR mediates vitamin K2-activated transcription of extracellular matrix-related genes and collagen accumulation in osteoblastic cells. J Biol Chem. 2006;281:16927–16934. DOI: 10.1074/jbc.M600896200.

[27] Ohta K, Lupo G, Kuriyama S, Keynes R, Holt CE, Harris WA, Tanaka H, Ohnuma S: Tsukushi functions as an organizer inducer by inhibition of BMP activity in cooperation with chordin. Dev Cell 2004;7:347–358. DOI: 10.1016/j.devcel.2004.08.014.

[28] Wagener R, Ehlen HW, Ko YP, Kobbe B, Mann HH, Sengle G, Paulsson M: The matrilins – adaptor proteins in the extracellular matrix. FEBS Lett. 2005;579:3323–3329. DOI: 10.1016/j.febslet.2005.03.018.

[29] Roman-Roman S, Garcia T, Jackson A, Theilhaber J, Rawadi G, Connolly T, Spinella-Jaegle S, Kawai S, Courtois B, Bushnell S, Auberval M, Call K, Baron R: Identification of genes regulated during osteoblastic differentiation by genome-wide expression analysis of mouse calvaria primary osteoblasts in vitro. Bone. 2003;32:474–482. DOI: 10.1016/S8756-3282(03)00052-8.

[30] Filipp D, Alizadeh-Khiavi K, Richardson C, Palma A, Paredes N, Takeuchi O, Akira S, Julius M: Soluble CD14 enriched in colostrum and milk induces B cell growth and differentiation. Proc Natl Acad Sci U S A. 2001;98:603–608. DOI: 10.1073/pnas.98.2.603.

[31] Manabe N, Kawaguchi H, Chikuda H, Miyaura C, Inada M, Nagai R, Nabeshima Y, Nakamura K, Sinclair AM, Scheuermann RH, Kuro-o M: Connection between B lymphocyte and osteoclast differentiation pathways. J Immunol. 2001;167:2625–2631. DOI: 10.4049/jimmunol.167.5.2625.

[32] Azuma K, Casey SC, Ito M, Urano T, Horie K, Ouchi Y, Kirchner S, Blumberg B, Inoue S: Pregnane X receptor knockout mice display osteopenia with reduced bone formation

and enhanced bone resorption. J Endocrinol. 2010;207:257–263. DOI: 10.1677/JOE-10-0208.

[33] Konno Y, Kodama S, Moore R, Kamiya N, Negishi M: Nuclear xenobiotic receptor pregnane X receptor locks corepressor silencing mediator for retinoid and thyroid hormone receptors (SMRT) onto the CYP24A1 promoter to attenuate vitamin D3 activation. Mol Pharmacol. 2009;75:265–271. DOI: 10.1124/mol.108.051904.

[34] Konno Y, Moore R, Kamiya N, Negishi M: Nuclear xenobiotic receptor PXR-null mouse exhibits hypophosphatemia and represses the Na/Pi-cotransporter SLC34A2. Pharmacogenet Genomics. 2010;20:9–17. DOI: 10.1097/FPC.0b013e328333bb28.

[35] Azuma K, Casey SC, Urano T, Horie-Inoue K, Ouchi Y, Blumberg B, Inoue S: Pregnane X receptor knockout mice display aging-dependent wearing of articular cartilage. PLoS One. 2015;10:e0119177. DOI: 10.1371/journal.pone.0119177.

[36] Oury F, Sumara G, Sumara O, Ferron M, Chang H, Smith CE, Hermo L, Suarez S, Roth BL, Ducy P, Karsenty G: Endocrine regulation of male fertility by the skeleton. Cell. 2011;144:796–809. DOI: 10.1016/j.cell.2011.02.004.

[37] Beulens JW, Bots ML, Atsma F, Bartelink ML, Prokop M, Geleijnse JM, Witteman JC, Grobbee DE, van der Schouw YT: High dietary menaquinone intake is associated with reduced coronary calcification. Atherosclerosis. 2009;203:489–493. DOI: 10.1016/j.atherosclerosis.2008.07.010.

[38] Gast GC, de Roos NM, Sluijs I, Bots ML, Beulens JW, Geleijnse JM, Witteman JC, Grobbee DE, Peeters PH, van der Schouw Y: A high menaquinone intake reduces the incidence of coronary heart disease. Nutr Metab Cardiovasc Dis. 2009;19:504–510. DOI: 10.1016/j.numecd.2008.10.004.

[39] Shea MK, O'Donnell CJ, Hoffmann U, Dallal GE, Dawson-Hughes B, Ordovas JM, Price PA, Williamson MK, Booth SL: Vitamin K supplementation and progression of coronary artery calcium in older men and women. Am J Clin Nutr 2009;89:1799–1807. DOI: 10.3945/ajcn.2008.27338.

[40] Oury F, Khrimian L, Denny CA, Gardin A, Chamouni A, Goeden N, Huang YY, Lee H, Srinivas P, Gao XB, Suyama S, Langer T, Mann JJ, Horvath TL, Bonnin A, Karsenty G: Maternal and offspring pools of osteocalcin influence brain development and functions. Cell. 2013;155:228–241. DOI: 10.1016/j.cell.2013.08.042.

[41] Shatenstein B, Kergoat MJ, Reid I: Poor nutrient intakes during 1-year follow-up with community-dwelling older adults with early-stage Alzheimer dementia compared to cognitively intact matched controls. J Am Diet Assoc. 2007;107:2091–2099. DOI: 10.1016/j.jada.2007.09.008.

[42] Lee NK, Sowa H, Hinoi E, Ferron M, Ahn JD, Confavreux C, Dacquin R, Mee PJ, McKee MD, Jung DY, Zhang Z, Kim JK, Mauvais-Jarvis F, Ducy P, Karsenty G: Endocrine

regulation of energy metabolism by the skeleton. Cell. 2007;130:456–469. DOI: 10.1016/j.cell.2007.05.047.

[43] Ferron M, Wei J, Yoshizawa T, Del Fattore A, DePinho RA, Teti A, Ducy P, Karsenty G: Insulin signaling in osteoblasts integrates bone remodeling and energy metabolism. Cell. 2010;142:296–308. DOI: 10.1016/j.cell.2010.06.003.

[44] Shiba S, Ikeda K, Azuma K, Hasegawa T, Amizuka N, Horie-Inoue K, Inoue S. γ-Glutamyl carboxylase in osteoblasts regulates glucose metabolism in mice. Biochem Biophys Res Commun. 2014;453:350–355. DOI: 10.1016/j.bbrc.2014.09.091.

Section 2

Vitamnin K2 Rich Food – Impact of K2 on Energy Metabolism

Chapter 3

Vitamin K_2 Rich Food Products

Muhammad Yasin, Masood Sadiq Butt and
Aurang Zeb

Additional information is available at the end of the chapter

http://dx.doi.org/10.5772/63902

Abstract

Naturally, vitamin K exists in two bioactive forms mainly phylloquinone (vitamin K_1) and menaquinones (vitamin K_2). Phylloquinone is mostly found in green leafy vegetables such as kale, spinach, broccoli, and vegetable oils. However, menaquinones abundantly occurs in fermented vegetable products as menaquinones-7 (MK-7) and in animal-based products as menaquinone-4 (MK-4). Diverse concentrations of menaquinones are present in various dietary sources such as fermented pulses and milk-based products, cheese, meat, and animal organs. Presently, MK-7 and MK-4 contribute about 24 and 7%, respectively, of the total vitamin K dietary intake in the population consuming fermented products regularly. However, about 10% of menaquinones are pooled in the liver out of total intake of vitamin K. Conclusively, fermented soybean products and fermented milk-based products such as cheese and soured milk contain ample amount of MK-7, whereas animal organs, meat, fish, and egg contain appreciable amount of MK-4.

Keywords: vitamin K_2, menaquinones, fermented soybean, fermented milk, meat, cheese, poultry products

1. Introduction

Vitamin K is an indispensable anti-hemorrhagic fat soluble nutrient important for posttranslational modification of the proteins. Generally, these proteins are called vitamin K–dependent or Gla proteins. These coagulation proteins are produced in liver and played an active role in blood-clotting cascade. Moreover, during γ-carboxylation process, vitamin K hydroquinone is oxidized to its epoxide. When vitamin K is insufficient in blood-circulating system, carboxylation of vitamin K–dependent proteins is hampered and synthesis of undercarboxylated proteins is high that affects the Coagulation *cascade* parameters such as bleed and clotting times, and

INTECH
open science | open minds

© 2017 The Author(s). Licensee InTech. This chapter is distributed under the terms of the Creative Commons Attribution License (http://creativecommons.org/licenses/by/3.0), which permits unrestricted use, distribution, and reproduction in any medium, provided the original work is properly cited. (cc) BY

prothrombin and partial prothrombin times. Additionally, inadequacy of vitamin K also affects the bone mineral density (BMD) that leads toward higher bone fracture rate among population especially postmenopausal women and increased the risk of coronary artery diseases [1, 2].

Vitamin K deficiency is one of the alarming dilemmas among newborns, teenagers, and postmenopausal women. Neonates and their nursing mothers are 33.3% and 65% deficit in vitamin K, respectively [3]. Moreover, gastrointestinal disorders, unnecessary use of antibiotics, and alcoholics are some potential causes of vitamin K deficiency that hinder the fat incorporation in the body, and thus absorption of this vitamin is reduced [4]. Similarly, continuous utilization of broad-spectrum antibiotics suppresses the synthesis of vitamin K in the intestinal gut, whereas antagonist drugs badly effect to vitamin K functioning that delays the blood coagulation and necessary modifies certain protein which are indispensable for bone health [5]. Likewise, inflammatory bowel disease individual also showed vitamin K deficiency prevalence that declined the bone mineral density possibly due to its malabsorption [6].

The ingestion of vitamin rich dietary sources is rational approach to manage the vitamin K deficiency population [7]. Additionally, supplementation of vitamin K in the diet is also improved the serum vitamin K concentration [8, 9]. In this context, pharmacological dose of vitamin K (45 mg/day) is effective to ameliorate problem of bone fractures [10]. Therefore, more osteocalcin carboxylation is attained via consumption of well above recommended dietary requirement amount of vitamin K [11].

Vitamin K predominantly, naturally, presents in the form of phylloquinone (vitamin K_1) and menaquinone (vitamin K_2). Primarily, phylloquinone is present in green leafy vegetables such as spinach, kale, broccoli, and certain vegetable oils [12]. However, menaquinones are present in the fermented soybean and animal products [8].

Fermented soybean (natto) is one of the richest sources of menaquinones principally menaquinone-7 (MK-7) (882–1034 μg/100 g) among other fermented plant and animal-based food products [8, 13]. Menaquinone-7 containing natto is prepared by the fermentation process with *Bacillus subtilis*. Raw soybean is converted into slimy and sticky food predominantly owing to glutamic residues [14]. This menaquinone is also available in other food products such as cheese, meat and meat-based products, and fermented milk. Menaquinones are also synthesized by intestinal microflora such as *Escherichia coli, Bacteroides vulgatus,* and *Bacteroides fragilis* [15].

From total vitamin K dietary intake, menaquinone-4 (MK-4), menaquinone-7 (MK-7), and phylloquinone contribute around 7, 24, and 60%, respectively [16]. Humans and laboratory animals have ability to transform phylloquinone into MK-4 and MK-7 by rearrangement of integral side chain that has the ability to strengthen the bone [17]. In both phylloquinone and menaquinones, total vitamin K is about 1.16 μg/kg body weight is required for normal functioning. Menaquinone-7 normally stored around 10% in the liver. However, only 5–25% of ingested vitamin K is catabolized to MK-4 followed by the conversion of menaquinones in the liver via prenylation [10].

2. Fermented soybean food products

Traditionally, *Bacillus subtilis* fermented soybean food products are consumed in different parts of world. Among the fermented soybean products, natto is attained special attention which is consumed in Japan, whereas similar products such as kinema is produced in India, Satlyangser and Bari in Bhutan, Thua-Nao in Thailand, Tao-si in Philippines, Douchi in China and Dawadawa in Nigeria, Pepok in Myanmar. Kinema is similar natto fermented soybean which is originated from south China and expanded in Nepal, India, Burma, Thailand, and Japan [18–20].

Soybean conquers special attention in the public owing to its protein and fat contents. Numerous cultivars are currently available for household consumption. Among, yellow, green, and black soybeans are commonly used after cooking or fermentation. Bacillus fermented products such as natto and cheonggukjang are also consumed [21, 22]. *B. subtilis* fermented soybean has slimy appearance, softer texture, and white goosy substance with distinctive rotten flavor. Moreover, fermentation also removes the beany odor of raw soybean completely and improves its hedonic response [19]. However, large number of population does not like its slimy appearance and powerful rotten odor of fermented soybean such as kinema, while other has enjoyable experience. Freshly prepared fermented soybean is consumed as a fried curry with boiled rice or mixed with vegetables or consumed with boiled rice, soup, and pickle [20].

Researchers are trying to identify the new strains that may have potential similar *B. subtilis* and have ability to produce ample amount of MK-7 in the soybean. In this context, *B. amyloliquefaciens KCTC11712BP* under its optimum conditions remarkably enhanced MK-7 concentration in cheonggukjang (fermented soybean) and has shorter production time than that of natto [23]. Menaquinones are produced by microorganisms during electron transport chain respiration process [24].

Moreover, natto is rich in menaquinones especially MK-7 followed by other menaquinones such as MK-5, MK-6, MK-8 which are ranged from 882–1034 μg/100 g, 7.1–7.8 μg/100 g, 12.7–14.8 μg/100 g, 78.3–89.8 μg/100 g, respectively [13]. Similarly, Berenjian et al. [25] also reported that MK-7 content in fermented soybean is varied from 800–900 μg/100 g. Moreover, Booth [26] also reported that natto contains 998 μg/100 g MK-7. Previously, Dajanta et al. explicated that low fat (15.6%), high protein (45.8%), and small seed size of soybean are ideal characteristics of soybean for nato preparation [22]. Moreover, fermentation with *B. subtilis* exerts better nutritional attributes and sensory acceptability of the developed natto. Earlier, *B. subtilis* fermented Pakistani soybean also improved the concentration of MK-7 ranged from 681.35 to 803.82 μg/100 g. Contrary, MK-7 was not present in freshly cooked or reconstituted spinach [2]. The details of menaquinones concentration in various fermented soybean products are summarized in **Table 1**.

Food	Type	Country	MK-4	MK-5	MK-6	MK-7	References
Natto	Fermented soybean	Netherlands	7.5	13.8	998	84.1	[13]
Natto	Fermented soybean	Japan	–	–	–	939	[12]
Hikiwari natto	Chopped natto	Japan	ND	–	–	827	[12]
Natto	Fermented black soybean	Japan	–	–	–	796	[12]
Natto	Fermented soybean	Japan	ND	–	–	87–102	[23]
Natto	Fermented soybean	Pakistan	–	–	–	668–881	[2]
Cheonggukjang	Fermented soybean	Korea	ND	–	–	112–461	[23]
Cheonggukjang	Fermented soybean	Korea	74–76	–	–	271–1171	[23]
Cheonggukjang	Fermented soybean extract	Korea	–	–	–	1674–3438	[23]
Sauerkraut	Fermented vegetables	Netherlands	0.4	0.8	1.5	0.2	[13]
Cotton tofu	Hard type	Japan	0.04	–	–	–	[12]

ND = not detected; (–) = unknown or not reported.

Table 1. Menaquinones content of various fermented soybean (μg/100 g or mL).

However, the level of MK-7 in natto is still less the recommended daily amount of 180 μg/day that needs the utilization of 20–22 g natto/day. Many consumers find natto unpalatable; therefore, the ingestion of sufficient MK-7 is impractical. In some countries such as Japan, China, Thailand, and India fermented soybean products are commonly consumed, whereas other consumers do not like the fermented soybean products due to unpalatable, slimy, and sticky in nature, so the ingestion of MK-7 is insufficient among the population. Furthermore, the digestion and utilization of MK-7 from fermented soybean products are less efficient in humans with aging. Hence, there is a need for production of concentrated, supplementary MK-7 in the diet [25]. The extracted MK-7 with oil is consumed in various functional foods to fulfill the recommended requirement of vitamin K.

2.1. Natto

Natto is a Japanese traditional food prepared by fermentation of soybean with *B. subtilis natto*. The sterilized soybeans are fermented at 40°C for 14–18 h till it was fully covered with a viscous, sticky, and string-like material on surface. The resultant fermented soybean has characteristic odor and musty flavor [28].

2.2. Thua nao

For thua nao preparation, washed soybeans are soaked in water at ambient temperature for 16 h. Subsequently, soaked soybeans are autoclaved at 121°C for 40 min and cooled to 55°C after removing water. Powder culture of *B. subtilis* is mixed and incubated at 42 °C about 24 h for fermentation. Afterwards, fermented soybeans are placed at 4°C to hamper the bacterial activity.

2.3. Kinema

Cleaned and pre-soaked soybeans are autoclaved at 121°C for 35 min and cooled to 30–35°C. The soybean is inoculated with active culture of *B. subtilis*. The inoculated soybeans are placed for fermentation at 37°C for 48 h at 85% RH. The activity of bacteria is reduced by using the refrigeration temperature for 12 h. Fresh kinema is immediately consumed because its shelf life is about 2–3 days in summer and 5–7 days in winter without refrigeration. After sun drying, it can be stored for several months [18].

2.4. Hawaijar

Generally, small-sized soybean seeds are used for the production of hawaijar. Soybean seeds are boiled without soaking and packed loosely in bamboo basket lined with fig or banana leaves. Natural fermentation process is completed within 3–5 days. Mixed microflora including *B. subtilis, B. cereus, B. licheniformis, Staphylococcus sciuri, Staphylococcus aureus, Providencia rettgeri, Alcaligenes* sp. are fermented organism present in the final product. Production of mucilage and ammonia flavor is the indicator of quality hawaijar [18].

2.5. Cheonggukjang

Washed soybean seeds are soaked in tap water for 12 h. They are autoclaved at 121°C for 30 min and cooled up to 40°C prior to inoculation with pre-culture cell suspension of *Bacillus amyloliquefaciens strain KCTC 11712BP*. General for the preparation of cheonggukjang inoculated soybean seeds were placed in fermented for 48 h [28].

3. Menaquinones synthesizing organisms

Menaquinones are commonly synthesized by the action of intestinal microflora *B. vulgatus, B. fragilis,* and *E. coli* in human large intestine during electron transport chain reaction of respiration. Commercially, MK-7 was produced by action of bacteria such as *E. coli, Flavobacterium,* and *B. subtilis natto*. Moreover, among the fermented microflora, *B. subtilis* is consider one of the best options for the production of MK-7 due to its safety aspects as well as potential to synthesis the range of menaquinones such as MK-4 to MK-8 [15, 24, 29]. These hydrolytic bacteria are significantly consumed and reduced the indigestible oligo- and polysaccharides. The *Bacillus* species also lowered the activity of anti-nutrients of the soybean that hamper the availability of nutrients such as proteins and bioactive molecules [30].

B. subtilis is rod shaped, aerobic, heat resistant, spore forming Gram-positive bacteria. It has potential to synthesis number of enzymes such as proteolytic, amylolytic, and lipolytic enzymes during fermentation. The colonies of *B. subtilis* are irregular with hair-like structure having somewhat sliming appearance. Food Drug Administration granted the status of generally *recognized* as safe (GRAS) to the *Bacillus subtilis* and its close relative strains due to the lack of pathogenicity [20].

Administration of high single dose of MK-7 (2000 mg/kg) did not impart any toxic effect in animal modeling study in both genders. Moreover, prolonged treatment of MK-7 is considered also safe for human consumption due to its non-toxic effect on biochemical, hematological, urinary, and histopathological parameters [31].

Recently, Indian researcher Puri et al. identified a *B. subtilis* strain (*Bacillus subtilis MTCC 2756*) for the biosynthesis of menaquinone-7 for commercial purposes [32]. They found that addition of Tween 80 in fermentation medium significantly boost the MK-7 yield. Likewise, for the production of MK-7, groundnut meal is also used as a substrate owing to its protein and oil quality with strong minerals and vitamins profile using *Flavobacterium* sp. *SP-L-01* [33].

4. Stability of menaquinones

The vitamin K_2 was quite stable at room temperature for a period of up to 3 years. However, its concentration in final product was little affected by storage timed. Therefore, it is recommended that the menaquinone-7 rich product should be stored not more than 15°C temperature in a dry, cool, and dark place away from humidity, high heat, and sunlight.

5. Nutritional profile of fermented soybeans products

Fermented soybean is demonstrated higher amount of protein than raw soybean. The higher content of protein in fermented soybean might be elevated due to the microbial synthesis of single cell protein or rearrangements of molecules or enzymes followed by mortification of other moieties [19]. Premarani et al. explicated the impact of fermentation on the proximate composition of hawaijar [34]. They inferred that natural fermented soybean had 62.1% moisture, 26.02% soluble protein 24.36% crude fat, 8.2% crude fiber, 1.42% ash, and 3.8% free amino acids content nonetheless, inoculation with bacteria increases these parameters. However, hawaijar prepared from different soybean varieties showed non-momentous differences in lipid, fatty acid, and amino acid contents [35]. The nutritional profile of some fermented soybean was listed in **Table 2**.

Food	Moisture	Crude protein	Crude fat	Crude fiber	Ash	Carbohydrate	References
Kinema (traditional) dry basis	7.2	44.0	21.8	4.2	4.7	29.7	[79]
Kinema (pure culture) dry basis	6.8	45.1	23.0	4.8	4.5	27.4	[79]
Natto fresh weight basis	62.72	13.19	7.61	1.17	1.37	13.89	[2]
Fermented soybean	62.1	26.02	24.68	8.2	1.42		[34]
B. natto Itobiki	59–61	36–42	19–23	–	4.38–4.97	31–34	[37]
B. Natto NRRL B-3393	61	39–44	25–27	-	4.72–4.86	25–29	[37]

Table 2. Nutritional profile of menaquinones rich fermented soybean (%).

Likewise, natto has 40% protein and 24.68% fat contents. Controlled fermentation process improves the nutritional composition of natto than that of natural fermentation process [36]. Earlier, Wei and Chang delineated that the natto was prepared from four different soybean cultivars using *B. natto "Itobiki"* for moisture, protein, lipid, ash and carbohydrates as 59–61, 37–43, 20–23, 4–5, and 31–34%, respectively [37]. However, the natto synthesis through *B. natto NRRL B-3393* exhibited the values ranged from 61–62, 39–44, 25–27, 4.72–4.86, and 25–29%, for these attributes.

Similarly, free amino acid contents were enhanced 60-fold in fermented products [38]. Free fatty acids impart significant role in the development of unique flavor in fermented soybean. Earlier, Kiuchi et al. reported that carbon length of fatty acids was momentously increased during fermentation [39]. Moreover, the amount of vitamin B complex such thiamine, riboflavin, and niacin is momentously improved during *B. subtilis* fermentation process of cooked soybeans. Additionally, total folate concentration was enhanced during fermentation process of Tempeh (fermented soybean). Fermented soybeans such as kinema and natto have appreciable amount of mineral contents such as iron, magnesium, copper, zinc, sodium, and calcium [40–42].

6. Antioxidant potential of fermented soybean

Fermented soybean had higher amount of total phenolic, antioxidant capacities, and flavonoid contents in thua nao produce by black soybeans by pure culture of *B. subtilis TN51* compared with yellow soybeans and cooked non-fermented soybeans. Fermented black soybean and yellow soybean with *B. subtilis* TN51 exhibited higher amount of total phenolics 27.05 and 35.88 mgGAE/g extract than that of non-fermented products. Whereas total flavonoids contents of fermented black and yellow soybean were 3.50 and 1.38 mg catechin eq/g methanolic extract. However, contents of total phenolics and flavonoids were significant less in non-fermented black and yellow cooked soybeans. Likewise, 2,2-diphenylpicrylhydrazyl (DPPH)-antiradical activity (IC_{50}, mg/mL), and lipid peroxidation inhibition potential were higher in fermented black and yellow soybean compared with non-fermented cooked soybean [22].

Total phenol content of kinema was 144% higher than the raw soybean. Moreover, kinema has better free radical scavenger ability and metal chelating power and improved reducing power than that of non-fermented soybean. Therefore, it is suggested that kinema may has potential used as designer foods to alleviate oxidative stress [43]. Likewise, *B. subtilis* fermentation inclines to improve total polyphenolic and anthocyanins content about 10 and 250% during natto preparation from black soybeans, respectively. Moreover, DPPH radical scavenging activity has positive association with fermentation time and concentration of black soybean [19].

Fermentation increases about 58% content of phytosterol in kinema [43]. Later, Moktan et al. [44] reported that the kinema had 144%, 44%, 147%, and 92% higher total phenolics content, antioxidant activity, DPPH scavenging activity, and Fe^{2+}-chelating activity, respectively, compared to non-fermented cooked soybean. Similarly, better total phenolic contents and

DPPH activity were observed in Bacillus fermented soybean than that of soaked or cooked soybean [19].

During fermentation process of soybean, peptides are released by the hydrolysis of soybean proteins. Specific bioactive peptides such as glycinin and β-conglycinin are synthesized through the hydrolysis of soybean proteins. These bioactive peptides may act as regulatory compounds and have potential to minimize the physiological dysfunctions such anti-diabetic and anticancer activities [45].

Generally, fermented products are still widely synthesized by traditional methods. Consequently, it was recommended to develop standard operating producers and adhere with the good manufacturing practice (GMP) for individuals directly involved in its production for ensuring its safety [43].

7. Sensory response of fermented products

Sensory response including color, flavor, odor, texture, and overall acceptability is the main contributing parameters for the acceptance of products. Sensory evaluation of fermented soybean is mostly carried out using seven-point hedonic scale. The prepared natto obtained higher scores for color, appearance, taste, and viscosity than non-fermented cooked soybean [46]. Similarly, chungkukjang flavor, taste, and overall acceptability are evaluated using nine-point hedonic scale and reported that fermented soybean has high savory flavor and lower bitterness than traditional natto [47].

Moreover, soybean fermented with *B. subtilis TN51* has superior aroma than that of conventionally fermented soybean [27]. In another study, sensory traits of natto, *that are,* color, aroma, stickiness, bitterness, sweetness, sourness, and chewiness were determined through continuous linear intensity-scale of 10 cm with multiple demographic panelists such as Chinese, Japanese, and American. There were non-significant variations observed in sweetness or sourness and flavor in commercial available natto and laboratory prepared similar product [37].

8. Animal-based vitamin K_2 food products

8.1. Meat

Concerning the vitamin K, only plants and fermented food commodities are considered as a major natural source, but limited attention has been paid to meat for its menaquinones contents. Recently, Rødbotten et al. reported that cattle meat such as Jersey and Norwegian Red have better amount of menaquinones predominately MK-4 [48]. Jersey meat has higher amount of MK-4 in *M. longissimus dorsi* (LD), *M. biceps femoris* (BF), and *M. psoasmajor* muscles compared to Norwegian Red Cattle. Some traces of MK-6 and MK-7 were also detected in both types of cattle meats. Moreover, it is suggested that vitamin K_2 content has no association with

http://dx.doi.org/10.5772/63902

intramuscular fat and tenderness of meat. Contrary, Fujiwara et al. reported that poultry birds rely on natto did not improve the menaquinones content of meat [49]. Previously, Elder et al. [50] quantified the MK-4 in beef, chicken, fish, and liver of calf available in retail outlets or fast food restaurants of various cities of USA and reported that MK-4 was present in chicken in substantial amount ranged from 6.3 to 22.1 μg/100 g as compared to other tested animal meat and meat products (**Table 3**).

Meat	Type	Country	MK-4	MK-6	MK-7	Reference
Chicken	Breast meat	Netherlands	8.9	–	–	[13]
Chicken	Leg meat	Netherlands	8.5	–	–	[13]
Chicken	Thigh raw	Japan	ND	-	27 ± 15	[12]
Chicken	Leg and thigh ng/g	Japan	600	ND	ND	[12]
Chicken	Fresh ng/g	Japan	89.9	ND	ND	[12]
Beef	Chuck, raw	Japan	0.6	-	15	[12]
	Beef raw	Netherlands	1.1	–	–	[13]
Beef	Fresh raw ground beef	USA	4.9	–	–	[50]
Beef	Fresh raw ground beef (medium fat)	USA	8.1	–	–	[50]
Beef	Fresh raw ground beef (high fat)	USA	7.4	–	–	[50]
Bovine beef	Fresh beef (ng/g)	Japan	34.3	0.3	0.3	[12]
Cattle (Jersey)	M. Biceps Femoris	Norway	4.85	0.004	0.006	[48]
	M. psoas major	Norway	2.46	–	–	[48]
	M. longissimus dorsi	Norway	3.39	–	–	[48]
Cattle (Norwegian Reg)	M. Biceps Femoris	Norway	3.02	0.006	0.082	[48]
	M. psoas major	Norway	1.82	–	–	[48]
	M. longissimus dorsi	Norway	2.43	–	–	[48]
Beef	Minced meat	Netherlands	6.7	–	–	[13]
Beef Product	Salami	Netherlands	9.0	–	–	[13]
Pork	Thigh, raw	Japan	ND		6 ± 2	[12]
Pork	Fresh pork meat (ng/g)	Japan	9.4	0.3	0.3	[12]
Pork	Pork meat chop (ng/g)	Japan	31	ND	1.2	[12]
Horse	Fresh house meat ng/g	Japan	2.0	0.2	2.3	[12]
Luncheon	Meat	Netherlands	7.7	–	–	[13]
Hare	Leg meat	Netherlands	0.1	–	–	[13]
Deer	Back meat	Netherlands	0.7	–	–	[13]
Goose	Leg meat	Netherlands	31	–	–	[13]
Goose	Liver paste	Netherlands	369	–	–	[13]
Duck	Breast meat	Netherlands	3.6	–	–	[13]

Table 3. Menaquinones contents of meat of various fresh meat (μg/100 g).

Some studies indicated that thigh raw chicken meat contained the menaquinone-7 about 27 ng/g [21], whereas other only reported menaquinone-4 in both leg and breast meat of chicken ranged from 89.9 ng/g to 8.9 μg/100 g [13, 21]. Similarly, beef meat including raw, raw ground low, medium, and high fat meats from Japan, Netherlands, and USA contained menaquinone-4 ranged from 0.6 to 8.1 μg/100 g. However, Japanese origin beef meat contained menaquinone-7 as 15 μg/100 g. Likewise, cattle meat including Jersey and Norwegian *Biceps Femoris* meat merely contained MK-4, MK-6, and MK-7, whereas other parts did not contained MK-7, but their amount is very low [13, 21, 48, 50]. Additionally, pork and horse fresh meat contained the limited amount of MK-7 and all other tested meat such as luncheon, hare, deer goose, and duck contained only MK-4 (**Table 3**).

In various countries, people used the organs of animal as a source of meat. Therefore, concentration of vitamin K_2 is also very important to know in the commonly consumed organs such as liver, kidney, and heart. Hirauchi and coworkers reported that the organs meat of horse, chicken, and pork had significant amount of MK-4 compared with long-chain menaquinones (MK-7 to MK-13) and phylloquinone [51]. However, bovine liver was rich in MK-13 (215 ng/g) followed by MK-12 (215.6 ng/g), whereas the lowest concentration was noticed of MK-9 (15.3 ng/g). Other livers of various tested animal contained traces of higher menaquinones. The higher amount of long-chain menaquinones are possibly synthesized by gut microflora and stored in liver [52].

However, roasted beef contained 2–4 μg/100 g of MK-4, while other menaquinones such as MK-5, MK-7 and MK-8 were also present with low concentration [53]. Few publications are available regarding the vitamin K_2 content of meat (**Tables 4** and **5**). Previously, it was reported that beef meat has limited amount of vitamin K_2 without specifying the type of muscle and breeds [50, 53].

Meat	Type	Country	MK-4	MK-5	MK-6	MK-7	MK-8	Reference
Beef liver	Raw (μg/100 g)	USA	0.4	–	–	–	–	[50]
Beef liver	Pan-fried (μg/100 g)	USA	0.4	–	–	–	–	[50]
Beef liver	Braised (μg/100 g)	USA	1.9	–	–	–	–	[50]
Beef liver	Raw (ng/g)	USA	8.2	–	24.5	181.8	48.4	[51]
Beef liver	Raw (ng/g)	Finland	6.8	ND	9.44	25.6	13.8	[53]
Beef liver	Fresh heart (ng/g)	Japan	21.7	–	2.8	0.9	ND	[51]
Calf liver	Raw (μg/100 g)	USA	5.0	–	–	–	–	[50]
Calf liver	Pan-fried (μg/100 g)	USA	6.0	–	–	–	–	[50]
Calf liver	Braised (μg/100 g)	USA	1.1	–	–	–	–	[50]
Chicken Liver	Raw (ng/g)	Japan	39.6	–	0.3	ND	0.9	[51]
Chicken Liver	Raw (μg/100 g)	USA	14.1	–	–	–	–	[50]
Chicken Liver	Pan-fried (μg/100 g)	USA	12.6	–	–	–	–	[50]

Meat	Type	Country	MK-4	MK-5	MK-6	MK-7	MK-8	Reference
Chicken Liver	Braised (μg/100 g)	USA	6.7	–	–	–	–	[50]
Chicken heart	Fresh (ng/g)	Japan	142.6	–	0.1	ND	ND	[51]
Horse liver	Raw (ng/g)	Japan	2.1	–	1.0	2.3	1.2	[51]
Horse heart	Fresh heart (ng/g)	Japan	0.4	–	0.2	ND	ND	[51]
Pork heart	Raw (ng/g)	Finland	10.8	ND	ND	16	25	[53]
Pork liver	Fresh	Japan	1.2		0.2	1.1	ND	[51]
Pork	Raw	Netherlands	0.3	–	–	0.3	–	[13]
Pork	Raw Liver (ng/g)	Japan	5.9	–	0.4	6.1	5.6	[51]

ND = not detected; (–) = unknown or not reported.

Table 4. Vitamin K-2 content of different organ meats.

Food	Type	Country	MK-4	MK-5	MK-6	MK-7	Reference
Beef product	Hot dogs, regular fat	USA	5.7	–	–	–	[50]
Beef product	Ham roasted and pan broiled	USA	5.1	–	–	–	[50]
Beef product	Bacon (raw, pan-fried, microwaved, cooked and baked)	USA	5.6	–	–	–	[50]
Beef product	Beef meat roasted (ng/g)	Finland	28	1.2	ND	1.17	[53]
Beef product	Beef products	USA	1.7–8.1	–	–	–	[50]
Beef product	Roasted beef		2–4	–	–	–	[53]
Beef product	Broiled ground beef (low-fat steak)	USA	1.7	–	–	–	[50]
Beef product	Broiled ground beef (medium fat)	USA	7.2	–	–	–	[50]
Beef product	Broiled ground beef (high fat)	USA	5.1	–	–	–	[50]
Pork product	Loin (raw, broiled, pan-broiled, braised)	USA	0.9	–	–	–	[50]
Pork product	Meat franks, regular fat	USA	9.8	–	–	–	[50]
Pork product	Pork steak	Netherlands	2.1	–	–	0.5	[13]

ND = not detected; (–) = unknown or not reported.

Table 5. Vitamin K-2 content of different animal meat-based products (μg/100 g).

8.2. Animal-based sea foods

Fishes such as rainbow trout contained MK-4 (31 ng/g), MK-5 (0.9 ng/g), and MK-7 (2.0 ng/g), whereas MK-6 and MK-8 were not present (**Table 6**). Similarly, pike-perch also contained these menaquinones along with MK-6. Baltic herring and salmon only contained MK-4. Moreover, plaice and eel contained MK-4 as 0.2 and 1.7, MK-6 as 0.3 and 0.1, and MK-7 as 1.6 and 0.0, respectively. Horse mackerel from Netherlands and Japan only contained MK-4 content (0.4 and 0.6 μg/100 g). Furthermore, shrimp also had MK-4 (0.2 μg/100 g). Canned crab and tilapia fillets did not contain any form of menaquinones [13, 21, 50].

Food	Type	Country	MK-4	MK-5	MK-6	MK-7	MK-8	Reference
Fishes	Rainbow trout (ng/g)	Finland	31	0.9	ND	2.0	ND	[53]
	Pike-perch (ng/g)	Finland	1.9	0.49	0.52	4.9	ND	[53]
	Baltic herring (ng/g)	Finland	2.07	–	ND	ND	ND	[53]
Fish	Mackerel	Netherlands	0.4	–	–	–	–	[13]
	Plaice	Netherlands	0.2	–	0.3	0.1	1.6	[13]
	Eel	Netherlands	1.7	–	0.1	0.4	–	[13]
	Salmon	Netherlands	0.5	–	–	–	–	[13]
Fish	Horse mackerel, raw	Japan	0.6	–	–	ND	–	[12]
	Mackerel, raw	Japan	1.0	–	–	ND	–	[12]
Crab	Canned	USA	ND	–	–	–	–	[50]
Shrimp	Cooked and canned	USA	0.2	–	–	–	–	[50]
Salmon	Raw, Alaska wild	USA	0.3	–	–	–	–	[50]
Tilapia fillets	Raw and baked	USA	ND	–	–	–	–	[50]

ND = not detected; (–) = unknown or not reported.

Table 6. Menaquinones contents of various sea foods (μg/100 g).

8.3. Milk

Fresh milk having varied amount of fat also contained some amount of menaquinones especially MK-4. Sheep and cow whole milk contained about 17.4 and 8.60 ng/g of MK-4, respectively, while menaquinones were not detected in goat and donkey milk [54]. Moreover, milk having 1% fat had 0.4 μg/100 g of MK-4, whereas milk with higher amount of fat (2%) and whole milk showed more MK-4 contents as 0.5 and 1.0 μg/100 g which is available in retail outlets of USA [50].

There are various microorganisms such as *L. lactis* subsp. *cremoris*, *L. lactis* subsp. *Lactis*, and *Leuconostoc lactis* which have potential to produce long-chain menaquinones like MK-7 to

MK-10 about 230 nmol/g of dry cells. Moreover, these strains also have capacity to synthesis the ample amount of long-chain menaquinones in reconstituted non-fat dry milk and soy milk. Therefore, the milk-based fermented foods have significant amount of menaquinones and considered as an important dietary sources of vitamin K_2 [55]. Earlier, it was documented that the human milk only contained phylloquinone, but menaquinones were not detected [56].

Some of the fresh milk only contained MK-4 and other menaquinones are not detected or contained in the whole milk with varied concentration of fat from different animals (**Table 7**). Likewise, various creams and dressing also had MK-4 as source of vitamin K2. Interestingly, fermented milk and sourced milk contained higher amount of long-chain menaquinones such as MK-6, MK-7, MK-8, and MK-9 while MK-4 and MK-5 were not present. Similarly, butter milk also contained MK-4 to MK-8 but in limited quantity, butter contained only MK-4 (15 µg/100 g).

Food	**Type**	**Country**	**MK-4**	**MK-5**	**MK-6**	**MK-7**	**MK-8**	**MK-9**	**Reference**
Fresh Milk	1% fat	USA	0.4	–	–	–	–	–	[50]
Fresh Milk	2% fat (Regular and chocolate)	USA	0.5	–	–	–	–	–	[50]
Fresh Milk	Whole milk	USA	1.0	–	–	–	–	–	[50]
Fresh Milk	Cow 3.5% fat (µg/L)	Italy	8.60	–	–	–	–	–	[54]
Fresh Milk	Buffalo 5.0% fat (µg/L)	Italy	ND	–	–	–	–	–	[54]
Fresh Milk	Sheep 5.5% fat (µg/L)	Italy	17.4	–	–	–	–	–	[54]
Fresh Milk	Goat 5.0% fat (µg/L)	Italy	ND	–	–	–	–	–	[54]
Fresh Milk	Donkey 1.0% fat (µg/L)	Italy	ND	–	–	–	–	–	[54]
Fresh Milk	Whole milk	Japan	2.03	–	–	ND			[12]
Fresh Milk	Whole milk	Nether lands	0.8	0.1	–	–	–	–	[13]
Yoghurt	Yogurt plain (ng/g)	Finland	3.6	1.01	ND	ND	ND	ND	[53]
Yoghurt	Whole milk	Japan	0.6	0.1	0	0.2	–	–	[51]
Yoghurt	Skimmed milk	Japan	0	0	0	0.1	–	–	[51]
Yoghurt	Yogurt, plain (whole milk)	Japan	1.0	–	–	0.1	–	–	[12]
Yoghurt	Whole yoghurt	Nether lands	0.6	0.1	–	–	0.1		[13]
Yoghurt	Fortified MK-7	Nether lands		–	–	11.65	–	–	[57]
Cream	Ice cream Regular fat (vanilla and chocolate)	USA	2.6	–	–	–	–	–	[50]

Food	Type	Country	MK-4	MK-5	MK-6	MK-7	MK-8	MK-9	Reference
Cream	Cream	Japan	8	–	–	ND	–	–	[12]
Cream	Whipping cream	Netherlands	5.4	–	–		–	–	[13]
Dressing	Mayonnaise (whole egg type	Japan	17	–	–	ND	–	–	[12]
Dressing	Mayonnaise (egg yolk type	Japan	38			ND	–	–	[12]
Chocolate	Market (brand or type is no specified)	Netherlands	1.5	–	–	–	–	–	[13]
Buttermilk	Market (brand or type is no specified)	Netherlands	0.2	0.1	0.1	0.1	0.6		[13]
Butter	Market (brand or type is no specified)	Netherlands	15.0	–	–	–	–	–	[13]
Soured milk(ng/g)		Finland	5.7	2.93	1.7	4.1	20.1	47	[53]
Fermented milk	Mesophilic fermented milk (MFM) ng/g	France	–	–	1.3–4.9	1.2–6.1	7.237.9	29–145	[58]
Fermented milk	MFM (ng/g)	Germany	–	–	2.1–6	4.1–6.3	31–42	88.4–198.5	[58]
Fermented milk	MFM (ng/g)	Poland	–	–	0.5–11.9	3.2–10.9	7.1–89.3	17–414.2	[58]

ND = not detected; (–) = unknown or not reported.

Table 7. Menaquinones contents of various dairy products (μg/100 g).

Recently, Knapen et al. delineated that vitamin K fortified foods are healthy choice to increase the nutritional intake of MK-7 [57]. The fortified yoghurt drink containing MK-7 about 28 μg/ml has similar absorption pattern as the soft gel containing same amount of pure menaquinone-7. It is therefore suggested that to fortify food products that are ideal choice among the public to enhance the nutritional intake of menaquinones in the body.

8.4. Yoghurt

Thermophilic bacteria such as *Lactobacillus delbrueckii, Streptococcus thermophilus,* and *Bifidobacterium* which are mostly used a lactic acid bacteria starter culture not have ability to produce the menaquinones. Therefore, yogurt type milk-based products prepared with pure culture of thermophilic bacteria have limited or no menaquinones. However, among other milk-based fermented products, 60% contained some amount of vitamin K_2 (**Table 7**). The mesophilic lactic acid bacteria species which as used as starter culture for the fermentation have a capability to produce ample amount of vitamin K_2[53, 58]. However, yoghurt including plain, whole and skimmed milk contained higher amount of MK-4, whereas very low concentration of other menaquinones such as MK-5, MK-6 and MK-7 was present (**Table 7**).

Fortified yoghurt drink with MK-7 significantly improved serum concentration from 0.38 ng/ml to 2.00 ng/ml, whereas yoghurt supplemented with MK-7 along with other vitamins increased better serum MK-7 level as 2.17 ng/ml. Fortified MK-7 yoghurt and soft gel containing MK-7 showed statistically non-significantly variations [57].

8.5. Cheeses

Soft cheese as well as blue cheese have tremendous amount of vitamin K_2 as 1100 and 700 ng/g [58]. Earlier, menaquinones content of these cheeses has never been reported because soft and blue cheeses were not evaluated in respect of their vitamin K_2 content. These cheeses have higher amount of menaquinones possibly due to the activity of lactic acid bacteria particularly *Leuconostoc* species and yeasts or molds which are involved in ripening of cheeses. Most of the soft (from table name) and blue cheeses characterized had high contents of vitamin K_2 [59]. Generally, MK-9 was present in soft and blue cheese almost fourfold than that of MK-8. It is suggested that *lactococci* is responsible for the production of MK-9 in the dairy products. However, these bacteria also produced MK-8 comparatively constant ratio. Likewise, *propionibacteria* also synthesized the MK-9 as a key menaquinone in cheese. Menaquinone-9 concentration was higher in Norwegian Jarlsberg cheese trailed by Swiss Emmental cheese. However, Appenzeller or Gruye're cheeses had extremely low concentrations of MK-9. Additionally, Comte and Raclette cheeses contained lesser amount of MK-9 than both Jarlsberg and Emmental cheeses (**Table 8**).

Food	Type	Country	MK-4	MK-5	MK-6	MK-7	MK-8	MK-9	Reference
Cheeses	Hard cheeses (μg/100g)	Nether lands	4.7	1.5	0.8	1.3	16.9	-	[13]
	Soft cheese (μg/100g)	Nether lands	3.7	0.3	0.5	0.5	1.0	-	[13]
	Curd cheese (μg/100 g)	Nether lands	0.4	0.1	0.2	0.3	5.1	-	[13]
Semi-hard cheese	Semi-hard cheese (ng/g)	Nether lands	–	–	14.5–34.5	0–14.1	33.9–73.1	100–321	[58]
	Semi-hard cheese (ng/g)	Denmark	–	–	16.1–19.8	7.1–13.5	25–35.8	115.3–185.1	[58]
	Semi-hard cheese (ng/g)	Poland	–	–	9.8–15.8	ND	27.8–56.4	124.5–166.3	[58]
Soft cheese	Soft cheese (ng/g)	France	–	–	13.7–25.9	0–17.1	89.2–139.9	176.1–939.7	[58]
Cheese	Edam type (ng/g)	Finland	33	10.2	5.6	12.6	105	300	[53]
	Emmental type (ng/g)	Finland	52.3	ND	Traces	Traces	ND	Nd	[53]
Blue cheese	Blue cheese (ng/g)	France	–	–	14.4–35.4	24.6	59.8	189–230	[58]
	Blue cheese (ng/g)	England	–	–	96.7	223	103	301	[58]
Cheddar	Hard cheddar (ng/g)	England	–	–	8.7–29.9	0–23.1	10.5–61.8	0–66.9	[58]

Food	Type	Country	MK-4	MK-5	MK-6	MK-7	MK-8	MK-9	Reference
	Cheshire hard cheese (ng/g)	England	–	–	15.7	Nd	57.9	241	[58]
Leicester	Leicester hard cheese (ng/g)	France	–	–	20	21.5	47.6	162.4	[58]
Cheese	Appenzeller (ng/g)	Swiss	43–52	–	–	–	–	20	[59]
	Comte (ng/g)	France	55–84	–	–	–	–	52–60	[59]
	Emmental (ng/g)	Swiss	81–86	–	–	–	–	222–314	[59]
	Gruyere (ng/g)	Swiss	81–96	–	–	–	–	ND	[59]
	Jarlsberg (ng/g)	Norway	84	–	–	–	–	652	[59]
	Raclette (ng/g)	Swiss	50	–	–	–	–	47	[59]
Cheese	Chedder cheese (μg/100 g)	USA	10.2	–	–	–	–		[50]
	Swiss cheese (μg/100 g)	USA	7.8	–	–	–	–	–	[50]
	Mozzarella cheese (μg/100 g)		-	–	–	–	–	–	[50]
	Processed cheese ((μg/100 g)	Japan	5	–	–	0.3	–	–	[12]

Table 8. Menaquinones contents of various cheeses

Positive correlation was found in propionate concentration and viable propionibacterial cell count which is contributed toward the production of MK-9 in cheeses [59]. Earlier, different reports showed that components of menaquinones are varied among the types of cheeses. In this context, menaquinones concentration is better in Edam-type cheeses than Emmental-type cheeses-specific bacterial activity [53]. Starter culture has prime importance during the preparation of cheese with higher amount of menaquinones. Commonly, Swiss-type cheeses are prepared with *propionibacteria* and lactic acid bacteria, while Edam-type cheese are produced with the action of lactic acid bacteria only such as *Lc. lactis* ssp. *Cremoris* and *Lactococcus lactis* ssp. *Lactis* which are mainly responsible for the synthesis of MK-8 and MK-9 [60]. Moreover, *Propionibacterium freudenreichii* isolated from Swiss-type cheese has a potential equivalent to the *Bacillus subtilis* to produce menaquinone-9 in milk whey [61].

Long-chain menaquinones such as MK-6 to MK10 was not present in Comte hard cheese produced in France. Likewise, these menaquinones were not present in the Emmental hard cheese while some amount of MK-10 and MK-11 was detected. Interestingly, mozzarella cheese did not have any type of menaquinones because during its production process no fermentation is involved [58]. Accordingly, further research is required to evaluate the stability of menaquinones in cheese that are stored for a long time [59].

8.6. Egg

Similar to the other animal-based products, hen egg also contained considerable amount of MK-4, whereas MK-7 was not detected or not quantified by the researchers. Egg yolk contained greater concentration of the MK-4 (31.4–64 μg/100 g) than that of egg white (0.9–7 μg/100 g). Additionally, cooking also increased the MK-4 content in the egg might be due to loss of moisture content compared with whole fresh egg (**Table 9**). In this context, whole fried egg contained 9.0 μg/100 g and hard cooked whole egg contained 7.0 μg/100 g compared to fresh whole egg 5.6 μg/100 g [13, 21, 50].

Food	Type	Country	MK-4	MK-6	MK-7	Reference
Egg	yolk	Netherlands	31.4	0.7	–	[13]
Egg	albumen	Netherlands	0.9	–	–	[13]
Egg	Whole and raw	Japan	7	–	ND	[12]
Egg	Raw yolk	Japan	64	–	ND	[12]
Egg	White fresh raw	USA	0.4	–	–	[50]
Egg	Yolk fresh raw	USA	15.5	–	–	[50]
Egg	Whole and fresh	USA	5.6	–	–	[50]
Egg	Whole and fried	USA	9.0	–	–	[50]
Egg	Whole and hard cooked	USA	7.0	–	–	[50]

Other menaquinones were not detected or quantified.

MK-7 was not detected in fresh whole egg and raw egg yolk.

Table 9. Menaquinones contents of hen egg (μg/100 g).

8.7. Fast foods

Elder et al. reported that various fast food products including hamburgers, sandwiches, burrito, taco, pepperoni, and shakes contained MK-4 [50]. Regular hamburger contained lower amount of MK-4 which was subsequently increased by the addition of cheese or sauces or both in the hamburgers. Likewise, Chicken sandwich contained relatively higher amount of MK-4 as 2.7–10.6 μg/100 g than that of hamburger due to higher amount of chicken that possibly contained higher amount of MK-4. Burrito prepared with beans, beef, or chicken contained MK-4 ranged from 0.6 to 2.7 μg/100 g. Pepperoni contained almost similar amount MK-4 as present in the burrito. Shakes available in USA market including chocolate and vanilla also have some amount of MK-4 (**Table 10**).

Food	Type	Country	MK-4	Reference
Hamburger	Regular, with cheese, sauces, and both	USA	1.4–2.9	[50]
Sandwich	Prepared the various meat-based products chicken sandwich	USA	2.7–10.6	[50]
Sandwich	Fish sandwich	USA	0.3	[50]
Burrito	Burrito with bean, beef, and chicken	USA	0.6–2.7	[50]
Taco	Taco regular, with beef, chicken, or cheese	USA	1.0–4.5	[50]
Pepperoni	Pepperoni (regular, thin, and thick crust) or meat and vegetables	USA	1.9–2.1	[50]
Shakes	Shakes, chocolate, and vanilla	USA	3.4	[50]

Other menaquinones were not quantified in this study.

Table 10. Menaquinones contents of various fast foods (μg/100 g).

9. Bioavailability

The Food and Nutrition Board established the RDA level for vitamin K as 65 and 80 μg for adult women and men, respectively [62]. The adequate intake of vitamin K from food sources is relative higher about 120 μg/day and 90 μg/day for men and women, respectively [63, 64]. Neonates need approximately 2–2.5 μg/day of vitamin K that progressively increases up to 30–55 μg/day in children.

Both active forms of vitamin K, *that is,* phylloquinone and menaquinones, have similar absorption and assimilation pattern. However, relative few studies were carried out to estimate the absorption efficiency, including transportation, distribution, and cellular uptake of various menaquinones [10, 65]. Absorbed vitamin K is transported mainly through lymphatic system in chylomicrons to the hepatic tissue which is main storage organ for long-chain dietary forms of vitamin K [2].

Gut absorption of all dietary forms of vitamin K appears to occur through the common pathway like most of dietary lipids. Bile acids and pancreatic enzymes accelerate the solubility, emulsification, and assimilation of vitamin K into mixed micelles in digestive system. In enterocytes, vitamin K is attached with chylomicrons and enters in lymphatic circulation system. The bioavailability of vitamin K dietary forms is positively associated with dietary lipids and integrity of food matrix [66, 67].

It was reported that isoprenoid side chain length was changed during cellular uptake, transportation, and storage of long-chain menaquinones. Variations were observed in absorption and transportation of vitamin K dietary forms such as phylloquinone, menaquinones (MK-4 and MK-9) after equivalent amount administration of respective form. Postprandial plasma concentration and absorption of MKs are relatively less than phylloquinone, and its uptakes are more in tissues.

In contract to phylloquinone, which is principally fund in triglyceride-rich lipoproteins during postprandial as well as fasting condition, menaquinones are reallocated from triglycerides-rich lipoproteins to low-density lipoproteins (LDL) in and after postprandial consumption of vitamin K. Whereas shorter-chain menaquinones, *that is,* MK-4 were redistributed earlier and incorporated with high-density lipoproteins (HDLs) than MK-9. Moreover, only MK-4 is transported via high-density lipoproteins (HDLs) [10, 68].

The clearance of shorter-chain menaquinones was quick, while other was detected after days in plasma. Likewise, MK-7 has same plasma kinetics with higher half-life of 72 h than that of MK-4 and phylloquinone [9, 13, 69–71]. Nevertheless, no information of plasma kinetics is available of other long-chain menaquinones. High concentration of MK-4 was found in non-hepatic tissues of the body after the ingestion of phylloquinone. This might be due to the conversion of phylloquinone to MK-4. The exact phenomenon is still unclear; however, some researchers suggested that it was converted to other menaquinones via prenylation. In this context, deuterium labeled MK-4 was administrated to mice which are converted to MK-4 via integral side chain removal through prenylation. Likewise, ingestion of MK-7 also increases the serum MK-4 level considerably [2, 12, 17, 69, 71]. Contrary, in germ-free rats, MK-4 and phylloquinone content were increased in extrahepatic tissues through administration of their respective supplements after deficient condition. Moreover, MK-4 concentration was enhanced by phylloquinone administration. They also inferred that the conversion of phylloquinone in to MK-4 in extraheptic tissues did not require the intestinal bacterial population. This conversion is purely biochemical and remain unclear thus far [72]. Optimal daily vitamin K_2 especially MK-4 intake as well as sufficient serum concentration is required to activate Vitamin K-dependent proteins [27, 73].

10. Demographic study

Progressive administration of MK-7 momentously increased the level of plasma MK-7; nonetheless, MK-4 supplementation did not enhance the MK-4 concentration in healthy individuals. Therefore, lower dose of MK-7 (45–90 μg/day) is considered to be effective for ameliorating the physiological dysfunctions [69]. Prime circulating form of vitamin K is phylloquinone, whereas menaquinones (MK-9 to MK-13) are abundantly present in liver. Stored vitamin K is rapidly depleted from the body, and almost 60–70% of absorbed vitamin K is finally lost from body through urine (20%) and faces (40–50%) [74, 75].

Various demographic studies were carried out to estimate the level of circulating MK-7 level in normal and unhealthy subjects. In 1990, study was conducted in London to estimate the level of MK-7 in young and elderly normal subjects. They found that serum concentrations ranged from 0.293 to 0.328 ng/ml in healthy individuals. Jamal et al. also assessed the circulating concentration in patients with hip and vertebral fractures subjects and noticed less amount (0.039 and 0.148 ng/ml) compared to normal subjects [76]. Additionally, French young and elderly women had non-significant varied amount of MK-7 (0.221–0.241 ng/ml), whereas hip fracture old women had 0.120 ng/ml of MK-7.

Japanese healthy adults and vertebral fracture older women had 3.820 and 3.290 ng/ml of MK-7, whereas elderly normal women had significantly higher amount of MK-7 (6.260 ng/ml) might be due to higher consumption of natto which are rich with MK-7. However, postmenopausal women contained less amount of serum MK-7 (0.75–1.10 ng/ml) as compared to normal adults (1.214ng/ml). Moreover, postmenopausal women from Osaka Japan having lower bone mineral density (BMD) showed lesser amount of circulating MK-7 than that of normal BMD women [77].

Likewise, Kaneki et al. reported the serum levels of MK-7 in postmenopausal women from United Kingdom, Hiroshima, and Tokyo [78]. They inferred that Tokyo women had higher amount of MK-7 as 5.26 ng/ml followed by women lived in Hiroshima 1.221 ng/ml and the lowest concentration as 0.371 ng/ml was noticed in United Kingdom population. They also reported that natto intake has positive association with serum concentration of MK-7 in elderly women living in Tokyo, Japan. Serum MK-7 level was maximum (7.91 ng/ml) in women that consumed natto twice or more in a week, whereas MK-7 level was decreased as 2.81 and 0.873 ng/ml in women when the intake of natto was reduced once or less than once in a week, respectively.

Recently, Knapen et al. demonstrated that the intake of MK-7 fortified yoghurt momentously increased the plasma concentration of MK-7 from 0.28 to 1.66 ng/ml after 14 weeks of intervention in postmenopausal women and healthy men of aged 45–65 years from the Limburg, Netherlands [57]. The regular intake of MK-4 was momentously lowered (29%) cardiovascular problems in hemodialysis patients of Poland. Although lower serum value of MK-4 in hemodialysis subjects might be due to the less intake of vitamin K_2 or probably slow conversion rate of phylloquinone to menaquinones, MK-4 intake is positively related with the amount of protein and fat consumed [73].

In Japanese young women, average consumption of vitamin K was adequate 230.2/day and nearly 94% individuals consume adequate intake level of vitamin K. There mean daily intake of MK-4 and MK-7 was 16.9 and 57.4 μg/day, respectively. Both menaquinones, *that is,* MK-4 and MK-7 contribute almost 30% of the daily requirement of vitamin K in the body [12].

11. Conclusion

Vitamin K_2 is present in numerous in varied concentration of long-chain menaquinones and their types. Fermented soybean of all region of the world contained abundant concentration of MK-7 compared with other menaquinones. Fermentation process is facilitated by action of bacteria which attained the status of generally recognized as safe (GRAS) due to their non-pathogenicity. However, animal-based products such as fresh meat of caw, buffalo, other animals, milk, fish, and egg contained greater amount of MK-4 contents. Additionally, fermented milk-based products such as cheese, source milk, butter milk, and menophilic fermented milk contained ample amount of long-chain menaquinones and MK-4 content were limited in these products due to the bacterial action. Some non-fermented cheese did not contain any form of vitamin K_2. During the physiological functioning, MK-4 is converted in to MK-7 which is more effective to curtail the vitamin K deficiency–associated dilemma.

Acknowledgements

Authors are highly thankful to the Higher Education Commission (HEC), Pakistan for providing funds for the research project entitled "Nutritional and biochemical evaluation of vitamin K enriched dietary sources." We are also thankful to the Dr. Leon J Schurger (*Department of Biochemistry, Maastricht University the Netherlands*) for valuable suggestions during my studies regarding the vitamin K. Moreover, special thanks for the generous support of Pakistan Atomic Energy Commission for providing the facilities.

Author details

Muhammad Yasin[1*], Masood Sadiq Butt[2] and Aurang Zeb[1]

*Address all correspondence to: yasinft_uaf@yahoo.com

1 Food Science Division, Nuclear Institute for Food and Agriculture (NIFA) Peshawar, Pakistan Atomic Energy Commission, Pakistan

2 National Institute of Food Science and Technology, University of Agriculture, Faisalabad, Pakistan

References

[1] Yasin, M., Butt, M.S., Yasmin, A., Bashir, S., 2014. Chemical, antioxidant and sensory profiling of vitamin K-rich dietary sources. J. Korean Soc. Appl. Biol. Chem., 57(2), 153–160.

[2] Yasin, M., Butt, M.S., Anjum, F.M., Shahid, M., 2013. Nutritional and antioxidant profiling of vitamin K dietary sources. Pak. J. Nutr., 12(11), 996–1002.

[3] Data, S., Mwanga, J., Shearer, M., Harrington, D., Voong, K., Parlett, T., Wariyar, U., 2012. Prevalence and associated risk factors for vitamin K deficiency in mothers and their newborn babies in an East African setting. Arch. Dis. Child., 97, 90.

[4] McCann, J.C., Ames, B.N., 2009. Vitamin K, an example of triage theory: is micronutrient inadequacy linked to diseases of aging? Am. J. Clin. Nutr., 90, 889–907.

[5] Booth, S.L., Al-Rajabi, A., 2008. Determinants of vitamin K status in humans. Vitam. Horm., 78, 1–22.

[6] Kuwabara, A., Tanaka, K., Tsugawa, N., Nakase, H., Tsuji, H., Shide, K., Kamao, M., Chiba, T., Inagaki, N., Okano, T., Kido, S., 2009. High prevalence of vitamin K and D deficiency and decreased BMD in inflammatory bowel disease. Osteoporos Int., 20, 935–942.

[7] Booth, S.L., Martini, L., Peterson, J.W., Saltzman, E., Dallal, G.E., Wood, R.J., 2003. Dietary phylloquinone depletion and repletion in older women. J. Nutr., 133, 2565–2569.

[8] Schurgers, L.J., Teunissen, K.J., Hamulyák K, Knapen, M.H., Vik, H., Vermeer, C., 2007. Vitamin K containing dietary supplements: comparison of synthetic vitamin K_1 and natto-derived menaquinone-7. Blood, 109, 3279–3283.

[9] Novotny, J.A., Kurilich, A.C., Britz, S.J., Baer, D.J., Beverly, A., 2010. Vitamin K absorption and kinetics in human subjects after consumption of ^{13}C-labelled phylloquinone from kale. Br. J. Nutr., 104(6), 858–862.

[10] Shearer, M.J., Newman, P., 2008. Metabolism and cell biology of vitamin K. Thromb. Haemost., 100, 530–547.

[11] van Summeren, M.J., Braam, L.A., Lilien, M.R., Schurgers, L.J., Kuis, W., Vermeer, C., 2009. The effect of menaquinone-7 (Vitamin K_2) supplementation on osteocalcin carboxylation in healthy prepubertal children. Br. J. Nutr., 102(8), 1171–1178.

[12] Kamao, M., Suhara, Y., Tsugawa, N., Uwano, M., Yamaguchi, N., Uenishi, K., Ishida, H., Sasaki, S., Okano, T., 2007. Vitamin K content of foods and dietary vitamin k intake in Japanese young women. J. Nutr. Sci. Vitaminol., 53, 464–470.

[13] Schurgers, L.J., Vermeer, C., 2000. Determination of phylloquinone and menaquinones in food; effect of food matrix on circulating vitamin K concentrations. Haemostasis, 30, 298–307.

[14] Kwak, C.S., Lee, M.S., Park, S.C., 2007. Higher antioxidant properties of chungkookjang, a fermented soybean paste, may be due to increased aglycone and malonylglycoside isoflavone during fermentation. Nutr. Res., 27(11), 719–727.

[15] Narumi, S., Sasaki, M., Okudera, D., 1998. Postoperative abnormal prothrombinemia in patients with cefoperazone: report of two cases. Surg. Today, 28, 227–230.

[16] Iwamotoa, J., Satob, Y., Takedaa, T., Matsumotoa, H., 2009. High dose vitamin K supplementation reduces fracture incidence in postmenopausal women: a review of the literature. Nutr. Res. 29(4), 221–228.

[17] Okano, T., Shimomura, Y., Yamane, M., Suhara, Y., Kamao, M., Sugiura, M., 2008. Conversion of phylloquinone (Vitamin K_1) into menaquinone-4 (Vitamin K_2) in mice: two possible routes for menaquinone-4 accumulation in cerebra of mice. J. Biol. Chem., 283, 11270–11279.

[18] Tamang, J.R., Chettri, Rajen, Sharma, Rudra Mani, 2009. Indigenous knowledge of Northeast women on production of ethnic fermented soybean foods. Indian J. Tradit. Knowl., 8(1), 122–126.

[19] Hu, Y., Ge, C., Yuan, W., Zhu, R., Zhang, W., Due, L., Xue, J., 2010. Characterization of fermented black soybeans Natto inoculated with Bacillus Natto during fermentation. J. Sci. Food Agric., 90, 1194–1202.

[20] Shrestha, A.K., Dahal, N.R., Ndungutse, V., 2010. Bacillus fermentation of soybean: a review. J. Food Sci. Technol. Nepal, 6, 1–9.

[21] Juan, M.Y., Chou, C.C., 2010. Enhancement of antioxidant activity, total phenolic and flavonoid content of black soybeans by solid state fermentation with *Bacillus subtilis* BCRC 14715. Food Microbiol., 27, 586–591.

[22] Dajanta, K., Janpum, P., Leksing, W., 2013. Antioxidant capacities, total phenolics and flavonoids in black and yellow soybeans fermented by *Bacillus subtilis*: a comparative study of Thai fermented soybeans (thua nao). Int. Food Res. J. 20(6), 3125–3132.

[23] Wu, W.J., Ahn, B.Y., 2011. Improved menaquinone (Vitamin K2) production in cheonggukjang by optimization of the fermentation conditions. Food Sci. Biotechnol., 20(6), 1585–1591.

[24] Berenjian, A., Mahanama, R., Talbot, A., Regtop, H., Kavanagh, J., Dehghani, F., 2014. Designing of an intensification process for biosynthesis and recovery of menaquinone-7. Appl. Biochem. Biotechnol., 172, 1347–1357.

[25] Berenjian, A., Raja, M., Peter, V., Andrea, T., Ray, B., Hubert, R., John, K., Fariba, D., 2011. Extraction of menaquinone-7 for supplementation of food. In: Chemeca 2011: Engineering a Better World: Sydney Hilton Hotel, NSW, Australia, 18–21 September 2011. Barton, A.C.T.: Engineers Australia, pp. 1616–1625. Availability: <http://search.informit.com.au>.

[26] Booth, S.L., 2012. Vitamin K: food composition and dietary intakes. Food Nutr. Res., 56, 5505.

[27] Dajanta, K., Apichartsrangkoon, A., Chukeatirote, E., 2011. Antioxidant properties and total phenolics of Thua Nao (a Thai fermented soybean) as affected by Bacillus-fermentation. J. Microbial Biochem. Technol., 3, 56–59.

[28] Obeid A.E.F.E., Alawad, Aisha Mudawi, Ibrahim, Hanan Moawia, 2015. Isolation and characterization of bacillus subtillus with potential production of nattokinase. Int. J. Adv. Res., 3(3), 94–101.

[29] Sato, T., Yamada, Y., Ohtani, Y., Mitsui, N., Murasawa, H., Araki, S., 2001. Production of menaquinone (Vitamin K2)-7 by *Bacillus subtilis*. J. Biosci. Bioeng., 91(1), 16–20.

[30] Esteves, E.A., Martino, H.S.D., Olivieira, F.C.E., Bressan, J., Costa, N.M.B., 2010. Chemical composition of soybeans cultivar lacking lipoxygenases. Food Chem., 122, 238–242.

[31] Ravishankar, B., Dound, Y.A., Mehta, D.S., Ashok, B.K., de Souza, A., Pan, M-H., Ho, C-T., Badmaev, V., Vaidya, A.D.B., 2014. Safety assessment of menaquinone-7 for use in human nutrition. J. Food Drug Anal. doi:10.1016/j.jfda.2014.03.001.

[32] Puri, A., Iqubal, M., Zafar, R., Panda, B.P., 2015. Influence of physical, chemical and inducer treatments on menaquinone-7 biosynthesis by *Bacillus subtilis* MTCC 2756. Songklanakarin J. Sci. Technol., 37(3), 283–289.

[33] Tan, M., Liu, H., Li, Z., Sun, X., Zheng, Z., Zhao, G., 2016. Optimization medium composition for vitamin K2 by *Flavobacterium* sp. using response surface methodology and addition of *Arachis hypogaea*. Braz. Arch. Biol. Technol., 59, e16150343.

[34] Premarani, T., Chhetry, G.K.N., 2011. Nutritional analysis of fermented soybean (Hawaijar). Assam. Uni. J. Sci. Technol. 7(1), 96–100.

[35] Li, H., Feng, F.Q., Shen, L.R., Xie, Y., Li, D., 2007. Nutritional evaluation of different bacterial douche. Asia Pac. J. Clin. Nutr., 16(S1), 215–221.

[36] Jeff-Agboola, Y.A., Oguntuase, O.S., 2006. Effect of *Bacillus sphaericus* on proximate composition of soybean (*Glycine max*) for the production of soy iru. Pak. J. Nutr., 5(6), 606–607.

[37] Wei, Q., Chang, S.K.C., 2004. Characteristics of fermented natto products as affected by soybean cultivars. J. Food Process. Preserv., 28(4), 251–273.

[38] Ang C.Y.W., Liu K., Huang Y., 1999. Asian Foods Science and Technology. Technomic Publishing Company Inc., pp. 139–195.

[39] Kiuchi K., Ohta O., Itoh H., Takahayahsi T., Ebine H., 1976. Studies on lipids of Natto. J. Agric. Food Chem., 55, 271–272.

[40] Ginting E., Arcot J., 2004. High-performance liquid chromatographic determination of naturally occurring folates during tempeh preparation. J. Agric. Food Chem., 52, 7752–7758.

[41] Sarkar P.K., Morrison E., Tinggi U., Somerset S.M., Craven G.S., 1998. B-group vitamin and mineral contents of soybeans during Kinema production. J. Sci. Food Agric., 78(4), 498–502.

[42] Nikkuni, S., Karki, T.B., Vilkhu, K.S., Suzukl, T., Shindoh, K., Suzuk, C., Okada, N., 1995. Mineral and amino acid contents of kinema, a fermented soybean food prepared in Nepal. Food Sci. Technol. Int., 1(2), 107–111.

[43] Iwuoha, C.I., Eke, O.S., 1996. Nigerian indigenous fermented foods, their traditional process operation, inherent problems, improvements and current status. Food Res. Int., 29(5–6), 527–540.

[44] Moktan, B., Saha, J., Sparker, P.K., 2008. Antioxidant activities of soybean as affected by *Bacillus*-fermentation to cinema. Food Res. Int. 41(6), 586–593.

[45] Sanjukta, S., Rai, A.K., 2016. Production of bioactive peptides during soybean fermentation and their potential health benefits. Trends Food Sci. Technol., 50, 1–10.

[46] Kim, N.Y., Song, E.J., Kwon, D.Y., Kim, H.P., Heo, M.Y., 2008. Antioxidant and antigenotoxic activities of Korean fermented soybean. Food Chem. Toxicol., 46(3), 1184–1189.

[47] Lee, N.K., Sowa, H., Hinoi, E., Ferron, M., Ahn, J.D., 2007. Endocrine regulation of energy metabolism by the skeleton. Cell, 130, 456–469.

[48] Rødbotten, R., Gundersen, Thomas, Vermeer, Cees, Kirkhus, Bente, 2014. Vitamin K2 in different bovine muscles and breeds. Meat Sci., 97, 49–53.

[49] Fujiwara, K., Miyaguchi, Y., Feng, X.H., Toyoda, A., Nakamura, Y., Yamazaki, M., Nakashima, K., Abe, H., 2008. Effect of fermented soybean, "Natto" on the production and qualities of chicken meat. Asian-Australasian J. Anim. Sci., 21(12), 1766–1772.

[50] Elder, S.J., Haytowitz, D.B., Howe, J.R., Peterson, J.W., Booth, S.L., 2006. Vitamin K contents of meat, dairy, and fast food in the US diet. J. Agric. Food Chem., 54(2), 463–467.

[51] Hirauchi, K., Sakano, T., Notsumoto, S., Nagaoka, T., Morimoto, A., Fujimoto, K., Masuda, S., Suzuki, Y., 1989. Measurement of K vitamins in animal tissues by high-performance liquid chromatography with fluorimetric detection. J Chromatogr., 497, 131–137.

[52] Matschiner, J.T., Amelotti, J.M., 1968. Characterization of vitamin K from bovine liver. J. Lipid Res., 9(2), 176–179.

[53] Koivu-Tikkanen, T.J., Ollilainen, V., Piironen, V.I., 2000. Determination of phylloquinone and menaquinones in animal products with fluorescence detection after postcolumn reduction with metallic zinc. J. Agric. Food Chem., 48, 6325–6331.

[54] Gentili, A., Caretti, Fulvia, Bellante, Simona, Ventura, Salvatore, Canepari, Silvia, Curini, Roberta, 2013. Comprehensive profiling of carotenoids and fat-soluble vitamins in milk from different animal species by LC-DAD-MS/MS hyphenation. J. Agric. Food Chem., 61, 1628–1639.

[55] Morishita, T., Tamura, N., Makino, T., Kudo, S., 1999. Production of menaquinones by lactic acid bacteria. J. Dairy Sci. 82, 1897–1903.

[56] Canfielda, L.M., Hopklnsonb, Judy M., Limal, Anne F., Martin, Gall S., Sugimoto, Kyoto, Burr, Jeanne, Clark, Larry, McGeea, Daniel L., 1990. Quantitation of vitamin K in human milk. Lipids 25, 406–411.

[57] Knapen, M.H.J., Braam, L.A.J.L.M., Teunissen, K.J., van't Hoofd, C.M., Zwijsen, R.M.L., van den Heuvel, E.G.H.M., Vermeer, C, 2016. Steady-state vitamin K2 (menaquinone-7) plasma concentrations after intake of dairy products and soft gel capsules. Eur. J. Clin. Nutr., 1–6.

[58] Manoury, E., Jourdon, K., Boyaval, P., Fourcassié, P., 2013. Quantitative measurement of vitamin K2 (menaquinones) in various fermented dairy products using a reliable high-performance liquid chromatography method. J. Dairy Sci., 96(3), 1335–1346.

[59] Hojo, K., Watanabe, R., Mori, T., Taketomo, N., 2007. Quantitative measurement of tetrahydromenaquinone-9 in Cheese. Fermented by propionibacteria. J. Dairy Sci. 90, 4078–4083.

[60] Morishita, T., Tamura, N., Makino, T., Kudo, S., 1999. Production of menaquinones by lactic acid bacteria. J. Dairy Sci., 82, 1897–903.

[61] Furuichi, K., Hojo, K., Katakura, Y., Ninomiya, K., Shioya, S., 2006. Aerobic culture of *Propionibacterium freudenreichii* ET-3 can increase production ratio of 1,4-dihydroxy-2-naphthoic acid to menaquinone. J. Biosci. Bioeng. 101, 464–470.

[62] Institute of Medicine, Food and Nutrition Board, 2001. Dietary Reference Intake for Vitamin A, Vitamin K, Arsenic, Boron, Chromium, Copper, Iodine, Iron, Manganese,

Molybdenum, Nickel, Silicon, Vanadium and Zinc: A Report of the Panel on Micro-Nutrients. National Academies Press, Washington, DC.

[63] Food and Nutrition Board, Institute of Medicine, 2001. Vitamin K. In: Dietary Reference Intakes for Vitamin A, Vitamin K, Arsenic, Boron, Chromium, Copper, Iodine, Iron, Manganese, Molybdenum, Nickel, Silicon, Vanadium, and Zinc. National Academy Press, Washington, DC, pp. 162–196.

[64] Schurgers, L.J., Shearer, M.J., Hamulya, K., Stocklin, E., Vermeer, C., 2004. Effect of vitamin K intake on the stability of oral anticoagulant treatment: dose response relationships in healthy subjects. Blood, 104(9), 2682–2689.

[65] Shearer, M.J., Fu, X., Booth, S.L., 2012. Vitamin K nutrition, metabolism, and requirements: current concepts and future research. Adv. Nutr., 3, 182–195.

[66] Booth, S.L., Lichtenstein, A.H., Dallal, G.E., 2002. Phylloquinone absorption from phylloquinone-fortified oil is greater than from a vegetable in younger and older men and women. J. Nutr., 132, 2609–2612.

[67] Jones, K.S., Bluck, L.J., Wang, L.Y., Stephen, A.M., Prynne, C.J., Coward, W.A., 2009. The effect of different meals on the absorption of stable isotopelabelled phylloquinone. Br. J. Nutr., 102, 1195–1202.

[68] Schurgers, L.J., Vermeer, C., 2002. Differential lipoprotein transport pathways of K-vitamins in healthy subjects. Biochim. Biophys. Acta, 1570, 27–32.

[69] Sato, T., Schurgers, L.J., Uenishi, K., 2012. Comparison of menaquinone-4 and menaquinone-7 bioavailability in healthy women. Nutr. J., 11, 93.

[70] Schurgers, L.J., Teunissen, K.J., Hamulyak, K., Knapen, M.H., Vik, H., Vermeer, C., 2007. Vitamin K-containing dietary supplements: comparison of synthetic vitamin K1 and natto-derived menaquinone-7. Blood, 109, 3279–3283.

[71] Walther, B., Karl, J. Philip, Booth, Sarah L., Boyaval, Patrick, 2013. Menaquinones, bacteria, and the food supply: the relevance of dairy and fermented food products to vitamin K requirements. Adv. Nutr. 4, 463–473.

[72] Ronden, J.E., Vermeer, C., Thijssen, H.H.W., 1998. Intestinal flora is not involved in the phylloquinone–menaquinone-4 conversion in the rat. Biochem. Biophys. Acta, 1379, 69–75.

[73] Katarzyna, W., Żak-Gołąb, A., Łabuzek, K., Suchy, D., Ficek, R., Pośpiech, K., Olszanecka-Glinianowicz, M., Okopień Boguslaw, M., Więcek, A., Chudek, J., 2015. Daily intake and serum concentration of menaquinone-4 (MK-4) in hemodialysis patients with chronic kidney disease, Clin. Biochem. doi:10.1016/j.clinbiochem.2015.08.011.

[74] Shearer, M.J., McBurney, A., Barkhan, P., 1974. Studies on the absorption and metabolism of phylloquinone (vitamin K1) in man. Vitam. Horm., 32, 513–542.

[75] Usui, Y., Tanimura, H., Nishimura, N., Kobayashi, N., Okanoue, T., Ozawa, K., 1990. Vitamin K concentrations in the plasma and liver of surgical patients. Am. J. Clin. Nutr., 51, 846–852.

[76] Jamal, S.A., Browner, W.S., Bauer, D.C., et al., 1998. Warfarin use and risk for osteoporosis in elderly women. Study of Osteoporotic Fractures Research Group. Ann. Intern. Med., 128, 829.

[77] Kanai, T., Takagi, T., Masuhiro, K., et al., 1997. Serum vitamin K level and bone mineral density in post-menopausal women. Int. J. Gynaecol. Obstet., 56, 25.

[78] Kaneki, M., Hodges, S.J., Hosoi, T., Fujiwara, S., Lyons, A., Crean, S.J., Ishida, N., Nakagawa, M., Takechi, M., Sano, Y., Mizuno, Y., Hoshino, S., Miyao, M., Inoue, S., Horiki, K., Shiraki, M., Ouchi, Y., Orimo, H., 2001. Japanese fermented soybean food as the major determinant of the large geographic difference in circulating levels of vitamin K2: possible implications for hip-fracture risk. Nutrition, 17(4), 315–21.

[79] Shrestha A.K., Noomhorm A., 2001. Composition and functional properties of fermented soybeans flour (Kinema). J. Food Sci. Technol., Mysore, 38(5), 467–470.

Chapter 4

Menaquinones, Bacteria, and Foods: Vitamin K2 in the Diet

Barbara Walther and Magali Chollet

Additional information is available at the end of the chapter

http://dx.doi.org/ 10.5772/63712

Abstract

Vitamin K2 is a collection of isoprenologues that mostly originate from bacterial synthesis, also called menaquinones (MKs). Multiple bacterial species used as starter cultures for food fermentation are known to synthesize MK. Therefore, fermented food is the best source of vitamin K2. In the Western diet, dairy products are one of the best known and most commonly consumed group of fermented products.

Although intensive research on metabolism and the biological effect of vitamin K2 continues today, data about vitamin K2 production and content in foods remain scarce. Dietary recommendations are still based on the classic role of vitamin K as an enzyme cofactor for coagulation proteins and do not consider differences in bioavailability and bioactivity between the various MKs and the possibly higher requirements for health effects apart from coagulation, such as bone or cardiovascular health. Here, we provide a global view of foods rich in vitamin K2 and their interactions together with other nutrients in selected health effects such as bone and cardiovascular health.

Keywords: Menaquinones, Bacteria, Food, Dairy, Health

1. Introduction

Vitamin K occurs naturally in two biologically active forms. Vitamin K1, also called phylloquinone (PK), is abundant in leafy green vegetables, such as cabbage, spinach, and lettuce [1]. The other form, vitamin K2, is called menaquinone (MK) and is predominantly of microbial origin [2, 3]. Vitamin K2 is mainly present in fermented food such as cheese and natto (fermented soybeans), but gut microbiota are also able to synthesize vitamin K2 [4]. One exception, menaquinone-4 (MK-4), is formed in humans and animals by tissue-specific

INTECH
open science | open minds

© 2017 The Author(s). Licensee InTech. This chapter is distributed under the terms of the Creative Commons Attribution License (http://creativecommons.org/licenses/by/3.0), which permits unrestricted use, distribution, and reproduction in any medium, provided the original work is properly cited.

conversion of PK and/or menadione [5]. However, in the literature, all MKs are mostly grouped under the term vitamin K2 resulting in the assumption that all MKs are similar in origin and function. Moreover, despite the knowledge that MKs are present in the food supply, little is known about their individual synthesis, growth conditions, and interactions of the producing bacteria and the total amounts of the different MKs in fermented foods. Regarding the findings that MKs play an important role in health aspects beyond coagulation, study of the interaction of MKs with other nutrients may lead to a better understanding of the effect of different food items on health aspects, such as bone health or cardiovascular health.

Such a global view could be essential for guiding the development of dietary intake recommendations for vitamin K.

2. Structure of vitamin K

Both vitamin K forms have 2-methyl-1,4-naphthoquinone, also called menadione or vitamin K3, as a common ring structure. However, they differ from each other in the length and degree of saturation of the polyisoprenoid side chain attached to the 3-position.

Phylloquinone (vitamin K1) possesses a phytyl side chain, which consists of four isoprene units, and one of them is unsaturated. Phylloquinone is found primarily in plants in association with chlorophyll, whereas menaquinone (vitamin K2) is principally synthesized by bacteria. Menaquinone contains side chains of varying length, for most the part of a polymer of repeating unsaturated 5-carbon prenyl units. Depending on the microorganism by which the chain is synthesized, the chain length generally ranges from 4 to 13 prenyl units. Menaquinones are classified according to the number of prenyl units. The number of units is given in a suffix (-*n*), that is, menaquinone-*n* and often abbreviated as MK-*n* [2, 6, 7]. Some bacteria produce isoprenologues in which one or more of the prenyl units are saturated. The additional hydrogen atoms are indicated with the prefix dihydro-, tetrahydro-, and so on and are abbreviated MK-*n*(H2), MK-*n*(H4), etc. [8].

Vitamin K is fat soluble. The melting points of menaquinones vary from 35°C to 62°C depending on the length of the multiprenyl side chain. Menaquinones are stable to heat and air but are very sensitive to alkali and ultraviolet (UV) irradiation [9].

3. Functions and biosynthesis of menaquinones in bacteria

The distribution of isoprenoid quinones has been studied in 900 microbial strains, 56 mold strains, and 88 yeast strains. About half of the studied bacteria contain menaquinone, but no menaquinones have been found in molds and yeast [10]. Menaquinone and demethylmenaquinones (DMKs) are found in the cytoplasmic membrane of bacteria. MKs and DMKs function as a reversible redox component of the electron transfer chain [11]. Additionally, reduced MKs exhibit antioxidant properties and can play a role in protecting cellular membranes from lipid oxidation [12]. Menaquinones are also necessary for sporulation and proper regulation of cytochrome formation in some Gram-positive bacteria such as *Lactobacillus subtilis* [13]. In particular, the food industry uses various lactic acid bacteria (LAB) as starter cultures to produce fermented milk products, meat products, and vegetables. As many LAB lack a heme biosynthesis pathway, which results in an incomplete electron transport chain, the addition of menaquinone to the media facilitates aerobic growth, improves yield, and reduces production costs [14].

Menaquinone synthesis has mostly been described in *Escherichia coli, Mycobacterium phlei,* and *Bacillus subtilis*. In *E. coli,* chorismate from the shikimate pathway is converted into the naphthoquinone ring by six enzymes (MenFDHCEB) [11, 15]. The isoprenoid side chain is synthesized separately and is joined to the naphthoquinone ring to form demethylmenaquinone. Prenylation and methylation catalyzed by polyprenyltransferase (MenA) and methyltransferase (MenG) are the last steps of the synthesis of menaquinone [16, 17]. An alternative pathway, called the futalosine pathway, was described in microorganisms that lack *men* genes. In this pathway, chorismate is converted to menaquinone with four enzymes encoded by *mqnABCD* genes and unknown enzymes [17–19]. The majority of the bacteria containing the classical menaquinone pathway are obligately or facultatively aerobic, and the majority of menaquinone in anaerobic bacteria is synthesized via the futalosine pathway [17]. For example, the metabolic pathway of *Lactoccocus lactis,* which is used as a cheese starter, can function through aerobic and anaerobic reactions, and the *men* genes for the synthesis of menaquinone were detected in its genome.

Menaquinones have side chains of different sizes in different organisms and sometimes even within the same organism. Depending on the growing conditions, the basic structure can be modified by demethylation of the naphthoquinone ring to reform DMK or by saturation of the isoprenoid side chain [2, 19].

4. Non-dietary sources of menaquinones

Bacterially synthesized menaquinones that contribute to human vitamin K2 requirements may be produced by the gut microbiota or by bacteria present in food. In humans, the most important genera of intestinal flora are *Bacteroides* and *Bifidobacteria*. However, only *Bacteroides* can synthesize menaquinone. The major forms produced by *Bacteroides* are MK-10 and MK-11. MK-6 produced by *Eubacterium lentum,* MK-7 produced by *Veillonella,* and MK-8 produced by *Enterobacter* were also found in isolates from intestinal flora [2, 7, 8, 20]. Most menaquinones are present in the distal colon, but the most promising site of absorption is the terminal ileum, where there are menaquinone-producing bacteria and bile salts that are needed for solubilization of menaquinones [7, 21]. Therefore, although intestinal microflora synthesize large amounts of menaquinones, the bioavailability of bacterial menaquinone is poor, and diet is the major source of functionally available vitamin K2 [3, 7, 8]. Recent studies also showed that a short-term decrease in dietary vitamin K intake is not compensated by intestinal menaquinones [22–24].

5. Dietary sources of menaquinones

Vitamin K2 is mostly synthesized by bacteria; therefore, the highest number of long-chain menaquinones is found in fermented dairy products, such as cheese and fermented vegetables, such as natto and sauerkraut [16]. One exception is MK-4, which is formed by a realkylation step from menadione present in animal feed or as a product of tissue-specific conversion directly from dietary phylloquinone [5]. The extent of the conversion to MK-4 is estimated to range from 5% to 25% of the ingested phylloquinone [25].

Searching for information about the concentration of vitamin K2 in food is not very fruitful. Out of more than 70 national food databases, only 12 provide the vitamin K content of food items. Only three of these food databases (the United States, the Netherlands, Turkey) specifically report the vitamin K2 concentration; all others publish only phylloquinone (PK) or total vitamin K or give no further information about the vitamin forms included in the given values. The comparison of the provided concentration of MK in these three databases is not possible because the values are based on different specifications and different processes. The data given in the US database are for MK-4. However, the Dutch database includes several types of menaquinones, ranging from MK-4 to MK-10. For the data from the Turkish database, there is no information concerning the definition of vitamin K2 [16, 26]. In countries where animals are supplemented with menadione as practiced in the United States [27] and the Netherlands [28], the MK-4 concentration is normally higher in food of animal origin. The supplementation practice used in Turkey is unknown. Last, the process and the bacterial strains used in the production of fermented food determine the concentration and forms of MK in products [16].

Scanning the literature for publications that report the results of vitamin K measurements in food provides additional separate values for different menaquinones. However, information about longer-chain menaquinones (MK-5 to MK-10) is very limited. **Table 1** summarizes the values of vitamin K2 for animal products such as dairy, meat, fish and eggs, and fermented vegetable products such as bread, sauerkraut, and legumes (natto).

Menaquinone content (μg/100 g; mean ± SD or range)								
Food	**MK-4**	**MK-5**	**MK-6**	**MK-7**	**MK-8**	**MK-9**	**MK-10**	**Source**
Dairy								
Whole milk	0.7–0.9	0.0–0.1	nd	nd	nd	nd	nr	[6]
Whole milk	0.8–1.0	nr	nr	nr	nr	nr	nr	[27]
Whole milk	2 ± 0.3	nr	nr	nd	nr	nr	nr	[32]
Whole milk	0.4–1.0	nr	nr	nr	nr	0–2	nr	[29]
Milk 1% fat	0.3–0.4	nr	nr	nr	nr	nr	nr	[27]
Milk 2% fat	0.4–0.5	nr	nr	nr	nr	nr	nr	[27]
Whipped cream	5.2–5.6	nd	nd	nd	nd	nd	nr	[6]

Menaquinone content (μg/100 g; mean ± SD or range)								
Food	**MK-4**	**MK-5**	**MK-6**	**MK-7**	**MK-8**	**MK-9**	**MK-10**	**Source**
Cream	8 ± 3	nr	nr	nd	nr	nr	nr	[32]
Butter	13.5–15.9	nd	nd	nd	nd	nd	nr	[6]
	21 ± 7	nr	nr	nd	nr	nr	nr	[32]
Fermented milk								
Whole milk, sour	0.6 ± 0.02	0.3 ± 0.002	0.2 ± 0.03	0.4 ± 0.04	2.0 ± 0.1	4.7 ± 0.2	nd	[31]
Buttermilk	0.2–0.3	0.1–0.2	0–0.2	0.1–0.3	0.5–0.6	1.2–1.6	nr	[6]
Mesophilic	nr	nr	4.2	5	25.9	100.8	8.5	[34]
Thermophilic	nd	nd	nd	nd	nd	nd	nd	[34]
Yogurt								
Whole	0.4–1.0	nr	nr	nr	nr	0–2.0	nr	[29]
Whole	0.5–0.7	0–0.2	nd	nd	nd	nd	nr	[6]
Whole	1 ± 0.1	nr	nr	0.1 ± 0.2	nr	nr	nr	[32]
Plain	0.4 ± 0.03	0.1 ± 0.006	nd	nd	nd	nd	nd	[31]
Skimmed	nd	nd	nd	nd	0–0.2	nd	nr	[6]
Cheese								
Curd	0.3–0.6	0–0.2	0.1–0.3	0.2–0.5	4.8–5.4	18.1–19.2	nr	[6]
Curd	2–10	nr	nr	nr	nr	40–70	nr	[29]
Hard	4.2–6.6	1.3–1.7	0.6–1.0	1.1–1.5	14.9–18.2	45.3–54.9	nr	[6]
Semi-hard	nr	nr	1.9	1.1	3.9	17.5	4.7	[34]
Soft	3.3–3.9	0.2 –0.4	0.5–0.7	0.9 – 1.1	10.7–12.2	35.1–42.7	nr	[6]
Soft	nr	nr	1.7	1.2	7.0	27.3	2.9	[34]
Processed	5 ± 2	nr	nr	0.3 ± 0.1	nr	nr	nr	[32]
Blue cheese	nr	nr	4.9	12.4	7.7	19.3	2.9	[34]
Appenzeller	4.3–5.2	nr	nr	nr	nr	nr	nr	[33]
Caerphilly	nr	nr	1.6 ± 0.1	nd	1.6 ± 0.1	32.4 ± 0.8	nd	[34]
Cheddar	10.2	nr	nr	nr	nr	nr	nr	[27]
Cheddar	nr	nr	2.2	2.1	3.2	12.9	5.2	[34]
Cheshire	nr	nr	1.6± 0.2	nd	5.8 ± 0.2	24.2 ± 0.4	nd	[34]
Comté	5.5–8.4	nr	nr	nr	nr	nr	nr	[33]
Comté	nd	nd	nd	nd	nd	nd	nd	[34]
Edam	3.3 ± 0.2	1.0 ± 0.1	0.6 ± 0.1	1.3 ± 0.1	10.5 ± 0.8	30.0 ± 2.6	0.9 ± 0.1	[31]
Emmental	8.1–8.6	nr	nr	nr	nr	nr	nr	[33]
Emmental	nr	nr	nd	nd	nd	nd	4.0	[34]

Menaquinone content (μg/100 g; mean ± SD or range)								
Food	**MK-4**	**MK-5**	**MK-6**	**MK-7**	**MK-8**	**MK-9**	**MK-10**	**Source**
Aged 90 d	5.2 ± 0.1	nd	trace	trace	nd	nd	nd	[31]
Aged 180 d	6.1 ± 0.5	nd	trace	nd	nd	nd	nd	[31]
Gamalost	1.0 ± 0.0	0.6 ± 0.0	0.3 ± 0.0	0.9 ± 0.1	4.8 ± 0.7	42.3 ± 7.0	2.1 ±0.4	[35]
Jarlsberg	8.4	nr	nr	nr	nr	nr	nr	[33]
Gruyère	8.1–9.6	nr	nr	nr	nr	nr	nr	[33]
Leicester	nr	nr	2.0 ± 0.1	2.1 ± 0.1	4.8 ± 0.2	16.2 ± 0.3	4.4 ± 0.2	[34]
Mozzarella	3.1–4.0	nr	nr	nr	nr	nr	nr	[27]
Mozzarella	nd	nd	nd	nd	nd	nd	nd	[34]
Norvegia	5.1 ± 0.9	nd	0.3 ± 0.1	1.3 ± 0.2	5.3 ± 0.5	29.6 ± 3.6	nd	[35]
Raclette	5	nr	nr	nr	nr	nr	nr	[33]
Swiss cheese	6.2–8.8	nr	nr	nr	nr	nr	nr	[27]
Meat								
Salami	8.2–10.1	nd	nd	nd	nd	nd	nr	[6]
Calf liver	1.1–8.9	nr	nr	nr	nr	nr	nr	[27]
Beef liver	0.4 ± 0.4	nr	nr	nr	nr	nr	nr	[27]
Bovine liver	6.8 ± 1.03	nd	9.44 ± 0.118	25.6 ± 0.59	13.8 ± 0.55	9.8 ± 0.7	14±1.7	[31]
Beef liver	0.8	nr	2.5	18.2	4.8	1.5	6.6	[30]
Pork liver	0.3–0.4	nd	nd	nd	nd	nd	nd	[6]
Pork liver	10.8 ± 1.44	nd	nd	16 ± 2.7	25 ± 5.2	6 ± 1.8	8±2.9	[31]
Pork liver	0.6	nd	0.04	0.6	0.5	0.3	0.5	[30]
Chicken liver	14.1 ± 2.0	nr	nr	nr	nr	nr	nr	[27]
Chicken liver	4	nr	0.03	nd	0.09	0.04	0.03	[30]
Beef kidney	2.1	nr	0.08	0.2	0.01	nd	0.1	[30]
Pork kidney	1.3	nr	0.02	0.07	0.05	0.22	0.24	[30]
Chicken kidney	5	nr	nd	nd	nd	nd	nd	[30]
Beef muscle	3.4	nr	0.03	0.03	nr	nr	nr	[30]
Pork thigh	6 ± 2	nr	nr	nr	nr	nr	nr	[32]
Pork steak	1.7–2.4	nd	nd	0.4–0.7	0.9–1.2	nd	nd	[6]
Pork chop	3.1 ± 0.46	nd	nd	0.12 ± 0.035	nd	nd	nd	[31]
Pork muscle	0.9	nr	0.03	0.03	nr	nr	nr	[30]
Chicken breast	6.4–11.3	nd	nd	nd	nd	nd	nd	[6]
Chicken leg	5.8–10.5	nd	nd	nd	nd	nd	nd	[6]
Chicken thigh	27 ± 15	nr	nr	nd	nr	nr	nr	[32]

Menaquinone content (μg/100 g; mean ± SD or range)								
Food	**MK-4**	**MK-5**	**MK-6**	**MK-7**	**MK-8**	**MK-9**	**MK-10**	**Source**
Chicken meat, leg and thigh	60 ± 8.2	nd	nd	nd	nd	nd	nd	[31]
Chicken muscle	8.9	nr	nd	nd	nr	nr	nr	[30]
Fish								
Rainbow trout, cultivated	3.1 ± 0.2	0.09 ± 0.019	nd	0.2 ± 0.058	nd	nd	nd	[31]
Pike perch	0.2 ± 0.025	0.05 ± 0.0044	0.05 ± 0.0008	0.5 ± 0.13	nd	nd	nd	[31]
Baltic herring	0.21 ± 0.002	nr	nd	nd	nd	nd	nd	[31]
Horse mackerel	0.6 ± 0.1	nr	nr	nd	nr	nr	nr	[32]
Mackerel	1 ± 0.2	nr	nr	nd	nr	nr	nr	[32]
Mackerel	0.3–0.5	nd	nd	nd	nd	nd	nr	[6]
Salmon	0.2–0.3	nr	nr	nr	nr	nr	nr	[27]
Plaice	0.1–0.3	nd	0.2–0.3	0.0–0.1	1.3–1.8	nr	nr	[6]
Eel	1.4–2.1	nd	0.0–0.2	0.2–0.6	nd	nd	nr	[6]
Salmon	0.4–0.6	nd	nd	nd	nd	nd	nr	[6]
Eggs								
Egg yolk	29.1–33.5	nd	0.6–0.8	nd	nd	nd	nr	[6]
Egg albumen	0.8–1.0	nd	nd	nd	nd	nd	nr	[6]
Whole egg	7 ± 3	nr	nr	nd	nr	nr	nr	[32]
Egg white	1 ± 1	nr	nr	nd	nr	nr	nr	[32]
Egg yolk	64 ± 31	nr	nr	nd	nr	nr	nr	[32]
Whole egg	5.6	nr	nr	nr	nr	nr	nr	[27]
Egg white	0.4	nr	nr	nr	nr	nr	nr	[27]
Egg yolk	15.5	nr	nr	nr	nr	nr	nr	[27]
Bread								
Bread	0	nr	nr	nr	nr	0.9–2	nr	[29]
Buckwheat	nd	nd	nd	1.0–1.2	nd	nd	nr	[6]
Plant products								
Sauerkraut	0.3–0.5	0.6–1.0	1.4–1.6	0.1–0.3	0.6–0.9	0.9–1.3	nr	[6]
Natto	nd	7.1–7.8	12.7–14.8	882–1034	78.3–89.8	nd	nr	[6]
	2 ± 3	nr	nr	939 ± 753	nr	nr	nr	[32]
Hikiwari natto (chopped natto)	nd	nr	nr	827 ± 194	nr	nr	nr	[32]

Menaquinone content (μg/100 g; mean ± SD or range)								
Food	**MK-4**	**MK-5**	**MK-6**	**MK-7**	**MK-8**	**MK-9**	**MK-10**	**Source**
Black bean natto	nd	nr	nr	796 ± 93	nr	nr	nr	[32]

nd, not determined; nr, not reported.

Table 1. Representative ranges of measured menaquinone concentration in food.

Values for MK-4 to MK-10 are available. MK-4 is found in all reported products except buckwheat, hikiwari natto, and black bean natto [6, 27, 29–35]. In non-fermented dairy products and in eggs, hardly any longer-chain menaquinones have been reported [6, 27, 29, 32]. Long-chain menaquinones are also rare in the muscle meat of beef, pork, and chicken [6, 30–32]. However, in offal, such as the liver and kidney, small-to-moderate concentrations of MK-6 to MK-10 have been detected [6, 27, 30, 31]. In fish, vitamin K2 concentrations are in general very low, and menaquinones other than MK-4 have been found in only a few fish species [6, 27, 31, 32]. These small amounts of longer-chain menaquinones are said to originate from the bacteria in decomposing organic material that serves as food for fish that live at the bottom of the sea such as eel and plaice [36]. In sour milk and buttermilk and in curd and hard and soft cheese, MK-8 and MK-9 mainly account for the total concentration of vitamin K followed by MK-6 and MK-7 [6, 27, 29, 31–35]. Fermented plant products are characterized by a high concentration of MK-7 (up to 1000 μg/100 g) [6, 32].

Almost no data are available about the stability and changes in vitamin K concentrations during storage of food in general and during ripening of fermented food in particular.

6. Production of different menaquinones by microorganisms in food

Fermentation is traditionally used to increase shelf life, to inhibit pathogens, and to improve organoleptic properties [37]. Additionally, the microbial production of vitamins provides a very attractive approach for improving the nutritional composition of fermented foods. A number of MK-producing species are commonly used in industrial food fermentation applications (**Table 2**). The main microorganisms used in fermented dairy products are lactic acid bacteria, which transform lactose into lactic acid. *Lactococcus lactis* ssp. *cremoris*, *Lactococcus lactis* ssp. *lactis*, and *Leuconostoc lactis* are used as starter cultures in semihard and soft cheeses. It was reported that these species produce menaquinone and MK-7 to MK-9 in particular for *Lactococcus* and MK-7 to MK-10 for *Leuconostoc* [2, 38]. For example, the starter cultures CHN211 and CHN22 from Hansen, which contain these species, produce MK-4 to MK-10; MK-9 is the main menaquinone with 472.4 ± 22.6 μg/100 g cells and 390.3 ± 10.4 μg/100 g cells, respectively [35]. Accordingly, the highest amounts of MK were detected in semihard and soft cheese and in Caerphilly and Cheshire, a crumbly cheese specialty, known for higher numbers of *Lactococcus* species (**Table 1**). In semihard cheese, menaquinones in amounts up to 29.1 μg/100 g have been detected. The main quantified form of menaquinone in dairy is MK-9 (usually more than 50%), and the second major form is MK-8. Manoury and coauthors also found a correlation between MK-9 and MK-8. For most dairy, the MK-9 level was four times higher than that of MK-8, and the authors suggested that microorganisms that produce MK-9 could also produce MK-8. Astonishingly, the level of MK-9 was not dependent on the fat level of the dairy products. Moreover, the authors found no link between pH and the MK-9 content. The highest amounts of MK-10 are usually found in hard cheese, with the exception of one semihard cheese [34].

Species/subspecies	Food use
Lactococcus lactis ssp. *lactis* and *Lactococcus lactis* ssp. *cremoris*	Cheese, buttermilk, sour cream, cottage cheese, cream cheese, kefir, yogurt
Lactococcus raffinolactis	Cheese
Leuconostoc lactis	Cheese
Leuconostoc mesenteroides	Vegetables, dairy
Brevibacterium linens	Cheese
Brochothrix thermosphacta	Meat
Hafnia alvei	Cheese
Staphylococcus xylosus	Dairy, sausage
Staphylococcus equorum	Dairy, meat
Arthrobacter nicotianae	Cheese
Bacillus subtilis "natto"	Natto (fermented soybean)
Propionibacterium shermanii	Cheese
Propionibacterium freudenreichii	Cheese

Adapted from Walther et al. [16].

Table 2. Menaquinone-producing bacteria in fermented food.

In Swiss Emmental cheese, *Propionibacterium* strains are added to the milk to improve the formation of holes inside the cheese body. During propionic acid fermentation, lactic acid is transformed into propionic acid, acetic acid, and carbon dioxide. Various studies showed the ability of *Propionibacterium* to produce menaquinone MK-9(4H) in anaerobic conditions [39, 40]. The highest amount of MK-9(4H) has been detected in Swiss Emmental (up to 31.4 μg/100 g MK-9(4H)) and Norwegian Jarlsberg (65.2 μg/100 g MK-9(4H)); both cheeses have a high propionic acid concentration. Smaller amounts are also found in Appenzeller (up to 2 μg/100 g), Comté (up to 6.0 μg/100 g), and Raclette (4.7 μg/100 g) [33].

In contrast, dairy products fermented with thermophilic lactic acid bacteria, such as Comté cheese, mozzarella, or yogurt products, contain only small amounts of menaquinone or none (**Table 1**). These thermophilic species include *Streptococcus thermophilus, Lactobacillus delbrueckii*, and *Bifidobacterium*, and they are known to be non-vitamin K producers [2, 34].

In soft cheese, the average total menaquinones range 40.1 μg/100 g to 61 μg/100 g depending on the source, analytical method, and type of cheese (**Table 1**). Manoury and coauthors reported a soft cheese and a blue cheese with very high concentrations (up to 4.110 μg/100 g and 70 μg/100 g, respectively), but the researchers could not explain why these two cheeses are so rich in menaquinones [34].

One cheese with mold was also analyzed for menaquinone content. Gamalost, a Norwegian mold (*Mucor mucedo*) ripened autochthonous cheese, contains more menaquinone than Norvegia, a semihard Norwegian cheese, but the mold did not contribute to the production of vitamin K in Gamalost. The low pH in Gamalost and a higher fermentation rate may explain the differences in menaquinone content [35].

Some work has been conducted to improve the content of different menaquinones in dairy products. New research demonstrated that strains of *Lactobacillus fermentum* LC 272 isolated from raw milk could be a starter culture for fermented milk with a high level of vitamin K2 (MK-4) production [41]. This strain can produce 185 μg/L in Rogosa medium and 64 μg/L in reconstituted skim milk. Morishita and coworkers published a study in 1999 that showed the possibility of producing MK-8 and MK-9 with *Lactocouccus lactis* ssp. *cremoris* YIT2011 and MK-9 and MK-10 with *Lactococcus lactis* YIT 3001 (29–123 μg of menaquinone/L of the fermented medium) [38]. Additionally, several patents for *Lactococcus* capable of producing a significantly increased amount of vitamin K2 have been deposed.

In contrast to fermented animal products, fermented vegetable products contain mainly MK-7 (**Table 1**). Natto, a traditional Japanese food produced with *Bacillus subtilis* natto, contains the highest amount of menaquinone. The highest measured value is almost 1000 μg/100 g. *B. subtilis* natto is the key microorganism for industrial production of MK-7, and much work has been done to improve the production. Optimization of the fermentation medium, mutations of the strains, and biofilm formation have been described as means for improving the yield of MK-7 [42–46]. The use of organic solvents to extract vitamins is one of the major issues of the bulk production of MK-7. Berenjian and coworkers demonstrated that the addition of vegetable oil during a dynamic fermentation process could be a good process for producing an oil rich in MK-7. In that study, the oil contained 724 mg/L of MK-7, and they suggested using the oil in supplementary and dietary food products [47].

7. Dietary recommendations for menaquinone

Dietary recommendations for vitamin K are still based on knowledge of phylloquinone and its classic role as an enzyme cofactor for coagulation proteins. The recommendations do not consider the differences in bioavailability and bioactivity between the different forms of vitamin K or the possibly higher requirements for health effects apart from coagulation, such as bone or cardiovascular health [16].

Depending on country, sex, and age, the recommendations for vitamin K range from 50 to 120 μg per day for adults 19 years and older. These recommendations are generally presented as adequate intake or estimated values, and no tolerable upper intake level has been established for vitamin K [16, 25, 48]. Research for valuable biomarkers to measure the status of vitamin K in the population is ongoing. A recent study from Maastricht University compared the biomarkers for coagulation with those of bone and vascular health in 896 healthy volunteers. Whereas all coagulation proteins were completely carboxylated by vitamin K, and a high concentration of undercarboxylated Gla proteins (osteocalcin and matrix Gla protein) was found in the majority of the blood samples, indicating that most of the volunteers in this study had an inadequate supply of vitamin K [23]. As long as robust physiological endpoints are missing to differentiate the contribution of MKs to human health from that of PK, it is unlikely that specific dietary recommendations for MKs will be widely adopted in the near future. In the meantime, a preferred recommendation could be to consume a wide variety of foods which are good sources of PKs and MKs, respectively, such as green leafy vegetables and fermented dairy products [16, 49].

8. Dietary intake of menaquinones

As shown in **Table 1**, the most important sources of menaquinones are cheese, curd, offal, and fermented soybeans (natto). Based on regional differences in dietary patterns, the form and amount of specific menaquinones consumed may vary widely between populations. For example, in Japan, as a result of natto consumption, MK-7 is the most frequently consumed form of menaquinones. The contribution of MK-7 to total vitamin K intake is 25% among young women living in eastern Japan. Nearly all of the MK-7 intake originates from pulses, including fermented soybean natto [32]. The mean daily intake of MK-7 in this study was 57.4 μg with a range from 0 to 340 μg.

In countries with a traditional high intake of dairy products, such as the Netherlands, Germany, and the United Kingdom (UK), MK-7 to MK-10 contribute mostly to the menaquinone supply. Beulens and coauthors compiled the results from several European studies that estimated menaquinone intake using Food Frequency Questionnaires (FFQs). The self-reported mean daily intake of menaquinones in adults ranged from 20.7 μg for women in the Rotterdam Study to 43 μg in men in the UK National Dietary and Nutrition Survey. In all of these studies, cheese was the most important food source of menaquinones [49]. However, these data should be interpreted carefully because they were collected by FFQs that are designed to estimate the relative dietary intake of large populations but not to estimate absolute dietary intake. A seasonal survey in postmenopausal women in Tehran, Iran, used a monthly food record for 1 year. The researchers found a significantly higher intake of vitamin K in the spring, summer, and autumn compared to the winter. Unfortunately, these authors did not further specify vitamin K and did not provide any information about consumption of different food items containing vitamin K [50]. A study in older individuals to calculate the desired duration of a diet recording to estimate the individual vitamin K intake concluded that 13 24-hour recalls are ideal to record intraindividual variance. As this would not be realistic in most studies, the authors proposed a minimum of six nonconsecutive days of diet recording [51]. Another possible approach for estimating nutrient status is to use biomarkers. Biomarkers for menaquinones are undercarboxylated vitamin K-dependent proteins in the circulatory system. However, in addition to vitamin K availability, these biomarkers depend on the total amount of protein. To be sure that protein status does not confound vitamin K status, the measurements must be corrected for the total amount of the protein under study [52].

These limitations, together with the scarce and widely varying data on concentrations of different menaquinones in food items, show how fragmentary our knowledge of the supply of vitamin K2 in the general population remains.

9. Pharmacokinetics of menaquinones

Although the forms of vitamin K are classified as fat-soluble nutrients, the lipophilicity of the different forms changes with side-chain length. Whereas menadione is water soluble, phylloquinone and MK-4 are mildly lipophilic. Long-chain menaquinones are strongly lipophilic and soluble only in apolar organic solvents [36]. This lipophilicity also influences the absorption of vitamin K, which varies greatly depending on the food matrix. As long-chain menaquinones are found mainly in the fat fraction of dairy products, the absorption of these menaquinones is almost 100% in contrast to PK, where the poor uptake of only 5–10% from cooked vegetables can be improved only slightly by concomitant fat intake [6]. As a consequence, even the dietary intake of phylloquinone is much higher, menaquinones are equally important for vitamin K status, because of their better intestinal absorption. Independently of their form and origin, all K vitamins are transported to the liver, incorporated in triglyceride-rich lipoproteins. Unlike phylloquinone, which mostly remains in the liver to be used for clotting factor synthesis, menaquinones are released to the bloodstream incorporated in low-density lipoproteins and transported to the target tissue such as bone and arteries for Gla-protein carboxylation. Absorbability is further supported by a longer half-life, up to several days for long-chain menaquinones compared to phylloquinone, which normally disappears from the bloodstream after 8 hours. This longer postprandial presence in the bloodstream leads to a more constant circulating level of vitamin K2 and, as a consequence, longer availability of these long-chain menaquinones for uptake by extrahepatic tissues [36, 53]. Although there is some evidence that menaquinones with medium-chain length like MK-7 are better absorbed than short- (MK-4) or long-chain menaquinones (MK-8 and MK-9) [6], human data on the bioavailability, absorption, and kinetics of K2 vitamins from food are limited to MK-7 and MK-9 and have not been systematically tested for all menaquinones thus far [36, 49].

As researchers have found that MKs play an important role in health aspects beyond coagulation, the cooperation with other nutrients in vitamin K-rich food such as fermented dairy products may lead to a better understanding of the effect of different food items on health aspects, for example, bone health or cardiovascular health.

10. Bone health, menaquinones, and fermented dairy products

One of the most important research fields in the past and present is the study of the factors that influence the formation and conservation of strong bones. Osteopenia, including osteoporosis, is one of the most prevalent diseases in elderly individuals and is a large social, medical, and economic burden throughout the world. One out of three women and one out of five men older than 50 years are at risk of experiencing an osteoporotic fracture [54]. Low bone mineral mass is the main factor that causes osteoporotic fractures. Bone mass in later life is the result of the peak bone mass achieved during growth and the rate of age-related bone loss. Consequently, a high peak bone mass at maturity and a low bone loss during aging are the most promising factors in the prevention of osteoporosis and fractures. In addition to factors that influence bone health such as gender, age, body size, genetics, and ethnicity that are not changeable, other factors, especially lifestyle factors such as physical activity, smoking, alcohol consumption, and dietary patterns, can be modified [55]. Different dietary factors are known to positively influence bone health. They range from minerals (e.g., calcium, magnesium, phosphorus, potassium, and various trace elements) and vitamins (A, D, E, K, C, and certain B vitamins) to macronutrients such as proteins and fatty acids and finally to bioactive food components (e.g., peptides) that in recent years have been proposed to be beneficial for bone health [55]. All these elements are involved in bone metabolism. Currently, researchers are trying to identify and understand the mechanisms and interactions of these factors in relation to bone health [56].

Most studies that have investigated the relationship between dairy and bone health have shown a beneficial effect of dairy consumption, even if the reason for this link is still unclear [56, 57]. After many years of focusing on calcium as the beneficial element for bone health in dairy, recent evidence suggests that other macro- and micronutrients, as well as food components such as bioactive peptides, milk fat globule membrane, prebiotics, and probiotics present in milk and dairy products, play an important role in this health outcome [56]. Many of these nutrients support the bioavailability (phosphorus, vitamin D, magnesium, zinc, potassium), absorption (casein phosphopeptides, phosphorus, lactose, protein) and homeostasis (magnesium, potassium, vitamin D) of calcium and contribute to bone-building properties (phosphorus, magnesium, potassium, zinc, vitamin D, vitamin B12, and vitamin K) [56–58].

Most of these components are not or are positively affected by fermentation. That means their concentration remains the same in the fermented product compared with milk or even increases either by processing (i.e., fat-soluble vitamins in cheese) or by the activity of microorganisms (i.e., bioactive peptides, vitamin B12, or vitamin K2).

The role of vitamin K2 in bone health is strongly bound to osteocalcin (OC), a key regulator of calcium usage. This small Ca^{2+}-binding protein is involved in the mineralization of bones and teeth, and its potential to bind calcium is dependent on carboxylation with vitamin K2.

Only the fully carboxylated OC is able to strongly bind calcium and to consolidate calcification of the hydroxyapatite crystal lattice that requires a sufficient supply with vitamin K2 and other nutrients, such as retinoic acid and vitamin D, all involved in the regulation of osteocalcin production [59].

Fermented dairy products are vital for bone health because of their unique combination of various nutrients and microorganisms that support and maintain positive bone metabolism [57, 60]. Additionally, dairy matrix and nutrient composition may affect the delivery of menaquinones and improve vitamin K status [61].

11. Cardiovascular health, menaquinones, and fermented dairy products

Coronary artery calcification (CAC) is a predictor of cardiovascular disease (CVD) and mortality. Based on vitamin K's role in activating matrix Gla protein (MGP), a calcification inhibitor, vitamin K is proposed to play a preventive role in CAC and CVD [59, 62]. As recently reviewed, randomized controlled trials that examined the influence of vitamin K on the risk of cardiovascular disease are scarce [63]. The results of observational studies have shown an association between higher dietary menaquinone consumption and less calcification [64], decreased risk of coronary heart disease (CHD), CHD mortality, and all-cause mortality [65–67]. The results of a Dutch prospective cohort study suggested that of all MKs the long-chain menaquinones (MK-7 to MK-9) have the most beneficial effects on cardiovascular disease [67]. Although these results are promising, they must be interpreted with caution, because validated biomarkers for single MK intake are missing [16].

Complex milk fatty acid chemistry and several minerals, such as calcium, magnesium, phosphorus, and potassium provided in relevant concentrations, have been proposed to be involved in the complex mechanism of dairy products and their support to reduce CVD risk [68]. Among the high number of different fatty acids in dairy products, trans-palmitoleic acid, stearic acid, lauric acid, myristic acid, and oleic acid have been associated with beneficial effects on blood lipids and serum lipoprotein levels [56]. These assumptions are supported by the inverse association observed between CHD risk and the consumption of milk, cheese, and meat as the richest sources of MKs in the Western diet [6, 67].

12. Conclusion

Our knowledge of the consumption of menaquinones should be improved with weighed and extended food records [51] in combination with (multiple) biomarkers in the blood for vitamin K status [52] and the content of the various menaquinones in food items such as cheese, which contribute most to the supply with this vitamin.

As different lactic acid bacteria strains used in cheese production influence the expression of various MKs, analysis of a wide variety of different cheeses may be necessary for a representative overview of the vitamin K2 content in this food group. Although results from well-designed clinical trials investigating the association between menaquinones and bone health, as well as cardiovascular health, are rare, dairy products seem to be predestined to play a major role in the Western diet because of their nutrient density and matrix properties that improve the bioavailability of vitamin K2.

Author details

Barbara Walther* and Magali Chollet

*Address all correspondence to: barbara.walther@agroscope.admin.ch

Federal Department of Economic Affairs, Education and Research EAER Agroscope, Institute for Food Sciences IFS, Berne, Switzerland

References

[1] Booth SL. Vitamin K: food composition and dietary intakes. Food & Nutrition Research. 2012;56. DOI: 10.3402/fnr.v56i0.5505:1–5

[2] Collins MD, Jones D. Distribution of isoprenoid quinone structural types in bacteria and their taxonomic implication. Microbiological Reviews. 1981;45(2):316–354.

[3] Suttie JW. The importance of menaquinones in human nutrition. Annual Review of Nutrition. 1995;15:399–417.

[4] Shearer MJ. Vitamin K. Lancet. 1995;345(8944):229–234.

[5] Okano T, Shimomura Y, Yamane M, Suhara Y, Kamao M, Sugiura M, et al. Conversion of phylloquinone (vitamin K1) into menaquinone-4 (vitamin K2) in mice: two possible routes for menaquinone-4 accumulation in cerebra of mice. Journal of Biological Chemistry. 2008;283(17):11270–11279.

[6] Schurgers LJ, Vermeer C. Determination of phylloquinone and menaquinones in food – effect of food matrix on circulating vitamin K concentrations. Haemostasis. 2000;30(6): 298–307.

[7] World Health Organization. Vitamin and mineral requirements in human nutrition, second edition. In: Report of a Joint FAO/WHO Expert Consultation, Bangkok, Thailand: Switzerland World Health Organization:, 2004, 341p

[8] Shearer MJ, Newman P. Metabolism and cell biology of vitamin K. Thrombosis and Haemostasis. 2008;100(4):530–547.

[9] Vandamme EJ, editor. Biotechnology of Vitamins, Pigments and Growth Factors: London and New York: Elsevier; 2012. 438 p. DOI: 10.1007/978-94-009-111-6

[10] Tani Y, Asahi S, Yamada H. Vitamin K2 (menaquinone): screening of producing microorganisms and production by Flavobacterium meningosepticum. Journal of Fermentation Technology. 1984;62:321–327.

[11] Bentley R, Meganathan R. Biosynthesis of vitamin K (menaquinone) in bacteria. Microbiological Reviews. 1982;46(3):241–280.

[12] Nowicka B, Kruk J. Occurrence, biosynthesis and function of isoprenoid quinones. Biochimica et Biophysica Acta (BBA) – Bioenergetics. 2010;1797(9):1587–1605. DOI: 10.1016/j.bbabio.2010.06.007

[13] Farrand SK, Taber HW. Changes in menaquinone concentration during growth and early sporulation in *Bacillus subtilis*. Journal of Bacteriology. 1974;117(1):324–326.

[14] Pedersen MB, Gaudu P, Lechardeur D, Petit MA, Gruss A. Aerobic respiration metabolism in lactic acid bacteria and uses in biotechnology. Annual Review of Food Science and Technology. 2012;3:37–58. DOI: 10.1146/annurev-food-022811-101255

[15] Meganathan R. Ubiquinone biosynthesis in microorganisms. FEMS Microbiology Letters. 2001;203(2):131–139. DOI: 10.1111/j.1574-6968.2001.tb10831.x

[16] Walther B, Karl JP, Booth SL, Boyaval P. Menaquinones, bacteria, and the food supply: the relevance of dairy and fermented food products to vitamin K requirements. Advances in Nutrition: An International Review Journal. 2013;4(4):463–473. DOI: 10.3945/an.113.003855

[17] Zhi X-Y, Yao J-C, Tang S-K, Huang Y, Li H-W, Li W-J. The futalosine pathway played an important role in menaquinone biosynthesis during early prokaryote evolution. Genome Biology and Evolution. 2014;6(1):149–160. DOI: 10.1093/gbe/evu007

[18] Hiratsuka T, Furihata K, Ishikawa J, Yamashita H, Itoh N, Seto H, et al. An alternative menaquinone biosynthetic pathway operating in microorganisms. Science. 2008;321 (5896):1670–1673. DOI: 10.1126/science.1160446

[19] Kurosu M, Begari E. Vitamin K2 in electron transport system: are enzymes involved in vitamin K2 biosynthesis promising drug targets? Molecules. 2010;15(3):1531–1553. DOI: 10.3390/molecules15031531

[20] Conly JM, Stein K. The production of menaquinones (vitamin K2) by intestinal bacteria and their role in maintaining coagulation homeostasis. Progress in Food & Nutrition Science. 1992;16(4):307–343.

[21] Conly JM, Stein K. Quantitative and qualitative measurements of K vitamins in human intestinal contents. American Journal of Gastroenterology. 1992;87(3):311–316.

[22] Iwamoto J, Sato Y, Matsumoto H. Vitamin K2 improves femoral bone strength without altering bone mineral density in gastrectomized rats. Journal of Nutritional Science and Vitaminology. 2014;60(2):71–77.

[23] Theuwissen E, Magdeleyns EJ, Braam LA, Teunissen KJ, Knapen MH, Binnekamp IA, et al. Vitamin K status in healthy volunteers. Food & Function. 2014;5(2):229–234. DOI: 10.1039/c3fo60464k

[24] Urano A, Hotta M, Ohwada R, Araki M. Vitamin K deficiency evaluated by serum levels of undercarboxylated osteocalcin in patients with anorexia nervosa with bone loss. Clinical Nutrition. 2015;34(3):443–448. DOI: 10.1016/j.clnu.2014.04.016

[25] Shearer MJ, Fu X, Booth SL. Vitamin K nutrition, metabolism, and requirements: current concepts and future research. Advances in Nutrition: An International Review Journal. 2012;3(2):182–195. DOI: 10.3945/an.111.001800

[26] Walther B, Schmid A. Effect of fermentation on vitamin content in food. In: Frías J, Martínez-Villaluenga C, Peñas E, editors. Fermented Foods in Health and Disease Prevention. Elsevier Science, San Diego; 2016. 880 p.

[27] Elder SJ, Haytowitz DB, Howe JR, Peterson JW, Booth SL. Vitamin K contents of meat, dairy, and fast food in the US diet. Journal of Agricultural and Food Chemistry. 2006;54(2):463–467.

[28] EFSA FEEDAP Panel. Scientific Opinion on the safety and efficacy of vitamin K3 (menadione sodium bisulphite and menadione nicotinamide bisulphite) as a feed additive for all animal species. EFSA Journal. 2014;12(1):1–29. DOI: 10.2903/j.efsa. 2014.3532

[29] Vermeer C, Knapen MH, Schurgers LJ. Vitamin K and metabolic bone disease. Journal of Clinical Pathology. 1998;51(6):424–426.

[30] Hirauchi K, Sakano T, Notsumoto S, Nagaoka A, Morimoto A, Fujimoto K, et al. Measurement of K vitamins in animal tissues by high-performance liquid chromatography with fluorimetric detection. Journal of Chromatography B: Biomedical Sciences and Applications. 1989;497:131–137.

[31] Koivu-Tikkanen TJ, Ollilainen V, Piironen VI. Determination of phylloquinone and menaquinones in animal products with fluorescence detection after postcolumn reduction with metallic zinc. Journal of Agricultural and Food Chemistry. 2000;48(12): 6325–6331.

[32] Kamao M, Suhara Y, Tsugawa N, Uwano M, Yamaguchi N, Uenishi K, et al. Vitamin K content of foods and dietary vitamin K intake in Japanese young women. Journal of Nutritional Science and Vitaminology. 2007;53(6):464–470.

[33] Hojo K, Watanabe R, Mori T, Taketomo N. Quantitative measurement of tetrahydromenaquinone-9 in cheese fermented by propionibacteria. Journal of Dairy Science. 2007;90(9):4078–4083. DOI: 10.3168/jds.2006-892

[34] Manoury E, Jourdon K, Boyaval P, Fourcassie P. Quantitative measurement of vitamin K2 (menaquinones) in various fermented dairy products using a reliable high-performance liquid chromatography method. Journal of Dairy Science. 2013;96(3):1335–1346. DOI: 10.3168/jds.2012-5494

[35] Qureshi TM, Vermeer C, Vegarud GE, Abrahamsen RK, Skeie S. Formation of biogenic amines and vitamin K contents in the Norwegian autochthonous cheese Gamalost during ripening. Dairy Science & Technology. 2013;93(3):303–314. DOI: 10.1007/ s13594-013-0123-8

[36] Vermeer C. Vitamin K: the effect on health beyond coagulation – an overview. Food & Nutrition Research. 2012;56:5329. DOI: 10.3402/fnr.v56i0.5329

[37] Bourdichon F, Casaregola S, Farrokh C, Frisvad JC, Gerds ML, Hammes WP, et al. Food fermentations: microorganisms with technological beneficial use. International Journal of Food Microbiology. 2012;154(3):87–97. DOI:10.1016/j.ijfoodmicro.2011.12.030

[38] Morishita T, Tamura N, Makino T, Kudo S. Production of menaquinones by lactic acid bacteria. Journal of Dairy Science. 1999;82(9):1897–1903.

[39] Schwartz AC. Terpenoid quinones of the anaerobic *Propionibacterium shermanii*. Archives of Microbiology. 1973;91(3):273–279.

[40] Furuichi K, Hojo K, Katakura Y, Ninomiya K, Shioya S. Aerobic culture of propionibacterium freudenreichii ET-3 can increase production ratio of 1,4-dihydroxy-2-naphthoic acid to menaquinone. Journal of Bioscience and Bioengineering. 2006;101(6): 464–470. DOI: 10.1263/jbb.101.464

[41] Lim SD, Kim KS, Do JR. Physiological characteristics and production of vitamin K2 by *Lactobacillus fermentum* LC272 isolated from raw milk. Korean Journal for Food Science of Animal Resources. 2011;31(4):513–520. DOI: 10.5851/kosfa.2011.31.4.513

[42] Sato T, Yamada Y, Ohtani Y, Mitsui N, Murasawa H, Araki S. Production of menaquinone (vitamin K2)-7 by *Bacillus subtilis*. Journal of Bioscience and Bioengineering. 2001;91(1):16–20.

[43] Sato T, Yamada Y, Ohtani Y, Mitsui N, Murasawa H, Araki S. Efficient production of menaquinone (vitamin K2) by a menadione-resistant mutant of *Bacillus subtilis*. Journal of Industrial Microbiology and Biotechnology. 2001;26(3):115–120.

[44] Berenjian A, Chan NL, Mahanama R, Talbot A, Regtop H, Kavanagh J, et al. Effect of biofilm formation by Bacillus subtilis natto on menaquinone-7 biosynthesis. Molecular Biotechnology. 2013;54(2):371–378. DOI: 10.1007/s12033-012-9576-x

[45] Song J, Liu H, Wang L, Dai J, Liu Y, Liu H, et al. Enhanced production of vitamin K2 from *Bacillus subtilis* (natto) by mutation and optimization of the fermentation medium. Brazilian Archives of Biology and Technology. 2014;57(4):606–612. DOI: 10.1590/S1982-88372014000100019

[46] Singh R, Puri A, Panda BP. Development of menaquinone-7 enriched nutraceutical: inside into medium engineering and process modeling. Journal of Food Science and Technology. 2015;52(8):5212–5219. DOI: 10.1007/s13197-014-1600-7

[47] Berenjian A, Mahanama R, Talbot A, Regtop H, Kavanagh J, Dehghani F. Designing of an intensification process for biosynthesis and recovery of menaquinone-7. Applied Biochemistry and Biotechnology. 2014;172(3):1347–1357. DOI: 10.1007/s12010-013-0602-7

[48] Ferland G. The discovery of vitamin K and its clinical applications. Annals of Nutrition and Metabolism. 2012;61(3):213–218. DOI: 10.1159/000343108

[49] Beulens JW, Booth SL, van den Heuvel EG, Stoecklin E, Baka A, Vermeer C. The role of menaquinones (vitamin K2) in human health. British Journal of Nutrition. 2013;110(8): 1357–1368. DOI: 10.1017/S0007114513001013

[50] Mansour A, Ahadi Z, Qorbani M, Hosseini S. Association between dietary intake and seasonal variations in postmenopausal women. Journal of Diabetes and Metabolic Disorders. 2014;13:52. DOI: 10.1186/2251-6581-13-52

[51] Presse N, Payette H, Shatenstein B, Greenwood CE, Kergoat MJ, Ferland G. A minimum of six days of diet recording is needed to assess usual vitamin K intake among older adults. Journal of Nutrition. 2011;141(2):341–346. DOI: 10.3945/jn.110.132530

[52] Shea MK, Booth SL. Concepts and controversies in evaluating vitamin K status in population-based studies. Nutrients. 2016;8(1):1-25.DOI: 10.3390/nu8010008

[53] Schurgers LJ, Vermeer C. Differential lipoprotein transport pathways of K-vitamins in healthy subjects. Biochimica et Biophysica Acta (BBA). 2002;1570(1):27–32. DOI: 10.1016/S0304-4165(02)00147-2

[54] International Osteoporosis Foundation. [Internet]. Available from: http://www.iofbonehealth.org/osteoporosis. [Accessed: 2016-04-07].

[55] Cashman KD. Diet, nutrition, and B one health. Journal of Nutrition. 2007;137(11): 2507S–2512S.

[56] Hess JM, Jonnalagadda SS, Slavin JL. Dairy foods: current evidence of their effects on bone, cardiometabolic, cognitive, and digestive health. Comprehensive Reviews in Food Science and Food Safety. 2016;15(2):251–268. DOI: 10.1111/1541-4337.12183

[57] Heaney RP. Dairy and bone health. Journal of the American College of Nutrition. 2009;28(S1):82S–90S.

[58] Price CT, Langford JR, Liporace FA. Essential nutrients for bone health and a review of their availability in the average North American diet. Open Orthopaedics Journal. 2012;6:143–149. DOI: 10.2174/1874325001206010143

[59] Kidd PM. Vitamins D and K as pleiotropic nutrients: clinical importance to the skeletal and cardiovascular systems and preliminary evidence for synergy. Alternative Medicine Review. 2010;15(3):199–222.

[60] Gaucheron F. Milk and dairy products: a unique micronutrient combination. Journal of the American College of Nutrition. 2011;30(5S1):400S–409S.

[61] Knapen MH, Braam LA, Teunissen KJ, Van't Hoofd CM, Zwijsen RM, van den Heuvel EG, et al. Steady-state vitamin K (menaquinone-7) plasma concentrations after intake of dairy products and soft gel capsules. European Journal of Clinical Nutrition. 2016; DOI: 10.1038/ejcn.2016.3

[62] Pizzorno L. Vitamin K2: optimal levels essential for the prevention of age-associated chronic disease. Longevity Medicinal Review. 2011:1–23. Available from: http://

www.lmreview.com/articles/view/Vitamin-K2-Essential-for-Prevention-of-Age-Associated-Chronic-Disease/[Accessed:2016-05-18]

[63] Hartley L, Clar C, Ghannam O, Flowers N, Stranges S, Rees K. Vitamin K for the primary prevention of cardiovascular disease. Cochrane Database Systematic Review. 2015;9:1–26. DOI: 10.1002/14651858.CD011148.pub2

[64] Shea MK, Holden RM. Vitamin K status and vascular calcification: evidence from observational and clinical studies. Advances in Nutrition. 2012;3(2):158–165. DOI: 10.3945/an.111.001644

[65] Geleijnse JM, Vermeer C, Grobbee DE, Schurgers LJ, Knapen MHJ, van der Meer IM, et al. Dietary intake of menaquinone is associated with a reduced risk of coronary heart disease: the rotterdam study. Journal of Nutrition. 2004;134(11):3100–3105.

[66] Beulens JWJ, Bots ML, Atsma F, Bartelink ML, Prokop M, Geleijnse JM, et al. High dietary menaquinone intake is associated with reduced coronary calcification. Atherosclerosis. 2009;203(2):489–493. DOI: 10.1016/j.atherosclerosis.2008.07.010

[67] Gast GCM, de Roos NM, Sluijs I, Bots ML, Beulens JWJ, Geleijnse JM, et al. A high menaquinone intake reduces the incidence of coronary heart disease. Nutrition, Metabolism and Cardiovascular Diseases. 2009;19(7):504–510. DOI: 10.1016/j.numecd.2008.10.004

[68] Rice BH. Dairy and cardiovascular disease: a review of recent observational research. Current Nutrition Reports. 2014;3(2):130–138. DOI: 10.1007/s13668-014-0076-4

Chapter 5

The Impact of Vitamin K2 on Energy Metabolism

Mona Møller, Serena Tonstad, Tone Bathen and
Jan Oxholm Gordeladze

Additional information is available at the end of the chapter

http://dx.doi.org/10.5772/67152

Abstract

Environmental and behavioral adaptations introduced during the last decades have synergistically enhanced man's lifespan, but also paved the ground for disease states involving impairment of multiple organs, which are both modulating and depending on homeostatic calorie "accounting."

Diabetes, obesity, and/or bone brittleness now occur frequently in our society, inducing ailments affecting overall health and well-being. Therefore, an improved comprehension of how organs (e.g. bone and adipose tissue) may provide homeostasis and sound strategies to treat these diseases, thus improving health and life quality of most age categories, should be sought. The steroid and xenobiotic receptor (SXR) (pregnane X receptor = PXR) is a nuclear steroid-like hormone receptor, stimulated by hormones, steroids, drugs, and xenobiotic compounds. SXR exhibits a versatile ligand binding domain, serving as a xenobiotic sensor, regulating xenobiotic clearance from the liver and intestine. However, new and interesting functions of SXR in the regulation of inflammatory processes, cholecalciferol and bone metabolism, lipid and energy homeostasis, and cancer therapy, have emerged.

Hence, the discovery and pharmacological development of new PXR modulators, like vitamin K2, represent an interesting and innovative therapeutic approach to combat various diseases, of which glucose and lipid metabolism (i.e., energy metabolism and adiposity) should be emphasized.

Keywords: vitamin K2, energy metabolism, PXR, SXR

1. Introduction

In the past years, we have seen a plethora of research reports illuminating the link between bone physiology and energy metabolism, even though the central and sympathetic nervous

INTECH
open science | open minds

© 2017 The Author(s). Licensee InTech. This chapter is distributed under the terms of the Creative Commons Attribution License (http://creativecommons.org/licenses/by/3.0), which permits unrestricted use, distribution, and reproduction in any medium, provided the original work is properly cited. (cc) BY

systems, but also the gastrointestinal and pancreatic axes serve essential functions related to the systemic regulation of energy expenditure. It may be asserted that the fat tissue is more prominent due to its basic implication in storing and dissipating energy. During the last 25 years, a major progress has been communicated within the medical societies, featuring a modern understanding of the origin of fat tissues, their specialized characteristics and functioning, as well as the pathophysiological consequences of their impairment. These advances consequently lead to the discovery that adipose tissue metabolism is heavily coupled to the homeostasis of the skeleton.

Fat tissue is able to deposit and releases energy rich compounds during feeding and fasting, respectively, and it modulates the energy homeostasis in a versatility of organs via its endocrine potential. Fat cells (adipocytes) may accumulate energy in the form of triglycerides, as well as burning it by degrading fatty acids via so-called β-oxidation. Additionally, adipocytes produce and release so-called adipokines, among which leptin and adiponectin are the most important ones. These hormones regulate both the ingestion of calories, as well as the body's sensitivity to insulin. The functional multiplicity of adipose tissue is sustained by various subtypes of adipocytes in depot storing fat. Mitochondria-sparse white adipose tissue (WAT), characterized as visceral and subcutaneous fat, stores energy as triglycerides for which the level is regulated by the body's sensitivity to insulin and the overall metabolism of glucose in the liver and skeletal muscle cells, respectively. Contrastingly, mitochondria-enriched brown adipose tissue (BAT), which in adults is positioned as discrete entities localized in the neck and other regions of the trunk of the body [1], serves to dissipate energy to sustain adaptive thermogenesis [2]. This process is facilitated by the action of uncoupling protein 1 (UCP1), stimulating a leakage of protons, in order to uncouple respiration from ATP synthesis, thus favoring heat production. The thermogenesis sustained by BAT is mastered via the central nervous system, the impact of both the signaling systems comprised by catecholamines (i.e., β-adrenergic signaling), as well as deiodinase 2 (Dio2)-facilitated thyroid hormone conversion from T4 (thyroxine) to T3 (triiodothyronine). As an extra feature, compared to its role in adaptive thermogenesis, BAT is also protecting against obesity, as well as against insulin resistance and development of diabetes [5–8]. Finally, it is worth noticing that genetic ablation of BAT in small experimental animals is resulting in diet-induced obesity, diabetes mellitus with insulin resistance, as well as enhanced blood lipids [3].

It has been asserted that BAT may originate from two sources. The classical or ordinary, preformed BAT stems from Myf5-positive dermomyotomal progenitor cells which may also yield skin and muscle, as well as functioning in so-called nonshivering thermogenesis [4]. On the other hand, Myf5-negative progenitor cells may differentiate into white adipocytes which play a role in energy storage, or to BAT-like or "beige" fat cells. The latter adipocytes demonstrate both brown and white fat cell characteristics [5]. Of major importance is the fact that BAT-like adipocytes can be transformed into WAT-like adipocytes through a plethora of mechanisms, of which a few deserves mentioning; cold exposure, endocrine action of FGF21 [6] or irisin [7], and via transcriptional regulators including FoxC2 [8], PRDM16 [9], and PPARc [10], which lead to SirT1-mediated deacetylation of the PPARc protein [11]. Beige fat possesses powerful antiobesity and antidiabetic activities. An overexpression of BAT-specific transcription factors, i.e., either FoxC2 or PRDM16 in WAT adipocytes, has been shown to protect mice from diet-induced obesity and metabolic dysfunction [9]. Furthermore, ablation of beige fat cells by adipocyte-only deletion of the transcriptional modulator PRDM16, yields experimental animals, which become prone to diet-induced obesity and ensuing insulin resistance [12].

In line with the cited article published by Lecka-Czernik et al. in Archives of Biochemistry and Biophysics, 2014 [13], we performed the following experiment: Human adipose stem cells and 3T3-L1 preadipocytes, the latter with an activated mutation (ref) in the $G_{i2\alpha}$-protein were differentiated into fully mature adipocytes as indicated by coloration with Alizarin Red. Interestingly, both adipocyte species accumulated triglycerides upon differentiation to mature adipocytes; however, exposure to vitamin K2 (MK-7) diminished their ability to turn into fully mature adipocytes. Furthermore, both adipocyte species were tested for expression of various genes differing white adipocytes from beige adipocytes. From the gene expression profile, featuring genes like the beta-adrenergic receptor β3-AR, Foxc2, PGC1α, PPARα, Dio2, UCP1, Adipoq, and Leptin, it was quite obvious that the preadipocytes in question differentiated more in the "direction" of beige, lipid (or fatty acid) metabolizing adipocytes than into white, triglyceride-storing adipocytes. Hence, it could be concluded or hypothesized that vitamin K2 (MK-7), in fact, was able to direct preadipocytes into becoming an energy-dissipating, rather than energy-storing adipocyte phenotype (see **Figure 1**, left and right panels). A putative model featuring this development, referring to a suggested mechanism, is given in **Figure 2**.

Here, we have demonstrated that vitamin K2 affects the differentiation of preadipocytes to mature adipocytes, ensuring that the "end-point" phenotype may be tilted in the direction of the beige, energy "dissipating" phenotype. In their paper from 2014, Lecka-Czernik and coworkers [13] assert that an impairment in fat function correlates with a reduced bone mass and an increased incidence of fractures. But, the question posed by the authors is: does accumulation of bone marrow adipose tissue (BMAT) exert a detrimental effect on bone structure and or mineralization, and will a diminished bone mass stimulate the accumulation of BMAT? Historically, BMAT was construed as a dormant or inert type of fat that accumulated within the bone marrow in order to pose as empty space filling subsequent to involution of hematopoietic tissue. Hence, its relationship with a lower bone mass was construed as circumstantial. However, novel evidence asserted that marrow adipogenesis should be construed as a process, which is tightly bound to the differentiation of osteoblasts, since they "share" common precursor cells, and they are subjected to same modulatory signaling patterns. This yields, however, opposite end results where a positive correlation with an overall fat metabolism indicates that BMAT, in fact, actively induces bone mass loss and inferior quality.

Both, over and malnutrition represent systemic changes in energy metabolism, affecting bone fat volume and bone mass. Despite the fact that adipose (overweight) individuals present with a higher body weight, the more often than not demonstrate a lower BMD. In obese older men and women, an enhanced percentage of body fat and low amount of lean (muscle) mass predicts a lower BMD and thus an enhanced frailty risk [14]. The enhanced fracture risk incurred by postmenopausal women and older men [15] is mainly due to enhanced circulatory levels of adipose tissue derived in adiponectin and proinflammatory cytokines, with a concomitant lower secretion of leptin and IGF-1. Furthermore, one will often detect lowered blood levels of 25-hydroxyvitamin D and higher serum parathyroid hormone levels in these "patients" [15].

Q-PCR of biomarkers differentiating between white and brown/beige types of fat cells/tissues

Human adipose stem cells or 3T3-L1 cells with activating mutations in the G-protein $Gi_{2\alpha}$ were differentiated towards adipocytes in the absence or presence of vitamin K2. Isolated mRNA levels were analysed by Q-PCR, and presented as percent ratio between cells grown in control medium or medium supplemented with MK-7.

	β3-AR	FoxC2	PGC1α	PPARα	PPARγ	Dio2	UCP1	Adipoq	Leptin
hASCs + K2	193±29	122±17	483±22	388±31	321±19	332±25	358±40	37±9	16±6
[R179E]$Gi_{2\alpha}$ 3T3-L1 + K2	221±22	143±33	563±51	395±43	355±28	387±49	488±56	55±12	38 ±11

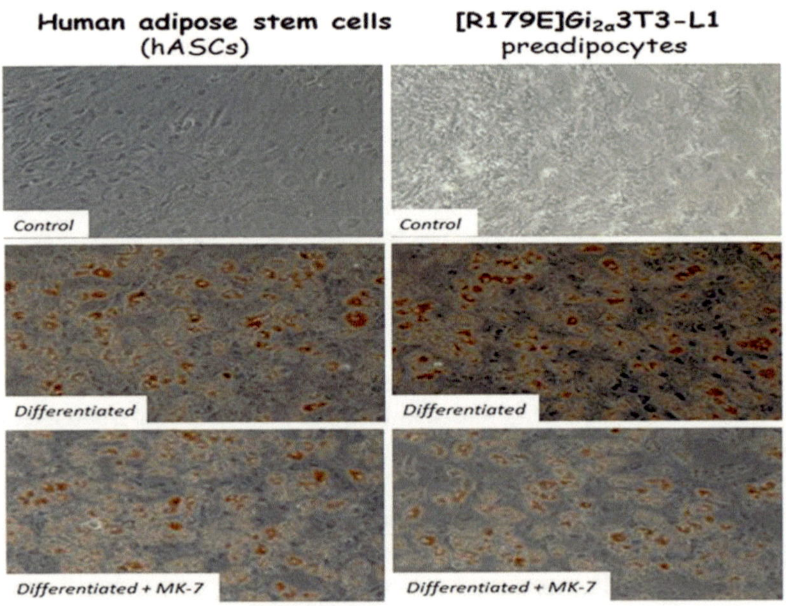

Figure 1. Human adipose stem cells and mutated 3T3-L1 preadipocytes were differentiated towards mature adipocytes. *Top panel*: Control cells are not manipulated with differentiating medium show lack of ability to store lipids (no Alizarin Red accumulation). Cells are differentiated in the presence of insulin, as well as IBMX (an inhibitor of phosphodiesterase) accumulate lipids, while cells are also exposed to vitamin K2 (MK-7) and lose their ability to produce and store triglycerides. *Bottom panel*: Percentage modulation relative to the "control" stage seen with exposure of human adipocytes and mutated 3T3-L1 cells to differentiating medium versus exposure to vitamin K2 (MK-7). Cells treated with vitamin K2 exhibit significant increments in genes characterizing beige adipocytes, where the alterations in PGC1α, PPARα, PPARγ, Dio2, and UCP1 expression were more prominent.

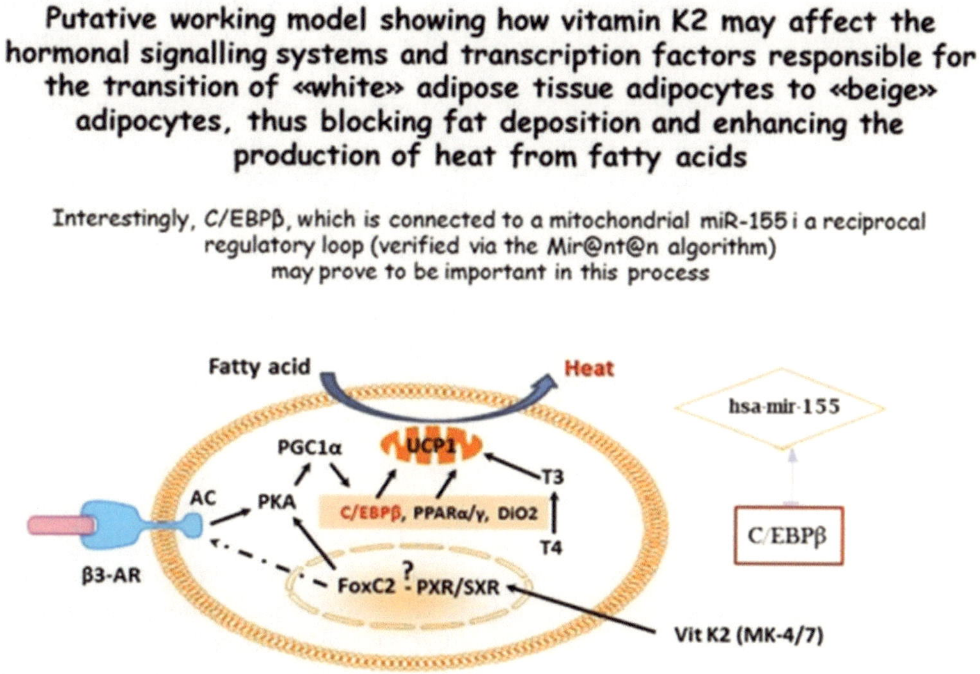

Figure 2. Putative model system featuring the mechanism behind the impact of vitamin K2 (MK-7). Based on previously published consensus material, our new findings of the effect of vitamin K2 (MK-7) shows that K2, via binding to the transcription factor PXR/SXR affects several signaling molecules and/or pathways, via the β3-adrenergic system, impinging on signaling molecules like PKA, PGC1α, C/EBPβ, and Dio2, which eventually affects the activity of the uncoupling protein UCP1, which produces heat (and not ATP) from fatty acids. Not mentioned in the text is the reciprocal regulatory loop consisting of the microRNA species hsa-mir-155 and the transcription factor C/EBPβ, which can be manipulated to reinforce the "beige" phenotype of the adipocyte after differentiation from the stem cells or preadipocytes.

2. Vitamin K2 and its mechanism of action – beyond xenobiotic metabolism

The steroid and xenobiotic receptor (SXR), which is synonymous with the pregnane X receptor = PXR, is characterized as a nuclear hormone receptor that is stimulated by a plethora of hormones, dietary steroids, pharmaceutically active agents, as well as xenobiotic compounds. SXR exhibits a binding domain, which diverges across mammalian species. SXR is construed as a xenobiotic sensor, facilitating xenobiotic clearance in the liver and intestine. However, newly published experiments unravel novel roles for SXR in modulating phenomena like inflammation, bone turnover, metabolism of vitamin D, lipid, and energy homeostasis, as well as cancer. The characterization of SXR as conveyer of hormone-like signals has now been recognized as a key instrument for the study of novel mechanisms, through which diet may ultimately affect health and disease. The discovery and pharmacological development of new PXR modulators, like vitamin K2, might represent an interesting and innovative therapeutic approach to combat various ailments and diseases.

Without elaborating on the details of how different natural products modulate the activity SXR to affect gene expression, it should be noted that St. John's wort (hypericum perforatum), vitamin E (tocopherols and tocotrienols), as well as sulforaphane and *Coleus forskohlii*, the latter producing forskolin which is able to stimulate adenylate cyclase activity, suffice to say that K2 emerges as an interesting "player" on the scene featuring major metabolic pathways accounting for a plethora of actions to be reckoned with [16]. One good example is the effect of SXR activation on the metabolism of cholesterol and lipid turnover, a second one is its impact on the Fox transcription factors (i.e., FoxO1 and FoxA2), and their influence on energy homeostasis [16]. FoxO1 and FoxA2 are both members of the "forkhead" family tree of transcription factors, serving critical roles in both lipid metabolism, as well as gluconeogenesis in the liver [17]. FoxO1 stimulates hepatic gluconeogenesis during fasting through the activation of gluconeogenic genes, such as PEPCK1 (phosphoenolpyruvate carboxykinase 1), G6P (glucose-6-phosphatase), as well as and insulin-like growth factor-binding protein 1.

FoxA2, on the other hand serves as a key switch, representing one of the regulatory factors of the breakdown of hepatic fatty-acids during calorie-restriction (i.e., fasting). Via mammalian cell-based two-hybrid screening, it was feasible to identify FoxO1 as a coactivator of both CAR (constitutive androstane receptor)-mediated, as well as SXR-stimulated transcription [18]. FoxO1 may directly associate with CAR and SXR as a hormone-ligand-receptor complex and stimulate their transcriptional ability. And, both CAR and SXR function as corepressors of FoxO1, suppressing the FoxO1-mediated transcription by counteracting its association with its response elements within the susceptible genes. As well as obliterating the FoxO1 activity, drug-stimulated SXR and CAR species were also shown to downregulate HNF4α transcriptional power via suppression of PGC1α, thus suppressing the transcription of both PEPCK1 and G6P [19]. This indicates that metabolic turnover of both drugs and glucose, the two major functions of the liver which are regulated independently, happens to be reciprocally coregulated via communication between xenobiotic sensors on one hand, and transcription factors in the liver, on the other. When blood sugar concentrations are rendered low due to fasting or subsequent to periods of exercise, the liver funnels energy rich molecules to extra-hepatic tissues and peripheral organs via either β-oxidation or the production of ketone bodies [20]. FoxA2 is stimulating both ketogenesis and β-oxidation via enhancement of the transcriptional activity of a plethora of genes, such as mitochondrial 3-hydroxy-3 methylglutarate-CoA synthase 2 (HMGCS2) and carnitine palmitoyltransferase 1A (CPT1A) in energy-depleted conditions (fasting), or subsequent to periods where extensive exercise prevails [21].

FoxA2 is phosphorylated and thus inactivated by the Akt pathway, and serves to decrease lipid turnover in response to insulin. Treatment with other drugs such as barbiturates has been shown to suppress lipid metabolism in an insulin-independent fashion [22]. Furthermore, it was reported that SXR may crosstalk with FoxA2 in order to induce the repression of lipid turnover in livers of fasting mice. By applying wild-type and SXR-/- animals, it was demonstrated that treatment with PCN (pregnenolone-16d-carbonitrile) diminished the steady-state mRNA levels of HMGCS2 and CPT1A in control animals, but not in SXR-deficient mice. When conducting biochemical and cell-based analyses, it was shown that SXR markedly downregulates the ability of FoxA2 to bring about activation of the HMGCS2 and CPT1A genes. SXR that serves as the ligand binding domain will associate directly with the DNA-binding area of

FoxA2, and the ensuing interaction will halt the "coupling" between FoxA2 and its response in DNA elements. The communication between SXR and FoxO1/FoxA2 signifies that SXR, apart from being a modulator of drug metabolism in the liver, is also serving as an important regulator in hepatic in glucose and energy homeostasis. Hence, since vitamin K2 binds to and enhances SXR-mediated transcription, it may serve as the natural "drug" to ingest, when one is aiming to treat insulin resistance and type II diabetes mellitus. The impact of vitamin K2 via SXR on the metabolic functions in general, is described in more detail in the forthcoming paragraphs ("SXR in glucose handling" and "SXR in lipid turnover") [22].

In **Figure 3**, we show our own experiments indicating that the adipocyte phenotype, characterized by the expression of the transcription factor PPARγ, is heavily dependent on the presence of FoxO1 and FoxA3, respectively, since siRNA against either of them completely obliterate the stimulatory effect of vitamin K2 obtained through its binding to SXR. However, the literature in general advocates a plethora of PIK3/Akt-stimulated transcription factors of the FoxA and FoxO families in the cascade of insulin/IGF-1 mediated signaling in general. However, the present data show, for the first time that vitamin K2 (i.e., the MK-7 variant) directly stimulates FoxO1 and FoxA3, short-cutting the insulin/IGF-1 activation cascade. The assertion to be drawn is simply: Vitamin K2 may directly fortify the action of insulin, thus ensuring a better glucose homeostasis, as well as protection from a detrimental turnover of lipids and protein structures of the body.

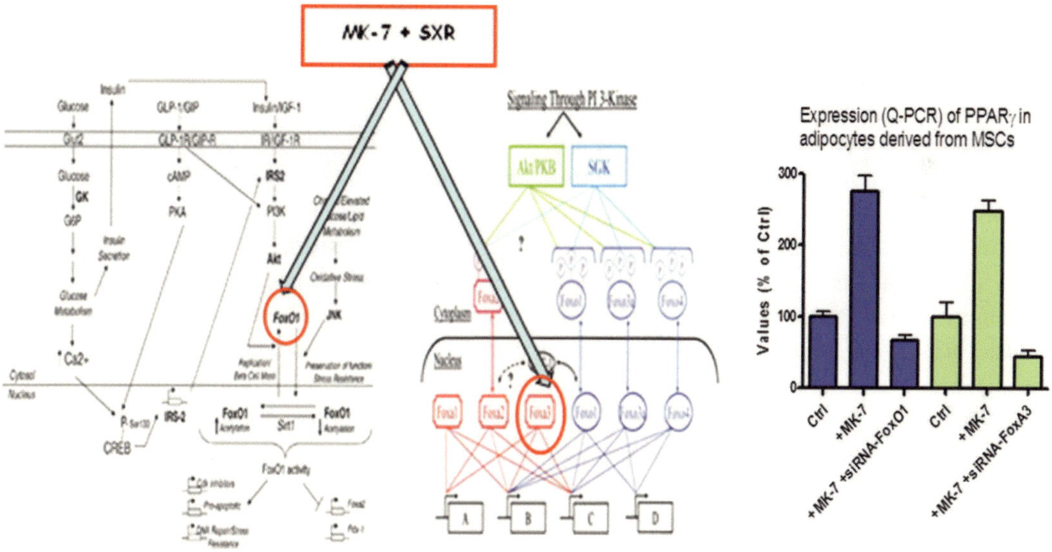

Figure 3. The impact of vitamin K2 (in the form of MK-7) on the signaling mechanism of insulin, via the PIK3/Akt/FoxO/FoxA cascade. *Left panel*: MK-7, bound to the transcription factor SXR activates FoxO1 and FoxA3, with ensuing stimulation of Sirt1. *Right panel*: Expression (mRNA analyses) of the transcription factor PPARγ in adipocytes generated from stem cells, either stimulated by vitamin K2 (MK-7) or inhibited in the presence of siRNA against FoxO1 or FoxA3. Reference to the figures: Left – DN Gross et al. (Oncogene, 2320–2336, 2008); Michael P. Czech (PNAS, 11198–11200, September 30, 2003).

3. SXR cross talks with biological signals in biological responses: implications in health and disease

3.1. The molecular mechanisms of SXR-mediated gene repression

Currently, SXR has been described as a repressor of gluconeogenic gene expression, some of which are glucose-6-phosphatase (G6Pase) and phosphoenolpyruvate carboxykinase 1 (PEPCK1), thus implicating vitamin K2 in metabolic (energy-related) reactions taking place in the liver [23–25]. The SXR-vitamin K2 complex may therefore interfere directly with transcription factors and thus be rendered responsive to both insulin and glucagon. Consequently, one would observe a release of transcription factors with their coactivators from target genes, which would lead to a general subactivation of gene transcription.

Furthermore, it has been asserted that SXR represses the CYP-genes through interference with the vitamin D receptor, VDR [26]. Hence, in tissues short of vitamin D3, liganded SXR would associate directly to vitamin D response elements (VDREs) and modulate transcription. Therefore, it may be asserted that vitamin K2 and vitamin D, as well as vitamin A (via VDR and RXA, respectively), and many other transcription factor like molecules (e.g., PPARs, FXR, LXRα, LRH-1 = NR5A2, RXR), when associated with their ligands, may act synergistically on gene transcription in general [16, 27–29]. Hence, it is not straightforward to predict the net results of a certain combination of liganded transcription factors on biological processes. However, there are several excellent reports on the impact of vitamin K2, in association with the nuclear factor SXR, on cellular metabolism.

3.2. Involvement of SXR in metabolic functions

Xenobiotics are able to enhance SXR-mediated expression of xenobiotic-metabolizing enzymes in both the liver and intestines. Even though such a modulation normally serves to detoxify the xenobiotics in question, these "alien molecules" enhance the production of intermediates, which confer harmful attack of tissues on the body [30, 31]. Additionally, SXR affects the balance of endobiotics (e.g., steroid hormones, cholesterols, and bile acids) aided by the same biochemical pathways. Hence, an SXR activation will consequently stimulate a plethora of physiological responses, i.e., in the liver, which plays an important role in the processing of, among many substances, glucose and lipids. Disruption of their metabolic fate may result in diseases, of which type II diabetes (T2DM) and obesity are most frequently encountered. Newly published studies of SXR-KO and SXR-humanized animals clearly shed light on the metabolic functions of SXR in man.

3.3. SXR and its role in energy metabolism of the liver

The liver provides energy sources to the rest of the body, i.e., carbohydrates and lipids are catabolized in order to fuel both central and peripheral tissues and organs. The effect of SXR on hepatic energy turnover was discovered with the aid of SXR-KO mice [25, 32, 33].

3.4. SXR in glucose handling

The pancreatic hormones (insulin and glucagon) reciprocally regulate the blood glucose level via transcriptional processes, where rate-limiting enzymes like G6Pase and PEPCK1 in the glucose metabolism play a decisive role [34–36]. The glucose-6-phosphatase dephosphorylates glucose-6-phospate (G6P), constituting the endpoint of both the glucose-forming and glucose-utilizing reactions, while PEPCK1 converts oxaloacetate to phosphoenolpyruvate (PEP) in the gluconeogenic reaction. These enzymes are instrumental in controlling blood glucose levels. During fasting and/or prolonged exercise, glucagon dominates glucose metabolism by activating the cAMP/PKA signaling pathway [35, 37, 38]. When phosphorylated by PKA, the cAMP-response element-binding protein (CREB) stimulates both the G6Pase and PEPCK1.

However, insulin acts in an opposite manner, and in response to high blood glucose levels, insulin is secreted and activates the phosphoinositide 3-kinase (PI3K)/Akt signaling pathway [39]. Thereafter, Akt phosphorylates and inactivates the transcription factor forkhead box O1 (FoxO1). FoxO1, being a key regulator of glucose turnover, subsequently stimulates the insulin response sequence (IRS)-bearing genes G6Pase and PEPCK1 [40]. When phosphorylation by Akt, FoxO1 cannot longer translocate to the nucleus, it is rapidly acetylated and therefore loses its activity [41, 42].

But, when using SXR-KO and SXR-humanized animals, it has been demonstrated that SXR is an important regulator of xenobiotic-dependent glucose turnover in the liver. Treatment of mice with potent SXR activators consistently leads to lowered blood glucose levels in laboratory animals [32]. Furthermore, it was shown that there existed so-called "cross-talk" between SXR and FoxO1 as a molecular mechanism underlying the downregulation of glucose metabolism [18]. It was reported by Kodama and Negishi that liganded SXR directly interacts with phosphorylated CREB in primary hepatocytes [25], and that SXR disturbs the binding of CREB to CRE with an ensuing repression of the CREB-mediated transcription of G6Pase and PEPCK1 genes. Looking at hitherto available information, it can be asserted that SXR, by "targeting" a plethora of factors modulated by insulin and glucagon, leads to an activation of many genes being functional in the intrinsic regulatory machinery maintaining serum glucose levels within "healthy" limits.

3.5. SXR in lipid turnover

The liver provides lipid-derived energy-rich compounds to different parts of the body. Hepatic lipid metabolism is controlled by the net influence of the "reciprocal" hormones insulin and glucagon, as well as by nutritional conditions. Some 10 years ago, Kodama and Negishi, along with Nakamura [25, 32], published that treatment with PCN (an activator of SXR) decreased the mRNA levels of carnitine palmitoyltransferase 1a (CPT1a) and 3-hydroxy-3-methylglutarate-CoA synthase 2 (Hmgcs2) in livers of starved wild-type mice, but not in SXR-KO mice [32]. CPT1a is instrumental in the overall mitochondrial β-oxidation by funneling long-chain fatty acids into mitochondria [43], and the mitochondrial enzyme HMGCS2 facilitates the initial reaction of ketogenesis [44].

Additionally, the stimulation of SXR by PCN has been demonstrated to enhance the mRNA steady state level of stearoyl-CoA desaturase 1 (Scd1) in hepatic tissue of starved wild-type experimental animals. SCD1, which serves as a key enzyme in hepatic lipogenesis, facilitates the rate-limiting step in the synthesis of unsaturated fatty acids [45]. The plasma concentrations of 3-OH-butylate were decreased, while the hepatic level of triglyceride (TG) was increased by the PCN treatment in wild-type mice during assay conditions. However, neither TG nor cholesterol levels in the blood were altered in those animals, despite the fact that there was a significant rise in TG accumulated in their liver. Hence, as a means of survival during fasting, SXR is thought to slow down hepatic lipid turnover by repressing β-oxidation and ketogenesis, while stimulating the transcription of lipogenic enzymes, in much the same way as induced by insulin.

The Akt-regulated forkhead transcription factor FoxA2 which serves as a facilitator of insulin-dependent modulation of β-oxidation and ketogenesis, enhances expression of both the CPT1a and HMGCS2 gene, respectively [21, 46]. It is well known that Insulin activates the PI3K/Akt signaling pathway to phosphorylate FoxA2, in order to translocate it from nucleus to cytosol, thereby downregulating both the genes. And, it has been asserted that a direct interaction between SXR and FoxA2 serves as the mechanism, by which SXR represses the transcription of CPT1a and HMGCS2 in the liver [32].

A plethora of transcription factors and coregulators have been asserted to serve as modulators of hepatic lipid metabolism, e.g., the peroxisome proliferator-activated receptors (PPARs), the liver X receptor α (LXRα), as well as the sterol regulatory element-binding proteins (SREBPs) [47]. The expression of SREBP1c, which is construed as the dominant regulator of hepatic lipogenesis, is under the control of LXRα, and mediates the insulin- and fatty acids-dependent responses of lipogenic genes such as fatty acid synthase (FAS), acetyl-CoA carboxylase 1 (ACC1), stearoyl-CoA-desaturase-1 (SCD1), and fatty acid elongase (FAE). SXR is believed to upregulate lipogenesis in the liver, independently of SREBP1c action, and it is not deemed to be associated with the steady state expression levels of both the Fas and Acc1 genes. Among the cluster of lipogenic genes, Cd36 (cluster of differentiation 36) is deemed to be serve as a direct target of SXR in the liver. And, upon stimulation by ligands, the receptor is believed to become recruited to a DR3-type SXR response element within its promoter region of the liver of experimental animals [48]. Furthermore, SXR has been asserted to serve as a link, facilitating the upregulation of the Pparγ-gene, which functions as a strong regulator of lipid-synthesizing enzymes [48]. Such a cross-talk involving nuclear receptors should confer a significant impact on the body's lipid homeostasis. Our data are in line with the published literature, however, it should be asserted that SXR probably affects a larger spectrum of FoxO and FoxA species than those presented in this review. In this way, one might speculate that SXR is able to recruit a "moving" representation of these transcription factors simultaneously, and that the net effect on various cell phenotypes depends on: (1) the distribution of FoxOs and FoxAs at any time within the cell or tissue, as well as (2) the epigenetic machinery or "make up" at any time within the same cells or tissues.

4. The effect of vitamin K2 on other genes related to metabolic processes in the cell

In 2009, Slatter [49] and coworkers published a paper, featuring oligonucleotide microarrays with the intention to reveal the heterogeneity of drug metabolism associated gene expression in liver tissue from healthy humans. Their intention was to define clusters of so-called "absorption, distribution, metabolism, and excretion" = ADME genes to define subgroups of coregulated genes. When analyzing the gene sets, they discovered distinct patterns of "parallel" gene expressions featuring gene "clusters", which proved to be modulated by the nuclear receptor SXR. So called "fold range metrics and frequency distributions" were applied in order to reveal the variability of solitary PKDM genes. The most variable gene entities chiefly correlated to: (1) drug metabolism, (2) intermediary metabolism, (3) inflammation, and (4) cell cycle control. Unique expression patterns of these genes allowed for a further correlation with a parallel expression of a plethora of other genes. Of major interest was the identification of SXR responsive genes.

A comprehensive list of these genes can be found in the article, however, quite a few of which are related to metabolic processes in the cell. The genes are the following (in alphabetical order): CLOCK, DUSP7, GCDH, IGFBP2, MAP2K2, NUCB2, OGT, PFKB1, PTPN11, and SLC16A2. By "looking up" current descriptions of the genes in "Gene-Cards", the following features of these SXR-sensitive genes were obtained (of which parts of their description is cited as presented):

Name of gene	Description of cellular function(s) http://www.genecards.org/cgi-bin/carddisp.pl?gene=NR1I2
CLOCK	Clock circadian regulator (The protein encodes a transcription factor, and serves as DNA-binding histone acetyl transferase). *Interpretation:* Polymorphisms in this gene may be associated with obesity and metabolic syndrome. CLOCK normally regulates gene products (proteins) in an optimal fashion, adapted to diurnal demands on the body related to food ingestion, physical activity and recreation/sleep.
DUSP7	Dual-specific phosphatase (DUSPs) constitutes a large subgroup of cysteine-base protein-tyrosine phosphatases characterized by their ability to dephosphorylate both tyrosine and serine/threonine residues. *Interpretation:* DUSP7 may function as a modulator of cellular exposure to insulin and growth factors, ensuring energy homeostasis within optimal limits.
GCDH	The protein encoded by this gene belongs to the acyl-CoA dehydrogenase family. It catalyzes the oxidative carboxylation of glutaryl-CoA to crotonyl-CoA and CO_2 in the degradative pathway of L-lysine, L-hydroxylysine, and L-tryptophan metabolism. *Interpretation:* GCDH is involved in the stability of mitochondria, and hence energy metabolism in general.
IGFBP2	Insulin-like growth factor binding protein, type 2. This protein inhibits IGF-mediate growth. *Interpretation:* A reduction in IGFBP2 may be responsible for organ hyperplasia and the development of neoplasia (cancer).
MAP2K2	This MAP-kinase catalyzes the concomitant phosphorylation of a threonine and a tyrosine residue in a Thr-Glu-Tyr sequence located in MAP-kinases. It activates the ERK1 and ERK2 MAP-kinases (by similarity). *Interpretation:* This kinase may block tumorigenesis and normalize energy metabolism via MEK1/2 and the FoxO- and FoxA-family of transcription factors.
NUCB2	Anorexigenic peptide; seems to play an important role in hypothalamic pathways regulating food intake and energy homeostasis, acting in a leptin-dependent manner. *Interpretation:* Appetite regulator fortifying the effect of leptin, but independent of the size of fat depots.

OGT	Glycosylates a substantial and diverse amount of proteins, encompassing species like histone H2B, AKT1 and PFK (phosphofructokinase). It can modulate their cellular processes through cross-talk between processes like glycosylation and phosphorylation, or via proteolytic processing. Involved in insulin sensitivity in muscle cells and adipocytes by glycosylating components of insulin signaling, blocks phosphorylation of AKT1, stimulates IRS1 phosphorylation, as well as attenuating insulin signaling. *Interpretation:* Modulator of insulin and IGF-1 signaling/sensitivity to maintain a healthy muscle tissue fat mass and distribution.
PFKFB1	Encodes a member of the family of the bifunctional 6-phophofructo-2 kinase: fructose-2, 6-bisphosphatase enzymes. These enzymes form homodimers, which catalyze the synthesis, as well as the degradation of fructose 2, 6-bishosphate, via independent catalytic domains. Fructose-2, 6-bisphosphate serves as the activator of the glycolytic pathway, and as the inhibitor of the gluconeogenetic pathway. *Interpretation:* Regulating fructose-2,6-bisphosphate levels through the activity of this enzyme is thought to regulate glucose homeostasis. Multiple alternatively spliced transcript variants have been found for this gene.
PTPN11	PTP (protein tyrosine phosphatase) is a member a large family of phosphatases and plays a regulatory role in various cell signaling events that are important for a diversity of cell functions, such as mitogenic activation, metabolic control, transcription regulation, and cell migration. *Interpretation:* Because of the activating effects of PTPN11 on ERK (extracellular signal regulated kinase), a lack of PTPN11 activation may lead to adiposity, diabetes, and hyperleptinemia.
SLC16A2	Very active and specific thyroid hormone transporter molecule. Stimulates cellular uptake of thyroxine (T4), triiodothyronine (T3), reverse triiodothyronine (rT3), and diiodothyronine. *Interpretation:* Lack of SLC16A2 activation may lead to a reduction in uptake and biological functions of T4 and T3, which is associated with adiposity, diabetes, and hyperleptinemia.

Author details

Mona Møller[1*], Serena Tonstad[2], Tone Bathen[3] and Jan Oxholm Gordeladze[4]

*Address all correspondence to: mona.moller@kappabio.com

1 KappaBio, Oslo, Norway

2 Ullevål Hospital, Oslo, Norway

3 NTNU, Trondheim, Norway

4 University of Oslo, Oslo, Norway

References

[1] Virtanen KA, Lidell ME, Orava J, Heglind M, Westergren R, Niemi T, Taittonen M, Laine J, Savisto NJ, Enerback S, et al. Different metabolic responses of human brown adipose tissue to activation by cold and insulin. *Engl. N. Med. J.* **360** (2009) 1518–1525.

[2] Gesta S, Tseng YH, Kahn CR. Developmental origin of fat: tracking obesity to its source. *Cell.* **131** (2007) 242–256.

[3] Lowell BB, Susulic VS, Hamann A, Lawitts JA, Himms-Hagen J, Boyer BB, Kozak LP, Flier JS. Development of obesity in transgenic mice after genetic ablation of brown adipose tissue. *Nature.* **366** (1993) 740–742.

[4] Seale P, Bjork B, Yang W, Kajimura S, Chin S, Kuang S, Scime A, Devarakonda S, Conroe HM, Erdjument-Bromage H, et al. PRDM16 controls a brown fat/skeletal muscle switch. *Nature.* **454** (2008) 961–967.

[5] Wu J, Bostrom P, Sparks LM, Ye L, Choi JH, Giang AH, Khandekar M, Virtanen KA, Nuutila P, Schaart G, et al. Beige adipocytes are a distinct type of thermogenic fat cell in mouse and human. *Cell.* **150** (2012) 366–376.

[6] Fisher FM, Kleiner S, Douris N, Fox EC, Mepani RJ, Verdeguer F, Wu J, Kharitonenkov A, Flier JS, Maratos-Flier E, et al. FGF21 regulates PGC-1α and browning of white adipose tissues in adaptive thermogenesis. *Genes Dev.* **26** (2012) 271–281.

[7] Bostrom P, Wu J, Jedrychowski MP, Korde A, Ye L, Lo JC, Rasbach KA, Bostrom EA, Choi JH, Long JZ, et al. A PGC1-α-dependent myokine that drives brown-fat-like development of white fat and thermogenesis. *Nature.* **481** (2012) 463–468.

[8] Cederberg A, Gronning LM, Ahren B, Tasken K, Carlsson P, Enerback S. FOXC2 is a winged helix gene that counteracts obesity, hypertriglyceridemia, and diet-induced insulin resistance. *Cell.* **106** (2001) 563–573.

[9] Seale P, Conroe HM, Estall J, Kajimura S, Frontini A, Ishibashi J, Cohen P, Cinti S, Spiegelman BM. Prdm16 determines the thermogenic program of subcutaneous white adipose tissue in mice. *J. Clin. Invest.* **121** (2011) 96–105.

[10] Ohno H, Shinoda K, Spiegelman BM, Kajimura S. PPARγ agonists induce a white-to-brown fat conversion through stabilization of PRDM16 protein. *Cell Metab.* **15** (2012) 395–404.

[11] Qiang L, Wang L, Kon N, Zhao W, Lee S, Zhang Y, Rosenbaum M, Zhao Y, Gu W, Farmer SR, et al. Brown remodeling of white adipose tissue by SirT1-dependent deacetylation of Pparγ. *Cell.* **150** (2012) 620–632.

[12] Cohen P, Levy JD, Zhang Y, Frontini A, Kolodin DP, Svensson KJ, Lo JC, Zeng X, Ye L, Khandekar MJ, et al. Ablation of PRDM16 and beige adipose causes metabolic dysfunction and a subcutaneous to visceral fat switch. *Cell.* **156** (2014) 304–316.

[13] Lecka-Czernik B, Stechschulte LA. Protein Phosphatase PP5 Controls Bone Mass and the Negative Effects of Rosiglitazone on Bone through Reciprocal Regulation of PPARγ (Peroxisome Proliferator-activated Receptor γ) and RUNX2 (Runt-related Transcription Factor 2). *Arch. Biochem. Biophys.* **561** (2014) 124–129.

[14] Aguirre L, Napoli N, Waters D, Qualls C, Villareal DT, Armamento-Villareal R. Increasing adiposity is associated with higher adipokine levels and lower bone mineral density in obese older adults. *Clin. J. Endocrinol. Metab.* **99**(9) (2014) 3290–3297.

[15] Compston J. Obesity and bone. *Curr. Osteoporos. Rep.* **11** (2013) 30–35.

[16] Zhou C, Verma S, Blumberg B. The steroid and xenobiotic receptor (SXR), beyond xenobiotic metabolism. *Nucl. Recept. Signal.* **7** (2009) 1–21.

[17] Montminy M, and Koo SH. PGC-1 promotes insulin resistance in liver through PPAR-alpha-dependent induction of TRB-3. Diabetes: outfoxing insulin resistance? *Nature.* **432** (2004) 958–9.

[18] Kodama S, Koike C, Negishi M. and Yamamoto Y. Nuclear receptors CAR and PXR cross talk with FOXO1 to regulate genes that encode drug-metabolizing and gluconeogenic enzymes. *Mol. Cell. Biol.* **24** (2004) 7931–40.

[19] Miao J, Fang S, Bae Y. and Kemper JK. Functional inhibitory cross-talk between constitutive androstane receptor and hepatic nuclear factor-4 in hepatic lipid/glucose metabolism is mediated by competition for binding to the DR1 motif and to the common coactivators, GRIP-1 and PGC-1alpha. *J. Biol. Chem.* **281** (2006) 14537–46.

[20] Eaton S, Bartlett K. and Pourfarzam M. Mammalian mitochondrial β-oxidation. *Biochem J.* **320** (Pt 2) (1996) 345–57.

[21] Wolfrum C, Asilmaz E, Luca E, Friedman JM. and Stoffel M. Foxa2 regulates lipid metabolism and ketogenesis in the liver during fasting and in diabetes. *Nature.* **432** (2004) 1027–32.

[22] Kiyosawa N, Tanaka K, Hirao J, Ito K, Niino N, Sakuma K, Kanbori M, Yamoto T, Manabe S. and Matsunuma N. Molecular mechanism investigation of phenobarbital-induced serum cholesterol elevation in rat livers by microarray analysis. *Arch. Toxicol.* **78** (2004) 435–42.

[23] Bhalla S, Ozalp C, Fang S, et al. Ligand-activated pregnane X receptor interferes with HNF-4 signaling by targeting a common coactivator PGC-1a. Functional implications in hepatic cholesterol and glucose metabolism. *J. Biol. Chem.* **279** (2004) 45139–45147.

[24] Kodama S, Koike C, Negishi M, Yamamoto Y. Nuclear receptors CAR and PXR cross talk with FOXO1 to regulate genes that encode drug-metabolizing and gluconeogenic enzymes. *Mol. Cell. Biol.* **24** (2004) 7931–7940.

[25] Kodama S, Moore R, Yamamoto Y, Negishi M. Human nuclear pregnane X receptor cross-talk with CREB to repress cAMP activation of the glucose-6-phosphatase gene. *Biochem. J.* **407** (2007) 373–381.

[26] Konno Y, Kodama S, Moore R, et al. Nuclear xenobiotic receptor pregnane X receptor locks corepressor silencing mediator for retinoid and thyroid hormone receptors (SMRT) onto the CYP24A1 promoter to attenuate vitamin D3 activation. *Mol. Pharmacol.* **75** (2009) 265–271.

[27] Francis GA, Fayard E, Picard F, and Auwerx J. Nuclear Receptors and the Control of Metabolism. *Ann. Rev. Physiol.* **65** (2003) 261–311.

[28] Bügel S ,Vitamin K, and bone health in adult humans. *Vitam. Horm.* **78** (2008) 393–416.

[29] Ahmadieh H, Arabi A. Vitamins and bone health: beyond calcium and vitamin D. *Nutr Rev.* **69**(10); (2011) 584–598. doi: 10.1111/j.1753-4887.2011.00372.x.

[30] Cheng J, Ma X, Krausz KW, et al. Rifampicin-activated human pregnane X receptor and CYP3A4 induction enhance acetaminopheninduced toxicity. *Drug Metab. Dispos.* **37** (2009) 1611–1621.

[31] Guo GL, Moffit JS, Nicol CJ, et al. Enhanced acetaminophen toxicity by activation of the pregnane X receptor. *Toxicol. Sci.* **82** (2004) 374–380.

[32] Nakamura K, Moore R, Negishi M, Sueyoshi T. Nuclear pregnane X receptor cross-talk with FoxA2 to mediate drug-induced regulation of lipid metabolism in fasting mouse liver. *J. Biol. Chem.* **282** (2007) 9768–9776.

[33] Zhou J, Zhai Y, Mu Y, et al. A novel pregnane X receptormediated and sterol regulatory element-binding protein-independent lipogenic pathway. *J. Biol. Chem.* **281** (2006) 15013–15020.

[34] Hanson RW, Reshef L. Regulation of phosphoenolpyruvate carboxykinase (GTP) gene expression. *Annu. Rev. Biochem.* **66** (1997) 581–611.

[35] Jiang G, Zhang BB. Glucagon and regulation of glucose metabolism. *Am. J. Physiol. Endocrinol. Metab.* **284** (2003) E671–E678.

[36] van Schaftingen E, Gerin I. The glucose-6-phosphatase system. *Biochem.* J. **362** (2002) 513–532.

[37] Gonzalez GA, Montminy MR. Cyclic AMP stimulates somatostatin gene transcription by phosphorylation of CREB at serine 133. *Cell.* **59** (1989) 675–680.

[38] Herzig S, Long F, Jhala US, et al. CREB regulates hepatic gluconeogenesis through the coactivator PGC-1. *Nature.* **413** (2001) 179–183.

[39] Taniguchi CM, Emanuelli B, Kahn CR. Critical nodes in signalling pathways: Insights into insulin action. *Nat. Rev. Mol. Cell Biol.* **7** (2006) 85–96.

[40] Barthel A, Schmoll D, Unterman TG. FoxO proteins in insulin action and metabolism. *Trends Endocrinol. Metab.* **16** (2005) 183–189.

[41] Matsuzaki H, Daitoku H, Hatta M, et al. Acetylation of Foxo1 alters its DNA-binding ability and sensitivity to phosphorylation. *Proc. Natl. Acad. Sci. USA.* **102** (2005) 11278–11283.

[42] Schmoll D, Walker KS, Alessi DR, et al. Regulation of glucose-6-phosphatase gene expression by protein kinase Ba and the forkhead transcription factor FKHR. Evidence for insulin response unit-dependent and independent effects of insulin on promoter activity. *J. Biol. Chem.* **275** (2000) 36324–36333.

[43] Louet JF, Le May C, Pegorier JP, et al. Regulation of liver carnitine palmitoyltransferase I gene expression by hormones and fatty acids. *Biochem. Soc. Trans.* **29** (2001) 310–316.

[44] Hegardt FG. Mitochondrial 3-hydroxy-3-methylglutaryl-CoA synthase: A control enzyme in ketogenesis. *Biochem. J.* **338** (1999) 569–582.

[45] Flowers MT, Ntambi JM. Role of stearoyl-coenzyme A desaturase in regulating lipid metabolism. *Curr. Opin. Lipidol.* **19** (2008) 248–256.

[46] Wolfrum C, Asilmaz E, Luca E, et al. Foxa2 regulates lipid metabolism and ketogenesis in the liver during fasting and in diabetes. *Nature.* **432** (2004) 1027–1032.

[47] Weickert MO, Pfeiffer AF. Signalling mechanisms linking hepatic glucose and lipid metabolism. *Diabetologia.* **49** (2006) 1732–1741.

[48] Zhou J, Zhai Y, Mu Y, et al. A novel pregnane X receptormediated and sterol regulatory element-binding protein-independent lipogenic pathway. *J. Biol. Chem.* **281** (2006) 15013–15020.

[49] Slatter JG, Templeton IE, Castle JC, Kulkarni A, Rushmore TH, Richards K, He Y, Dai X, Cheng OJ, Caguyang M, Ulrich RG. Compendium of gene expression profiles comprising a baseline model of the human liver drug metabolism transcriptome. Xenobiotica. **36**(10-11) (2006) 938–962.

Section 3

The Impact of Vitamnin K2 on Bones and Teeth

Chapter 6

Vitamin K2 and Bone Health

Niels Erik Frandsen and Jan Oxholm Gordeladze

Additional information is available at the end of the chapter

http://dx.doi.org/10.5772/64876

Abstract

During the last 20 years, the main clinical effects of vitamin K2 on bone homeostasis have been investigated in both indirect and direct vitamin K treatment regimens. This chapter is mainly based on randomized clinical trials (RCT) lasting for more than 1 year. As for vitamin K1 (phylloquinone, indirect treatment) and vitamin K2 (menaquinone MK-4 and MK-7 direct treatment), respectively, the clinical trials have consistently shown decreased fracture rate incidents, however, mainly in Asian populations. In 2013, a major breakthrough was observed by Knapen et al. in the Netherlands, where menaquinone MK-7 supplementation of 180 μg/day for 3 years to healthy postmenopausal women significantly decreased the age-related decline in BMC (bone mineral contents) and BMD (bone mineral density) at the lumbar spine and femoral neck, but not at the total hip, as compared to placebo. Thus, MK-7 supplementation has shown a significant "double"-positive action through (1) increased bone building and (2) decreased bone resorption. We look forward to seeing the clinical effects on low bone mass and osteoporosis as well as other bone diseases.

Keywords: bone health, RCT trials, vitamin K1, vitamin K2 (MK-4, menatetrenone), menaquinone-7 (MK-7)

1. Introduction

Vitamin K2 (menaquinone-4, MK-4, or menatetrenone) is a very important vitamin K species serving special functions in several extrahepatic organs, like bone tissue, heart, blood vessels, kidneys, brain, and cartilage. MK-4 is a member of a sub-family with eliciting the same cellular reactions, but with different effects. MK-4 is deemed necessary for γ-carboxylation of proteins, and activation of the vitamin K-dependent proteins, i.e., Osteocalcin (bone-Gla-protein), matrix-Gla proteins (MGPs), Periostin, as well as protein S. Without these activated proteins, the body

INTECH
open science | open minds

© 2017 The Author(s). Licensee InTech. This chapter is distributed under the terms of the Creative Commons Attribution License (http://creativecommons.org/licenses/by/3.0), which permits unrestricted use, distribution, and reproduction in any medium, provided the original work is properly cited.

is not able to regulate the process of calcium uptake and bone mineralization. To date, we know of 19 different vitamin K-dependent proteins. The process also ensures the access of energy for a plethora of cellular reactions. The most common of the K vitamins is phylloquinine-1 (K1). K1 is found in leafy green vegetables, olive oil, and soybean oil, while MK-4 is found in small amounts in egg yolks, butter, and hard cheese. Fermented soybeans, known as natto, a special Japanese food, contain another subfamily member menaquinone-7 (MK-7). It is the richest dietary source of menaquinones (yielding about 1100 μg/100 g). In Western diets, MK-7 in fermented cheeses amounts to only 62 μg/100 g, approximately. Very small amounts of MK-4 are produced by intestinal bacteria. The adequate intake (AI) for vitamin K1 from 2001 in the United States and Canada was recommended to be 90 μg/day for women, 120 μg/day for men. These values are based on the need for hepatic vitamin K1 to ensure sufficient clotting factor synthesis. To date, we know that the early AI is not enough to meet the request for a complete carboxylation of all the extrahepatic vitamin K-dependent proteins. When MK-4 ingestion is insufficient, the serum concentration of uncarboxylated MGP (ucMGP) or uncarboxylated Osteocalcin (ucOC) is increased. This serves as the major sign of a deficient intake of K1 or a deficient hepatic transformation of vitamin K1 to MK-4. High serum concentrations of MK-1 and MK-4 reflect only the recent intake of vitamin K, not the accumulation or steady state levels of a "chronic" intake. The bioavailability of K1 and MK-4 is very short, i.e., about 1.5–3 h in contrast to 72 h for MK-7. To date, K1, MK-4, and MK-7 are available as dietary supplements [1, 2].

Osteoporosis is, on a worldwide basis, the reason for many emerging, spontaneous fractures. On a global basis, osteoporosis causes more than 88.9 million fractures annually. The condition is estimated to affect 200 million women worldwide, i.e., approximately one-tenth of women aged 60, one-fifth of women aged 70, two-fifths of women aged 80, and two-thirds of women aged 90 years. Worldwide, one in three women over the age of 50, will experience osteoporotic fractures, as will one in five of men aged over 50 [3]. From Europe, the number of new fractures in 2010 was estimated at 3.5 million, comprising approximately 620,000 hip fractures, 520,000 vertebral fractures, 560,000 forearm fractures, and 1,800,000 of other fractures (types or sites). For women in the EU: approximately 50% of fractures related deaths, were due to hip fractures, 28% to clinical vertebral fractures, and 22% to other types of fractures. In men, the corresponding figures were 47, 39, and 14%, respectively [3]. From the USA Preventive Services Task Force (USPSTF) for 2013, it is therefore concluded that the current evidence is insufficient to assess the balance of the benefits and harm of combined vitamin D and calcium supplementation for the primary prevention of fractures In the treatment of pre- and postmenopausal women or in men, these indicate that new treatments of osteoporosis are wanted [3, 4].

In 1984, Hart et al. [5] revealed that the serum concentrations of vitamin K1 were very low among patients with hip fractures, and hypothesized that poor vitamin K status is associated with increased rates of osteoporotic fractures. The period of 1991–1993 was regarded as the beginning of the era of vitamin K examinations, since patients-based studies by Hodges et al. confirmed that low serum concentrations of vitamin K1 and K2 were associated with increased risk of spine and hip fracture. The same phenomena were also shown by Szulc et al. in a French study, yielding a positive association between ucOC and fracture risk [6–8]. Other countries, like the Netherlands, as shown by the group of Knaben et al. [9] and Finland, as shown by Luukinen et al. [10], clearly demonstrated that this association (ucOC versus fracture risk) point to the fact that vitamin K status is a predictor of bone health (fracture-free percentage of a population group). The correlation between low intake of vitamin K and increased fracture rate was also revealed by prospective analysis within the Nurses' Health Study cohort 1984. Here, the diet was assessed in 72,327 women, aged 38–63 years, with a food-frequency baseline-questionnaire. During the subsequent 10 years of follow-up, 270 hip fractures resulting from low or moderate trauma were reported. Results: women in quintiles 2–5 of vitamin K intake had a significantly lower age-adjusted relative risk (RR: 0.70; 95% CI: 0.53, 0.93) of hip fracture than women in the lowest quintile [11].

In Japan, a high dose of menaquinone-4 (MK-4) of 45 mg/day (15 mg × 3/day) was used as therapeutic treatment for osteoporosis. The principle cause-effect of vitamin K2 on osteoporosis is mainly the prevention of bone fractures due to its improvement of bone quality, and not the ensuing increasing bone mineral density. Due to variable contents of vitamin K in the diet, the Japanese Society of Osteoporosis included in 1995 vitamin K2 as Menatetrenone (MK-4), together with vitamin D, in the first line treatment of osteoporosis [12].

Osteoporosis, in its primary form, is characterized by bone loss, and the age of females upon onset or development of menopause, however, the disease also affects a smaller cohort of men. Genetics is the single most important cause for both sexes, however, changes of lifestyle, exercise levels, smoking habits, and low body weight, are important contributors, which may trigger the onset of the disease. Secondary forms of osteoporosis-related chronic diseases are rheumatoid arthritis, chronic lung disease, and anorexia. In addition, it is well known that the use of cortisone or prednisolone can reduce bone mineral density (BMD). Standard examination of bone mineral density (BMD) is the dual energy X-ray absorptiometry (DEXA). BMD values is not the same for all women and men but different due to race. In the last decade, research have combined DEXA scan with more specific geometric hip and vertebra measurements as femoral neck width (FNW) and hip-axis length (HAL) and vertebral fracture (VF) assessment (VFA).

However, it may be asserted to a certain extent, that proper bone 3D-architecture may "make up" for a marginal reduction in BMD-values in terms of predictive value of the BMD-levels *per se*.

The incidence of osteoporosis is lower in Japanese women, even though they are subjected to the same period with menopause as other women. However, in some regions, like in Tokyo, the intake of vitamin K as MK-7 in the special fermented food of natto, soybeans together with other treatments result in higher BMD-values and lower fracture rates than age-matched woman in the United States and Europe [3].

Of other health outcome studies featuring patients with an without osteoporosis, a few warrant special attention: observational studies of subjects displaying a low long-term vitamin K intake revealed a higher incidence of osteoarthritis in the hand and knee [13], dementia [14, 15]. From population-based studies of atherosclerosis, Jie et al. revealed that, in atherosclerotic women, vitamin K status is associated with lower bone mass. All these findings support our hypothesis that vitamin K status affects the mineralization processes in both bone and atherosclerotic plaques in a healthy manner [16]. And, from the population-based Rotterdam Study, the relation between low vitamin K status and development of coronary artery disease [17] is indisputable.

The importance of vitamin D for bone health has been known since it was used in the treatment of rachitis in the 1930s. Vitamin D and calcium supplements have been recommended as a pillar in the treatment of osteoporosis over the last three decades. Vitamin D and calcium supplementation increases spinal BMD in healthy, postmenopausal women [18, 19], and vitamin D is crucial in the process of mobilizing the Ca^{2+} ions into the bone tissue [20]. Interestingly, vitamin D and calcium alone are not able to rebuild bone tissue and infrastructure that are being lost. The synergistic effect of vitamin D and retinoic acid on osteoblast production of osteocalcin was shown in 1993 [21], where Hara et al. demonstrated that MK-4 was able to partly inhibit the bone resorption induced by inflammation, vitamin D loss and ensuing PTH induction, as seen in the calvariae and incubation medium in his mode *ex vivo* model system. This observation served later as the basis for the introduction and acknowledgement of menatetrenone (MK-4) supplementation in the clinic [22]. And finally, in 1995, Hara et al. [23] also showed that the inhibition of bone resorption was related to vitamin K2's long side chain.

2. Vitamin K-dependent proteins in the bone building process

Osteocalcin is produced by osteoblasts during bone formation, and serves as the most abundant protein in bone after collagen. Furthermore, it is crucial for bone mineralization. Activated osteocalcin is located within hydroxyapatite crystals and binds calcium strongly to facilitate mineralization of the hydroxyapatite crystal grid [24]. Osteocalcin production is regulated by a plethora of factors including retinoic acid (RA), estrogens, glucocorticoids, as well as vitamin D [25–27]. In 1995, Douglas et al. showed the percentage of carboxylated osteocalcin (cOC), as calculated from total osteocalcin, was found to be less than 60% in osteoporotic postmenopausal women compared to 70–80% in young, healthy adults [28]. The osteocalcin production is increased by vitamin D but also increased by both MK-4 and MK-7 in a synergistic fashion. From 2007, as used by Knapen et al., osteocalcin has been employed as a marker for the deficiency of vitamin K in bone. Today, the specific osteocalcin molecules are total Osteocalcin (tOC), ucOsteocalcin (ucOC), and cOsteocalcin (cOC) [28, 44]. A secondary action of MK-4 and MK-7 is the ensuing increased collagen production by cells of the osteoblastic lineage. Collagen should make the structural fundament, on which calcium and other minerals are accumulated within the bone matrix. Increased deposition of collagen makes the bone more flexible and this is very important for the attainment of "higher" or better bone quality [24]. In 2001, Yamaguchi et al. unraveled the stimulatory effect of MK-7 on osteoblastic bone formation *in vitro*, but they also discovered the suppressive effect of MK-7 on osteoclast-like cell formation and osteoclastic bone resorption in rat bone tissues *in vitro* [29]. Furthermore, in 2001, Yamaguchi and Ma [30] confirmed the dual effects of MK-7 but also a significant decrease number of osteoclasts. Finally, in 2011, Yamaguchi and Weilzmann showed that MK-7 reinforces the synthesis of various bone-specific proteins, mediated through the pathways of calcium-dependent protein kinase C signaling, as well as cyclic AMP-dependent signaling. MK-7 also antagonizes the "receptor activator of NF-κB (RANK) ligand (RANKL)" induced NF-κB activation on osteoclast precursors. This concept now makes up the basis for the search of novel antiosteoporotic medication regimens, mimicking the plethora of effects induced by MK-7 [31].

3. Clinical-related publications featuring K1 supplementation

In relation to the loss of bone associated by deficient intake of vitamin K1, many observational studies have been conducted. However, few randomized studies have been able to reveal a significant positive rebuilding of bone mass and increased BMD. This chapter deals only with randomized clinical trials with a duration exceeding 1 year.

Schaafsma et al. from the Netherlands showed in a 1-year long randomized study from 2000 [32] (featuring four groups with 400 IU vitamin D3; vitamin K1 supplementation 80 μg daily, vitamin K1 and D3, and placebo of Dutch postmenopausal women (with a patient total, n = 141) with either normal or low bone mineral densities (BMD). It was shown that women with low BMD had a lower %cOC at baseline than the women with normal BMD. However, this difference disappeared after 1 year of supplementation with vitamin K1 [(mean ± S.D.) 68 ± 11% (95% CI = 64.5 ± 71.2%) versus 72 ± 6% (95% CI = 70.1 ± 72.9%)], respectively. On the other hand, 1 year of supplementation with vitamin D3 showed maximum increases in 25(OH)D of 33 ± 29% (95% CI = 24.8 ± 41.8%) and 68 ± 58% (95% CI = 50.1 ± 84.6%) in women with normal and low BMD, respectively. No effect was observed on BMD [32].

In the Bram et al. study from 2003, three groups were examined; group K1D: the effect of vitamin K1 (1 mg/day) and vitamin D (8 μg/day including standard mineral supplementation), group D (vitamin D and minerals) and group 3, placebo on bone loss retardation in a randomized, double-blind, placebo-controlled 3-year intervention study. Of 181 healthy postmenopausal women between 50 and 60 years of age, 155 completed the observation period. The main outcomes of the study were significant changes in BMD-values of the femoral neck and lumbar spine after 3 years. The group receiving the supplement with additional vitamin K1 showed a decline in the bone loss from the femoral neck. The difference in femoral neck bone mass between the K1D group and the placebo group was 1.7% (95% CI: 0.35–3.44). The difference between the K1D group and the D group was 1.3% (95% CI: 0.10–3.41). No significant differences were observed among the three groups with respect to changes in BMD at the site of the lumbar spine. It was therefore concluded that the minerals and vitamin D, coadministered with the vitamin K1 supplement, substantially contributed to a significant reduction in postmenopausal bone loss at the site of the femoral neck [33].

In a systematic review and meta-analysis of 700 patients, Cockayne et al. showed in 2006 that the MK-4 intake in Japan yielded a powerful reduction in the incident of fractures. However, the authors would not recommend vitamin K supplementation until a new randomized clinical trial confirmed the results [34].

Boton-Smith et al. performed a 2-year, randomized, double-blind, and placebo-controlled study in 2007, scrutinizing the effect of dietary supplementation with either: (1) 200 μg vitamin K1 daily, (2) vitamin D 400 UI daily (3) and calcium 1000 mg daily, or (4) their combination on 244 healthy nonosteoporotic older women. Baseline and 6-month measurements included DXA bone mineral scans of the hip and wrist, markers of bone turnover, and vitamin status. The results reported were the following: the combined vitamin K1 with vitamin D plus calcium treatment was associated with only a modest but significant increase in BMC at the ultradistal radius, however not at other sites in the hip or radius [35].

The Booth et al.'s study was published 1 year later, in 2008. The goal of the present investigation was to pinpoint the effect of a consecutive 3-year administration of vitamin K1 on putative alterations in bone mineral density (BMD) of the femoral neck in elderly patients of both sexes, who presented upon inclusion with calcium and vitamin D repletion. In the present double-blind, controlled survey, 452 individuals (both men and women, 60–80 years of age) were evenly distributed, in a randomized fashion, each to receive a multivitamin containing either 500 μg/day of vitamin K1 or placebo, in addition to a daily, 600 mg elemental calcium with vitamin D (400 IU) supplement. Analyses of the femoral neck, spine (L2–L4), and total-body BMD, turnover of bone mass, and indigenous status of both vitamins K and D was checked every 6–12 months. Results were as follows: one could not find any distinction in BMD values localized to any pertinent body sites, when comparing the two patient groups. Furthermore, the population receiving the vitamin K1 supplement showed a markedly higher vitamin K1 level, as well as a substantially lower degree (percentage) of ucCO concentrations, when matched with the patients not ingesting K1. Neither of the additional biochemical variables measured differed between the patient treatment groups. Hence, the authors concluded that vitamin K1 supplementation, in a dose attainable in the diet, does not confer any additional benefit for bone health at the spine or hip when taken with recommended amounts of calcium and vitamin D [36].

The emerging questions to ask were then: Was the dose of vitamin K1 supplement too low, or the follow up period too short to elicit an increase of BMD, or was it possible to use other sub-family vitamin K members, such as menaquinone-4 (MK-4)? Or, could vitamin K supplementation potentially be harmful to the body?

Cheung et al. from Canada addressed the last of these questions in their 2–4-year study from 2008. In this trial, 440 postmenopausal women with osteopenia were randomized into a placebo-controlled double-blind trial, and it was conducted, mainly to determine whether daily high-dose vitamin K1 supplementation safely reduces bone loss, bone turnover, and fractures. The conclusions coming out of the study were: 5 mg of daily vitamin K1 supplementation for 2–4 years does not protect against age-related decline in BMD, but may protect against fractures and cancers in postmenopausal women with osteopenia. Overall fracture rate was reduced by 50% (9 versus 20, $P = 0.04$) versus placebo. Interestingly, cancer incidence was reduced by 75% with vitamin K1 (3 versus 12, $P = 0.02$). However, more studies are needed to further examine the effect of vitamin K on fractures and cancers [37].

Brinkley et al. from USA conducted a 1-year study in 2009 and found that low vitamin K status is associated with low BMD and an increased fracture risk. From the bulk of reports available at that time, it seemed that the menaquinones (menatetrenone: MK-4), might diminish the fracture risk incurred by the enrolled patients. Whether vitamin K is an important "by-player" in maintaining skeletal health in females situated in the northern part of the US remains an unsolved issue. Furthermore, different entities of K vitamins (i.e., phylloquinone (K1) and MK-4) may exert differing biological effects on the skeleton. The present study was designed to assess the efficacy of either vitamin K1 or MK-4 exposure on biomarkers of skeletal health and bone mineral density (BMD) in postmenopausal nonosteoporotic, North American women. In the present, placebo-controlled and double-blind investigation, a total of 381 postmenopausal females were given either vitamin K1 (1 mg/day), MK-4 (45 mg/day), or a placebo treatment. The whole observation period was 12 months. All enrolled patients/participants were given either Ca^{2+} or vitamin D_3 supplementation. Bone-specific BSALP (alkaline phosphatase) in blood samples, as well as the *n*-telopeptide of collagen, type 1 (NTX) were measured at before the onset of "medication," and subsequent to 1, 3, 6, and 12 months, respectively. Both lumbar spine and proximal femur BMD values, as well as proximal femur geometry were assessed by DXA, before the onset of the trial, and after 6 and 12 months of treatment. At the onset of the trial, all treatment groups showed identical demographic parameters. The patients' compliance rates related to the intake of either calcium, vitamin K1, or MK-4, were some 87–93%, respectively. Interestingly, K1 and MK-4 treatments both diminished the patient levels of serum ucCO, however, neither BSALP nor NTX levels were changed. Lastly, no effects of K1, or MK-4 on lumbar spine or proximal femur BMD or proximal femoral geometric parameters could be observed. This study does not support a role for vitamin K supplementation in osteoporosis prevention among healthy, postmenopausal, North American women, receiving calcium and vitamin D supplementation [38].

4. Clinical randomized controlled osteopenia/osteoporosis studies with menatetrenone-4 (MK-4) from 1 year duration, in countries with different background intake of vitamin K

In 1998, Orimo et al from Japan evaluated the effects of menatetrenone-4 (MK-4) on bone and calcium metabolism in osteoporosis patients in a 24-week double-blind placebo-controlled study, where 80 osteoporotic patients were included. Treatment was MK-4, 90 mg/day (n = 39) or placebo (n = 41). Bone density was assessed on X-ray films of the right, second metacarpal bone, using the microdensitometric method. In the MK-4 group, bone density increased by about 2.2 ± 2.5% from the baseline; in the placebo group it decreased by about 7.3 ± 3.7% (P = 0.037, K_2 treatment versus placebo). The excretion of γ-carboxyglutamic acid (Gla) to the urine was markedly enhanced (i.e., from 72.6 ± 4.1 nmol/mg of creatinine before initiation of "medication," to 88.4 ± 5.4 during the 24th week subsequent to the sustained MK-4 treatment (P = 0.008) period). In the group receiving placebo, no significant changes in the excretion of urinary Gla could be observed. However, during the 24-week long treatment period, the urinary ratio of calcium over creatinine in the K_2 treatment group was reduced from 0.14 ± 0.02 to 0.12 ± 0.02, respectively. However, in the placebo group it increased from 0.15 ± 0.02 to 0.19 ± 0.03. Accordingly, the 24-week levels shown by members of both the MK-4 and the placebo groups turned out to be significantly different ($P = 0.03$) with unpaired test. Finally, it should be noted that there were but a few adverse effects, being attributable to the vitamin MK-4 treatment. One patients increased hepatic enzymes of GOT, GPT, al-P, and γ-GTP evaluated as probable relationship. The results suggest that MK-4, at a dosage of 90 mg/day, is effective in maintaining peripheral cortical bone density and is safe in treatment of osteoporosis. The dose was increased in order to maintaining peripheral cortical density. Interference of diet was not observed. This study is one of very few, where side effects of the treatment were observed [39].

In 2000, Shiraki et al. conducted a 2-year study in Japan, to assess whether MK-4 effectively prevented the incidence of new fractures in osteoporotic patients. Two hundred forty-one osteoporotic women were enrolled in a 24-month, randomized, and open label study. The population constituted: a control group without treatment ($n = 121$) and an MK-4 group ($n = 120$), the latter receiving 45 mg/day. All patients received follow up measurements of lumbar bone mineral density (LBMD) analyzed by DXA and the occurrence of new clinical fractures had occurred. Both serum concentrations of Glu-osteocalcin (Glu-OC) (RIA, Takara Japan), as well as MK-4 were analyzed after termination of the follow-up period, while both the level of serum-OC (RIA, CIS, France) and excretion of deoxypyridinoline (DPD) to the urine were analyzed prior to and at the end of the treatment. The demographic data of the present groups did not differ significantly, and the results obtained run as follows: the clinical vertebral fracture incidence in the control group was 30, compared with 13 in the MK-4 treated group ($P = 0.027$). Furthermore, the percentage change from the initial LBMD value at 6–24 months after the initiation of the study ranged between −1.8 ± 0.6, and −3.3 ± 0.8% for the control group, and between 1.4 ± 0.7, and −0.5 ± 1.0% for the MK-4-treated group, respectively. The alterations in LBMD-values around each measure point turned out to be significantly different, when comparing the control group with the treated group ($P = 0.0010$ at 6 months, $P = 0.0153$ after a year, and $P = 0.0339$ after 2 years, respectively). The blood concentrations of Glu-OC at termination of the period of observation of the controls and the group receiving active "drug" were 3.0 ± 0.30 and 1.6 ± 0.10 ng/ml, respectively ($P < 0.0001$), while blood concentrations of OC (as analyzed by standard radioimmunoassay (RIA) methodology, gave a marked and significant rise (42.4 ± 6.9%) over basal value observed in the treated group at 24 months, but only 18.2 ± 6.1% for the individuals constituting the controls ($P = 0.0081$)). However, one did not find any significant change in the amount of DPD excreted into the urine of the treatment group members. This compilation of information indicates that MK-4 treatment is effective in reducing the incidence of additional fractures, even though the MK-4 treated individuals failed to show an enhancement of LBMD. Lastly, the study was able to show that MK-4 treatment elevates the levels of γ-carboxylated OC [40].

In 2000, Iwamoto et al. showed in 92 postmenopausal women, aged 55–81 years, completing a 2-year randomized controlled trial in four groups receiving either menatetrenone (MK-4, 45 mg/day), 1α-hydroxyvitamin D3 (0.75 μg/day), a combination of MK-4 and D3 (same dosage as above), or calcium lactate (2 g/day). The MK-4 and vitamin D3 groups experienced significant enhancements of their BMD-values (+0.91 and 0.38%), compared to the "Calcium group" (−0.79%), while the combined MK-4 and D treatment, being synergistic, significantly increased lumbar BMD by 1.5 %, $p < 0.001$. These findings indicate that combined administration of vitamin D3 and MK-4, compared with calcium administration alone, appears to be instrumental in increasing the BMD-values of the lumbar spine in postmenopausal women with osteoporosis [41, 42].

Ushiroyama et al. completing a randomized 2-year study in 2002, investigated the therapeutic effect of the combined use of menatetrenone MK-4 and vitamin D3 on vertebral bone mineral density in 172 postmenopausal women with low bone mass and osteoporosis. Four groups, each with 43 subjects received the following; either (a) MK-4, 45 mg/day, (b) standard vitamin D3 supplement, (c) combined MK-4 and vitamin D3 therapy, and (d) control group receiving dietary therapy alone. BMD (bone mineral density) was assessed before initiation of therapy and subsequent to 6–24 months of treatment, respectively. Analyzed biological markers of osseous metabolism constituted: serum type 1 collagen carboxyterminal propeptide (P1CP), intact or total osteocalcin, as well as urinary pyridinoline. Tests for potential of blood coagulation was done analyzing "activated-partial-thromboplastin-time" (APTT), as well as assessment of levels of antithrombin III (AT III), fibrinogen, and plasminogen, respectively. Conclusions summarized in the paper were combined therapy with MK-4 and vitamin D_3 given for 24 months significantly increased bone mineral density = BMD (4.92 ± 7.89%), $P < 0.001$, but also while MK-4 alone was significantly enhanced by 0.135 ± 5.44%, $P < 0.05$. A majority of the population (77.5%) increased their BMD values, while 22.5% experienced the opposite. In the MK-4 group, the marker of bone formation (P1CP) showed an increase by 20% after 6 months, and while thereafter returning to baseline. Urinary pyridinoline was significantly increased after 6 month, and peaked after 18 months (89.6 ± 112.3%), while slightly decreasing at 24 months to 53.4 ± 55.7%, $P < 0.05$. In the combined MK-4 and D3 group, P1PC was unchanged for the first 12 months, then it increased at 24 months to 24.2 ± 23.1%, $P < 0.05$. Urinary pyridinoline was increased throughout the 24 months to 84.5 ± 51.9%, $P < 0.01$. The MK-4 group at 24 months showed a significant positive correlation between changes of P1CP and changes of BMD, $P < 0.001$. In the MK-4 and D3 groups at 12 months, a significant positive correlation ($P < 0.03$) between changes of P1PC and changes of BMD could be seen. In the MK-4 and D3 groups, at 12 and 24 months, respectively, significant negative correlations between changes in pyridinoline and changes in BMD ($P < 0.001$ and $P < 0.004$, respectively) could be observed. The authors conclude that these findings indicate both concomitant enhancement of net bone formation, as well as some osseous resorption activities. Additionally, observed significant increments in the coagulation and fibrinolytic reaction pathways were seen. However, they were both restricted within normal physiological range, suggesting maintenance of a normalized balance within the fibrinolysis versus coagulation system, since no side effects were observed [43].

In 2007, Knapen et al. from the Netherlands presented the result of a 3-year randomized clinical intervention study of 325 healthy postmenopausal non- osteoporotic women receiving MK-4, 45 mg/day, or placebo. BMC and hip geometry were assessed by DXA, and bone strength indices were calculated from DXA-BMD, femoral neck width (FNW) and hip axis length (HAL). Computations showed that MK-4 significantly improved the hip bone strength, BMC and FNW, but in the placebo group, bone strength decreased significantly. However, MK-4 did not affect the DXA-BMD values. It was therefore speculated that the high vitamin K2 intake prevented the postmenopausal, nonosteoporotic bone loss. The importance of K vitamins for optimal bone health has been suggested on the basis of population-based analyses, however, intervention trials performed with DXA-BMD serving as measures of clinical endpoints have given contradicting conclusions. In contrast, BMC, compared with DXA-BMD does not take into account the geometry (size, thickness) of bone, which is construed as an independent contributor to and "interpreter" of bone resilience and fracture risk [44].

Jiang et al. conducted a 1-year study in China in 2014. In this randomized, double-blinded study, it was evaluated whether the efficacy of MK-4 is inferior to α-calcidol in Chinese postmenopausal women. Two hundred and thirty-six women were randomized to two groups: Group MK-4, receiving 45 mg/day or Group α-calcidol, receiving 0.5 µg/day, respectively. Furthermore, all enrolled individuals ingested 500 mg/day of calcium. Assessment of bone mineral density (BMD) post-treatment, onset or occurrence of new fractures, as well as serum OC and ucOC levels were matched with patient baseline values in both patient groups. The information obtained was as follows: 90.3% of the patients completed the investigation. Results showed that the BMD-values in the MK-4 group significantly increased from baseline by 1.2% at the lumbar spine, and 2.7% at the trochanter, respectively ($P = 0.001$). The percentage enhancement in BMD in the α-calcidol group and the MK-4 group was 2.2 and 1.8%, respectively ($P = 0.001$). No difference could be seen when comparing either treatment groups. And one could not spot any alterations in femoral neck BMD between the populations observed. However, one tibia and one femoral neck fracture in the MK-4 group, and three lumbar spine compressive and one forearm fracture in the α-calcidol group revealed the appearance of new fractures ($P < 0.05$). Within the MK-4 group, the concentration of OC and ucOC, respectively, fell from baseline levels by some 39 and 82%, respectively ($P < 0.001$). In the α-calcidol group, OC and ucOC fell by some 26 and 35%, respectively ($P < 0.001$), and the decreases in serum OC and ucOC were more obvious in the MK-4 group than in the calcidol group ($P < 0.001$). The safety profile of menatetrenone was similar to that of α-calcidol. It was therefore concluded that MK-4 is an effective and safe choice in the treatment of postmenopausal osteoporosis in Chinese women [45].

Bisphosphonates, combined with vitamin D and calcium are widely used around the world in the treatment of osteoporotic patients. Vitamin K apparently sustains the lumbar BMD, thus reducing the incidence of osteoporotic fractures it was natural to investigate and compare bisphosphonates and MK-4.

In 2001, Iwamoto et al. presented a 2-year preliminary randomized study, comparing the effects of etidronate (E) and MK-4 on forearm BMD-values and fracture incidence. Seventy-two osteoporotic women, all 5 years after menopause were enrolled. The E-group received 200 mg/day, 14 days per 3 months; $n = 25$, the MK-4 group received 45 mg/day; $n = 23$, and the placebo (C = control group) was given calcium lactate 2 g/day; $n = 24$. At baseline, there was no difference between the three groups. Significant results obtained were summarized as follows: mean percentage change in BMD in the E-group was a significant increase of 2.1%; $P < 0.01$. In the MK-4 group, it was −0.1%, which was not significant. In the C-group, there was a significant decrease in BMD; −1.7%; $P < 0.0001$. Incidence of new fractures was in the E-group; total 13 (nine vertebral and four lumbar), in the MK-4 group total 10 (seven vertebral and three lumbar), and in the C-group; total 12 (seven thoracal and five lumbar). No significant differences could be spotted between groups. It was therefore concluded that a maintained BMD seen in the MK-4 group could be construed as a positive effect, which should be evaluated together with etidronate treatment [46].

In 2003, Iwamoto et al. examined 98 postmenopausal osteoporotic women in a randomized, controlled, 2 years study, comprising four groups: (1) MK-4, 45 mg/day of vitamin K; (2) etidronate 200 mg/day for 14 days per 3 months; (3) etidronate and MK-4 combined; and (4) calcium supplement. End points assessments were: Forearm (distal radius) BMD measured by DXA using DTX-200 (Osteometer®), and incidence of vertebral fractures (level T4-L4). Results reported are summarized as follows: the forearm BMD in the calcium group was reduced from baseline, sustained by MK-4, increased significantly by etidronate, and increased even more in the MK-4 and etidronate group. The incidence of vertebral fractures in the MK-4 group was 8.0%, in the etidronate group 8.7%, in the MK-4 and etidronate group 3.8%, and in the calcium group 20.8%, respectively. The compelling conclusion is combined treatment with MK-4 and bisphosphonates showed significant different ($P < 0.01$) between other groups alone. The author concluded that combined treatment seems to have the greatest efficacy in prevention of new osteoporotic fractures, and this treatment should be recommended to treat osteoporotic women [47, 48].

Ishida and Kawai published in 2004 a 2-year long study, examining 396 postmenopausal women with osteoporosis, randomized to six equally sized groups: (1) hormone replacement therapy (conjugated estrogen 0.625 mg/day plus medroxyprogesterone 2.5 mg/day), (2) etidronate (2 weeks of treatment with 200 mg/day followed by a 10-week medication-free period), (3) ell calcitonin (20 IU/week), (4) α-calcidol 1 μg/day, (5) MK-4 45 mg/day, and (6) control group (i.e., no treatment). End point BMD, bone markers and incidence of vertebral fractures served as clinically relevant end point measures. The results (2-year mean changes in BMD) were 2% for the hormone replacement therapy, −0.5% for the etidronate treatment, 1.6% for the calcitonin treatment, −3.6% for the α-calcidol treatment, −1.9 for the MK-4 treatment, and −3.3 for the controls, respectively. Seventeen (26%) of the 66 control patients developed new vertebral fractures. Compared with controls, the relative risk of incurring vertebral fractures was 0.35 (95% CI: 0.14–0.83) in the hormone replacement therapy, 0.40 (95% CI: 0.17–0.92) in the etidronate group, 0.41 (95% CI: 0.17–0.93) in the calcitonin group, 0.56 (95% CI: 0.26–1.12) in the α-calcidol group, and 0.44 (95% CI: 0.20–0.99) in the MK-4 group, respectively. Hence, the conclusion: a substantial and significant reduction in the incidence of vertebral fractures was achieved due to either: (a) hormone replacement therapy, (b) etidronate, or (c) calcitonin medication. Significant improvements in BMD were obtained by the patients enrolled in the hormone replacement therapy group and in the calcitonin group [49].

Hirao et al. conducted a 1-year study in Japan in 2008. He enrolled and examined 48 postmenopausal women, but only 44 were followed up after 1 year. This randomized controlled study consisted of the following groups A, monotherapy of alendronate (5 mg/day) and group AK, alendronate plus MK-4 (5 mg/day an 45 mg/day, respectively). The take home message was clear: MK-4 decreased undercarboxylated osteocalcin significantly more than alendronate, which is known not to influence the degree of carboxylated osteocalcin. In the AK group, the femoral neck BMD was significantly increased. The small number of patients enrolled, and the short observation time undermines any conclusion drawn from this study. Further investigations using this combination therapy were recommended [50].

Je et al. from Korea published in 2011, a study evaluating the effects of MK-4 supplementation on bone mineral density and undercarboxylated osteocalcin (ucOC) in 78 postmenopausal Korean women 60-years-old plus. These women, not receiving any medical treatment, were randomized into two groups: (1) calcium and vitamin D, $n = 40$) or (2) vitamin K supplementation, $n = 38$, receiving 15 mg of MK-4 three times a day after every meal, calcium carbonate 315 mg twice daily, and active vitamin D3, 400 UI once a day. The dosage of calcium and vitamin D3 was the same in both groups. After 6 months of treatment, the members of the MK-4 group showed a decrease in ucOC (−1.6 ± 1.6 ng/dl versus −0.4 ± 1.1 ng/dl) with a P-value of 0.008. The vitamin K (MK-4) group members showed a significant improvement of L3 BMD-values, however, no significant enhancement of the BMD-values in L1, L2, and L4. Similarly, no significant increase was observed for femoral BMD, which remained unchanged in the women receiving vitamin D and calcium [51].

In 2012, Fang et al. presented a meta-analysis of randomized controlled trials published during the period of 1999–2009. The analysis examined the role of vitamin K on bone mineral density (BMD). The study revealed that vitamin MK-4 supplementation was associated with increased BMD at the lumbar spine, however not necessarily at the femoral neck. This untoward heterogeneity may be the result, may reside within different participant groups, different regions of the skeleton, gender, and type of vitamin K1 and MK-4 supplementation. Hence, further studies are deemed required to investigate and unravel the detailed effects of vitamin K2 sub"populations" or metabolites on BMD [52].

5. Clinical randomized controlled studies with menaquinone-7 (MK-7)

Japanese menopausal women have traditionally a lower fracture risk than women from the western world [3]. An association with the ingestion of natto (a processed food containing fermented beans) was evaluated by Katsuyama et al. [53] and Ikeda et al. [54] in the so-called Japanese population-based osteoporosis (JPOS) study. The dietary natto intake over 3 years was shown to significantly increase the changes of total hip BMD $P < 0.003$. The alleged prevention of postmenopausal bone loss through the effects of MK-7 is more abundant with natto than other soybean products. More randomized controlled studies are clearly warranted to substantiate this contention [54].

The Norwegian study published by Emaus et al. in 2010. features a 1-year randomized double-blind placebo-controlled study with 334 early menopausal, but otherwise healthy, women. The patient groups received either MK-7 (360 μg/day) or placebo. The summary of the present trial disclosed the following: in the MK-7 group, serum uncarboxylated Osteocalcin (uc-OC) was significantly reduced (from 4.14 to 2.22 ng/ml, respectively). In contrast, carboxylated Osteocalcin (c-OC) was significantly increased (from 13.5 to 19.1 ng/ml, respectively). However, there was no statistical difference in bone loss rate between the groups at the "total hip site," nor at

http://dx.doi.org/10.5772/64876

any other measurements sites. This was most probably due to the shorter follow-up time, i.e., after only 1 year of MK-7 exposure [55].

After solid organ transplantations, loss of bone mass often occurs and may cause substantial health problems. In a study (published in 2010) on such a patients group, Forli et al. from Norway looked at the effect of MK-7 on bone mass, measured as BMD of the lumbar spine. Despite the fact that the impact of MK-7 on the measured BMD was not conclusive, it was recommended that further studies over an extended period of time should be conducted. Here, we refer to the main findings in the study.

This study was the first in organ transplantation, featuring the effect of MK-7 on bone mass, 1 year after lung and heart transplantation. Postoperatively, 35 lung and 59 heart recipients were actively treated with MK-7 in a prospective and longitudinal study, receiving MK-7 supplement, 180 μg/day or placebo. The results reported were the following: 1 year after solid organ transplantation, the difference between MK-7 and placebo for the lumbar spine (L2–L4) BMD was 0.028 (SE 0.014) g/cm^2, $P = 0.055$, and for L2 to L4, BMC emerged as 1.33 (SE = 1.91) g/cm^2, $P = 0.5$. Scrutinizing the lung recipients separately, the difference for BMC was 3.39 g (SE = 1.65), $P = 0.048$. In the heart recipients, however, observed values were 0.45 (SE = 0.02) g, $P = 0.9$ subsequent to correcting for measures of baseline values.

In a stepwise linear regression analysis, alterations in the L2-L4 BMD, controlled for alleged confounding variables (which include the use of bisphosphonates), significant predictors turned out to be: (a) organ (if heart = 1, BMD = −0.065 g/cm^2 , $P = 0.001$) and (b) MK-7 versus placebo (BMD = 0.034 g/cm^2, $P = 0.019$). It so happened, that insufficient vitamin D status was frequent, and that PTH (parathyroid hormone) levels were augmented in the MK-7 group, indicating a more imminent need for ingestion of vitamin D. In conclusion, it turned out that 12 months of MK-7 ingestion generally suggests a positive effect on BMD of the lumbar spine, but with diverging responses in "cardio-pulmonary" recipients. Thus, the patients' vitamin D status would benefit from a closer monitoring during vitamin K supplementation [56].

Knapen et al., the Netherlands, published in 2013 the results of their 3 year study on the effect of low-dose MK-7 supplementation on bone loss in 244 healthy postmenopausal women. The study was a double-blind, randomized placebo controlled study, with two groups: (a) active low-dose vitamin K2 (MK-7, 180 μg/day), and placebo. Their main task was to investigate whether low-dose MK-7 supplements beneficially could affect bone health in general.

Secondary to an improved and favorable vitamin K status, MK-7 ingestion from supplements should have the possibility to significantly reduced age-related loss of bone mineral density and ensuing bone mechanical properties. Hence, low-dose MK-7 supplements should consequently result in preventing bone loss in postmenopausal women. In spite of contradictory data emanating from trials with vitamin K supplementation on the status of bone health, the European Food Safety Authorities (EFSA) has accepted the health claim on vitamin K's role in the maintenance of normal and healthy bone structure. In accordance with EFSA's opinion, it was clearly demonstrated that a 3-year high-dose of vitamin K1 and MK-4 supplementation improved bone health after menopause.

Because of the longer half-life, bioavailability, and greater potency of the long-chain MK-7, they also measured the effect of low-dose MK-7 supplementation on bone health, as reflected by bone mineral density (BMD) of lumbar spine, total hip, and femoral neck. The assessment of vertebral fractures was performed using DXA. Furthermore, blood levels of ucOC and cOC were also analyzed, and the ucOC/cOC ratio functions as an indicator of vitamin K "health" status. All analyses were performed at baseline, and subsequent to 1, 2, and 3 years of supplementation, respectively. A carboxylation rate of >50% was achieved during the first year of treatment, and it was maintained throughout the study period.

The main results obtained were as follows: MK-7 ingestion significantly enhanced vitamin K status and decreased the age-related reduction in bone mass, as well as both BMC and BMD at the level of lumbar spine and femoral neck. However, total hip BMC and BMD decline could not be "rescued." Bone strength also seemed to be favorably affected by MK-7 ingestion, significantly decreasing the loss of vertebral height of the lower thoracic region at the mid-site of the vertebrae. These results confirm the hypothesis that long term supplementation with MK-7 beneficially affects bone health. Whether these results can be extrapolated to other populations with osteoporosis, needs further investigation [57].

6. Anticoagulation and vitamin K-antagonist association with loss of bone mass

The need of vitamin K for the activation of clotting factors is lower than for the activation of extrahepatic vitamin K-dependent proteins. This is the plausible reason for deficient rebuilding of bone mass and increased calcification process in blood vessels. Long-term use of anticoagulants, likes warfarin, may potentially lead to the loss of bone, and/or an increased incidence of osteopenia or osteoporosis with and without fractures. The crucial question then is: What time span for patients on warfarin medication will suffice for the detection of bone loss?

In an observational study conducted in Japan by Namba et al. 2015, the biomarkers during warfarin use in a 1-year follow-up on 42 patients treated for atrial fibrillation were described. Twenty-four patients received warfarin (WF group) and 18 patients received non-warfarin treatment (Non-WF group). Results revealed an increased significant difference in ucOC in WF the group 10.3 ± 0.8 ng/ml, versus non-WF group 3.4 ± 0.9 ng/ml, $P < 0.01$. In cytokines, RANKL in WF group 0.6 ± 0.1 ng/ml versus non-WF group 0.4 ± 0.1 ng/ml, $P < 0.01$. After 1 year, DEXA scan showed no significant different between groups. It was concluded long-term use of warfarin might be associated with high risk of osteoporosis but also risk of ectopic calcification in blood vessels. Further randomized studies are needed to evaluate these patients [58].

Twenty years ago, clinical observations and research demonstrated that women, taking warfarin during the first trimester of their pregnancy, gave birth to children with punctate calcifications in the axial skeleton, proximal femurs, and calcanei. The presumed reason has since long been that prenatal vitamin K deficiency, induced by warfarin, was the reason for these calcifications [59].

Large clinical studies on bone mass have given different results, and in early observation studies, the evaluation of duration of warfarin use and other patients receiving treatment for osteoporosis was not included. However, two newer studies showed no further risk on bone mass of warfarin use in elderly patients: In the first one published by Woo et al. in 2008, in a large cohort of elderly community-dwelling men, no association was observed between current warfarin use and bone mass, bone loss or fracture risk. Although warfarin use was based upon a single assessment, the findings suggest that current warfarin use in older men does not appear to have clinically important effects on the skeleton [5, 60].

The second experience was summarized by Misra et al. in 2014, featuring long-term treatment of incident atrial fibrillation without prior history of fractures. Long-term warfarin use was defined in two ways: (1) warfarin use ≥1 year; (2) warfarin use ≥3 years. Event-score on warfarin users and nonusers were created to evaluate the association between long-term warfarin use and risk of hip, spine, and wrist fractures separately, as well as combined, using Cox-proportional hazards regression models. Among more than 20,000 participants with incident atrial fibrillation, the hazard ratios (HR) for hip fracture with warfarin use ≥1 and ≥3 years, respectively, were 1.08 (95% CI 0.87, 1.35) and 1.13 (95% CI 0.84, 1.50).

The conclusion of the present trial was as follows: long-term warfarin use among elders (i.e., >65 years of age) with atrial fibrillation was not associated with any increased risk of osteoporotic fractures and therefore does not appear to necessitate additional surveillance or prophylaxis [61]. These observational studies have focused on clinical fractures as endpoints below follow-up time at 5–10 years, but the thesis that warfarin-induced clinical fractures was not confirmed. This may be due to the beneficial effect of MK-7 on bone mass, which appears to stay unaffected by the impact of warfarin on vitamin K1, which again reinforces the notion that vitamin K2 status (measured as ucOC) per se is a good marker of bone homeostasis [58].

7. Chronic kidney disease and loss of bone mass (CKD-MBD)

The link between increased calcification of vessels and bone complications changes the definition of CKD-MBD to better describe the complexity of the syndrome [62]. The link between osteoporosis and cardiovascular morbidity is well described in postmenopausal women with intact renal function [63]. In chronic hemodialysis patients, a lower bone volume is associated with higher coronary calcification scores measured by multislice computed tomography, reflecting a higher risk of cardiovascular events [64]. This association between vascular calcification, arterial stiffness and bone mineral density in chronic kidney disease was also described in 2008 by Toussaint et al. [65]. Furthermore, the mortality of hemodialysis patients was evaluated in 2003 by Taal et al. among 88 hemodialysis patients over a 3.5 years follow-up period. Here, it appeared that mortality was associated with age, Ca-P product, lack of transplantation and a low bone mineral density measured at the hip. The leading cause of death (42.5%) appeared to be related to cardiovascular events [66].

The therapeutic options are few, since many women on renal replacement treatment did not accept reinstigation of hormone replacement therapy. However, the efficacy of hormone replacement was confirmed in a randomized trial in women on continuous dialysis, receiving hormone therapy (estradiol and cyclic norethisterone) for 1 year. At the end of the study, the active group showed an increase in bone mineral density at the lumbar spine. This difference between the active and control group (receiving cinacalcet) was significant at all measurement sites [67]. Active vitamin D analogs, calcimimetics administration and phosphate-binders are widely used to suppress iPTH and thus bone specific alkaline phosphatase, as a marker of enhanced bone turnover.

Kohlmeier et al. were the first to show an independent association between serum concentration of phylloquinone <1.2 nmol/l or less (poor vitamin K status) and an increased risk of bone fracture in patients with end-stage renal disease [68]. This observation was confirmed by Fusaro et al. [69] in 2013, showing that hemodialysis patients treated by warfarin for longer than 1 year had an increased risk of vertebral fractures, compared with patients not on warfarin. McCabe et al. enrolling 172, stage 3–5 CKD patients without dialysis treatment, showed that intake of vitamin K was insufficient in more than 50–60% of individuals on a given diet, if measures of ucOC were conducted (>20% ucOC), and 97% if evaluation was done by the prothrombin induced by vitamin K absence-II (PIVKA-II) assessment (>2 nmol/l) [70, 71]. After establishment of dialysis as a therapeutic intervention, Cranenburg et al. showed in a study of 40 chronic hemodialysis patients from 2012 that the dietary intake of vitamin K1 and K2, in general, was insufficient. This was reflected by analyses of plasma levels of desphospho-undercarboxylated (dp-uc) MGP (matrix-GLA protein), which was increased over the normal range by some 82.5% with elevated PIVKA-II values 3.81.4–12.4 ng/ml, reference value <2 ng/ml. [72]. Elevated dp-ucMGP levels suggest insufficient vitamin K2 levels on the vascular site, while high ucOC reflects insufficient vitamin K2 on bone or osseous sites.

A 6 weeks randomized controlled trial on hemodialysis patients evaluated the response of biomarkers of vitamin K status (dp-ucMGP, PIVKA-II and ucOC) to the ingestion of 45, 135, 360 μg/day of MK-7. The study confirmed that most patients displayed a functional deficiency at baseline, and that MK-7 supplementation decreased dp-ucMGP and PIVKA-II. However, only the highest doses brought about a significant decrease in ucOC [73].

In osteoporosis, the main treatment aims at inhibiting osteoclastic bone resorption. The osteoclast and osteoblast are functionally tightly coupled, and the mechanism of this reciprocal link is now very well known. By the discovery of MK-7, which is able to play a role in the prevention of bone loss from most sites of the skeleton, there is hope for efficient treatment. MK-7 has been shown to stimulate osteoblastic bone formation, as well as suppressing osteoclastic bone resorption *in vitro* and in humans, as showed by Knapen et al. [57]. MK-7 suppresses the activation of NF-κB signaling pathways in both osteoblasts and osteoclasts. These treatments have not yet been enrolled side by side with vitamin D analogs in CKD patients. Unfortunately, vitamin K2 is rare in Western diets, but in CKD patients, vitamin K2 levels are very low due to recommended restriction of potassium and phosphate in the diet.

New trials enrolling CKD and chronic dialysis patients treated with MK-7 supplementation are presently being conducted to fully evaluate the effect of MK-7 on atherosclerosis and bone mineral density.

8. Conclusion

In the near future, the dose, bioavailability and potency of the vitamin K2 subfamily member menaquinone MK-7, will most probably make it possible to improve on the bone building process, yielding enhanced bone strength and resilience in several bone-losing patient categories, such as those suffering from osteoporosis of different etiologies, and patients presenting with low bone mass (osteopenia).

Author details

Niels Erik Frandsen[1*] and Jan Oxholm Gordeladze[2]

*Address all correspondence to: nef@dadlnet.dk

1 Department of Nephrology, University Hospital Roskilde, Roskilde, Denmark

2 Section for Biochemistry, Department of Molecular Medicine, Institute of Basic Medical Science, University of Oslo, Oslo, Norway

References

[1] Shearer MJ, Fu X, Booth LS. *Vitamin K nutrition, metabolism, and requirements: current concepts and future research 2012 American Society for Nutrition.* Adv. Nutr., 2012. 3: 182–195.

[2] Price PA. *Vitamin K-dependent formation of bone Gla protein (Osteocalcin) and its function.* Vitam Horm, 1985. 42: 65–108.

[3] International-Osteoporosis Foundation. *Facts and Statistics.* 2013.

[4] USA Preventive Services Task Force (USPSTF). Clinical summary: *Vitamin D and Calcium to Prevent Fractures: Preventive Medication.* 2014.

[5] Hart JP, Catteral A, Dodds RA, Klenerman L, Shearer MJ, Bitensky L, Chayen J. *Circulating vitamin K1 levels in fractured neck of femur.* Lancet, 1984. 324: 283.

[6] Hodges SJ, Akesson K, Vergnaud P, Obrant K, Delmas PD. *Circulating levels of vitamin K1and K2 decreased in elderly women with hip fracture.* J Bone Miner Res, 1993. 8: 1241–1245.

[7] Hodges SJ, Pilkington MJ, Stamp TC, Catterall A, Shearer MJ, Bitensky L, Chayen J. *Depressed levels of circulating menaquinones in patients with osteoporotic fractures of the spine and femoral neck.* Bone, 1991. 12: 387–389.

[8] Szulc P, Chapuy MC, Meunier PJ, Delmas P D. *Serum undercarboxylated osteocalcin is a marker of the risk of hip fracture in elderly women.* J Clin Invest, 1993. 91(4): 769–1774.

[9] Knapen MH, Neuwenhuijzen Kruseman AC, Wouters RSME, Vermeer C. *Correlation of serum osteocalcin fractions with bone mineral density in women during the first 10 years after menopause.* Calcif Tissue Int, 1998. 63(5): 375–379.

[10] Luukinen H, Kakonen SM, Pettersson K, Koski K, Laippala P, Lovgren T, Kivela SL, Vaananen HK. *Strong prediction of fractures among older adults by the ratio of carboxylated to total serum osteocalcin.* J Bone Miner Res, 2000. 15(12): 2473–2478.

[11] Feskanich D, Weber P, Willett WC, Rockett H, Booth SL, Colditz GA. *Vitamin K intake and hip fractures in women: a prospective study.* Am J Clin Nutr, 1999. 69(1): 74–79.

[12] *Japanese 2011 guidelines for prevention and treatment of osteoporosis- executive summary.* Arch Osteporos, 2012. 7: 3–20.

[13] Neogi T, Booth SL, Zhang YQ, Jacques PF, Terkeltaub R, Aliabadi P, Felson DT. *Low vitamin K status is associated with osteoarthritis in hand and knee.* Arthritis Rheum, 2006. 54: 1255–1261.

[14] Sato Y, Honda Y, Kaji M, Asoh T, Hosokawa K, Kondo I, Satoh K. *Amelioration of osteoporosis by menatetrenone in elderly female Parkinson's disease patients with vitamin D deficiency.* Bone, 2002. 31: 114–118.

[15] Presse N, Shatenstein B, Kergoat MJ and Ferland G. Low *vitamin K intake in community-dwelling elders at an early stage of Alzheimer's disease.* J Am Diet Assoc, 2008. 108: 2095–2099.

[16] Jie KG, Bot ML, Vermeer JC, Witteman J C and Grobbee DE. *Vitamin K status and bone mass in women with and without aortic atherosclerosis: a population- based study.* Calcif Tissue Int, 1996. 59: 352–356.

[17] Geleijnse JM, Vermeer C, Grobbee DE, Schurges LJ, Knapen MH, van der Meer I M, Hofman A, Witteman JC. *Dietary intake of menaquinone is associated with a reduced risk of coronary heart disease: the Rotterdam Study.* J Nutr, 2004. 134: 3100–3105.

[18] Baeksgaard L, Andersen KP and, Hyldstrup L. *Calcium and vitamin D supplementation increases spinal BMD in healthy, postmenopausal women.* Osteoporos Int, 1998. 8(3): 255–260.

[19] Kidd PM. *Vitamin D and K as pleiotropic nutrients: clinical importance to the skeletal and cardiovascular systems and preliminary evidence for synergy.* Altern Med Rev, 2010. 15: 199–122.

[20] Ooms ME, Roos JC, Bezemer PD, van der Vijgh WJ, Bouter LM, Lips P. *Prevention of bone loss by vitamin D supplementation in elderly women: a randomized double-blind trial.* J Clin Endocrinol Metab, 1995. 80(4): 1052–1058.

[21] Olivia A, Della Ragione F, Fratta M, Marrone G, Palumbo R, Zappia V. *Effect of retinoic acid on osteocalcin gene expression in human osteoblasts.* Biochem Biophys Res Commun, 1993. 191(3): 908–914.

[22] Hara K, Akiyama Y, Tajima T, Shiraki M. *Menatetrenone inhibits bone resorption partly through inhibition of PGE2 synthesis in vitro.* J Bone Miner Res, 1993. 8: 535–542.

[23] Hara K, Akiyama Y, Nakamura T, Murota S, and Morita I. *The inhibitory effect of vitamin K2 (menatetrenone) on bone resorption may be related to its side chain.* Bone, 1995. 16: 179–184.

[24] Koshihara Y, Hosti K. *Vitamin K2 enhances osteocalcin accumulation in the extracellular matrix of human osteoblasts in vitro.* J Bone Miner Res, 1997. 12: 431–438.

[25] Shiraishi A, Higashi S, Ohkawa H, Kubodera N, Hirasawa T, Ezawa I, Ikeda K, Ogata E. *The advantage of alfacalcidol over vitamin D in the treatment of osteoporosis.* Calcif Tissue Int, 1999. 65(4): 311–316.

[26] Berkner KL. *Vitamin K-dependent carboxylation.* Vitam Horm, 2008. 78: 131–156.

[27] Ichikawa T, Horie-Inoue K, Ikeda K, Blomberg B, Inoue S. *Steroid and xenobiotic receptor SXR mediates vitamin-activated transcription of extracellular matrix-related genes and collagen accumulation in osteoblastic cells.* J Biol Chem, 2006. 281: 16927–16934.

[28] Douglas AS, Robins SP, Hutchison JD, Porter RW, Stewart A and Reid DM. *Carboxylation of osteocalcin in postmenopausal osteoporotic women following vitamin K and D supplementation.* Bone, 1995. 17: 15–20.

[29] Yamaguchi M, Sugimoto E, Hachiya S. *Stimulatory effect of menaquinone-7 (vitamin K2) on osteoblastic bone formation in vitro.* Mol Cell Biochem, 2001. 223: 131–137.

[30] Yamaguchi M and Ma ZJ. *Inhibitory effect of menaquinone-7 (vitamin K2) on osteoclast-like cell formation and osteoclastic bone resorption in rat bone tissues in vitro.* Mol Cell Biochem, 2001. 228: 39–47.

[31] Yamaguchi M and Weilzmann MN. *Vitamin K2 stimulates osteoblastogenesis and suppresses osteoclastogenesis by suppressing NF-kappaB activation.* Int J Mol Med, 2011. 27: 3–14.

[32] Schaafsma A, Muskiet FAJ, Storm H, Hofstede GJ, Pakan I, Van der Veer E. *Vitamin D3 and vitamin K1 supplementation of Dutch postmenopausal women with normal and low bone mineral densities: effects on serum 25-hydroxyvitamin D and carboxylated osteocalcin.* Eur J Clin Nutr, 2000. 54: 626–631.

[33] Braam LA, Knapen MHJ, Geusens P, Brouns F, Hamulyák K, Gerichhausen MJW, Vermeer C. *Vitamin K1 supplementation retards bone loss in postmenopausal women between 50 and 60 years of age.* Calcif Tissue Int, 2003. 73(1): 21–26.

[34] Cockayne S, Adamson J, Lanham New S, Gilbody S, Torgerson DJ. *Vitamin K and the prevention of fractures: systematic review and meta-analysis of randomized controlled trails.* Arch Intern Med, 2006. 166: 1256–1261.

[35] Bolton-Smith C, McMurdo ME, Paterson CR, Mole PA, Harvey JM, Fenton ST, Prynne CJ, Mishra GD, Shearer MJ. *Two-year randomized controlled trial of vitamin K1 (phylloquinone) and vitamin D3 plus calcium on the bone health of older women.* J Bone Miner Res, 2007. 22: 509–519.

[36] Booth SL, Dallal G, Kyla Shea M, Gundberg C, Peterson J W, and Dawson- Hughes B. *Effect of vitamin K supplementation on bone loss in elderly men and women.* J Clin Endocrinol Metab, 2008. 93(4): 1217–1223.

[37] Cheung AM, Tile L, Lee Y, Tomlinson G, Hawker G, Scher J, Hu H, Vieth R, Thompson L, Jamal S, Josse R. *Vitamin K supplementation in postmenopausal women with osteopenia (ECKO trial): a randomized controlled trial.* PLoS Med, 2008. 5(12): e247.

[38] Binkley N, Harke J, Krueger D, Engelke J, Vallarta-Ast N, Gemar D, Checovich M, Chappell R and Suttie J. *Vitamin K treatment reduces undercarboxylated osteocalcin but does not alter bone turnover, density, or geometry in healthy postmenopausal North American women.* J Bone Miner Res, 2009. 24(6): 983–991.

[39] Orimo H, Shiraki M, Tomita A, Morii H, Fujita T, Ohata M. *Effects of menatetrenone on the bone and calcium metabolism in osteoporosis: a double- blind placebo-controlled study.* J Bone Miner Metab, 1998. 16: 106–112.

[40] Shiraki M, Shiraki Y, Aoki C, Miura M . *Vitamin K2 (menatetrenone) effectively prevents fracture and sustains lumbar bone mineral density in osteoporosis.* J Bone Miner Res, 2000. 15: 515–521.

[41] Iwamoto J, Takeda T, Ichinura S. *Effect of combined administration of vitamin D3 and vitamin K2 on bone mineral density of the lumbar spine in postmenopausal women with osteoporosis.* J Orthop Sci, 2000. 5: 546–551.

[42] Iwamoto J, Takeda T, Ichmura S. *Treatment with vitamin D3 and/or vitamin K2 for postmenopausal osteoporosis.* Keio J Med, 2003. 52: 147–150.

[43] Ushiroyama T, Ikeda A, Ueki M. *Effect of continuous combined therapy with vitamin K_2 and vitamin D_3 on bone mineral density and coagulofibrinolysis function in postmenopausal women.* Maturitas, 2002. 41(3): 211–212.

[44] Knapen MH, Schurgers LJ, Vermeer C. *Vitamin K2 supplementation improves hip bone geometry and bone strength indices in postmenopausal women.* Osteoporos Int, 2007. 18(7): 963–972.

[45] Jiang Y, Zhang ZL, Zhang ZL, Zhu HM, Wu YY, Cheng Q, Wu FL, Xing XP, Liu JL, Yu W, Meng XW. *Menatetrenone versus alfacalcidol in the treatment of Chinese postmenopausal women with osteoporosis: a multicenter, randomized, double-blinded, double dummy, positive drug-controlled clinical trial.* Clin Interv Aging, 2014. 9: 121–127.

[46] Iwamoto J, Takeda T, Ichimura S. *Effect of menatetrenone on bone mineral density and incidence of vertebral fractures in postmenopausal women with osteoporosis: a comparison with the effect of etidronate.* J Orthop Sci. 2001; 6(6): 487–492.

[47] Iwamoto J, Takeda T, Ishmura S. *Combined treatment with K2 and bisphosphonate in postmenopausal women with osteoporosis.* Yonsei Med J, 2003. 44: 751-756.

[48] Iwamoto J. *Vitamin K2 therapy for postmenopausal osteoporosis.* Nutrients, 2014. 6: 1971–1980.

[49] Ishida Y and Kawai S. *Comparative efficacy of hormone replacement therapy, etidronate, calcitonin, alfacalcidol, and vitamin K in postmenopausal women with osteoporosis: The Yamaguchi Osteoporosis Prevention Study.* Am J Med, 2004. 117(8): 549–555.

[50] Hirao M, Hashimoto J, Ono WAT, Yoshikawa H. *Response of serum carboxylated and undercarboxylated osteocalcin to alendronate monotherapy and combined therapy with vitamin K2 in postmenopausal women.* J Bone Miner Metab, 2008. 26: 260–264. abstract.

[51] Je SH, Joo NS, Choi B h, Kim KM, Kim BT, Park SB, Cho DY, Kim KN, Lee DJ. *Vitamin K supplement along with vitamin D and calcium reduced serum concentration of undercarboxylated osteocalcin while increasing bone mineral density in Korean postmenopausal women over sixty years old.* J Korean Med Sci, 2011. 26(8): 1093–1098.

[52] Fanf Y, Hu C, Tao X, Wan Y, Tao F. *Effect of vitamin K on bone mineral density: a meta-analysis of randomized controlled trial.* J Bone Miner Metab, 2012. 30: 60–68.

[53] Katsuyama H, Ideguchi S, Fukunaga M, Saijoh K, Sunami S. *Usual dietary intake of fermented soybeans (Natto) is associated with bone mineral density in premenopausal women.* J Nutr Sci Vitaminol (Tokyo), 2002. 48: 207–215.

[54] Ikeda Y, Iki M, Morita A, Kajita E, Kagamimori S, Kagawa Y and Yoneshima H. *Intake of fermented soybeans, natto, is associated with reduced bone loss in postmenopausal women: Japanese population-based osteoporosis (JPOS) study.* J Nutr, 2006. 136: 1323–1328.

[55] Emaus N, Gjesdal CG, Almås B, Christensen M, Grimsgaard AS, Berntsen G KR, Salomonsen L, Fønnebø V. *Vitamin K2 supplementation does not influence bone loss in early menopausal women: a randomised double-blind placebo-controlled trial.* Osteoporos Int, 2010. 21: 1731–1740.

[56] Forli L, Bollerslev J, Simonsen S, Isaksen GA, Kvamsdal KE, Godang K, Gadeholdt G, Pripp AH, Bjortuft O. *Dietary vitamin K2 supplement improves bone status after lung and heart transplantation.* Transplantation, 2010. 89: 458–468.

[57] Knapen MHJ, Drummen NE, Smit E, Vermeer C, Theuwissen E. *Three-years low-dose menaquinone-7 supplementation helps decrease bone loss in healthy postmenopausal women.* Osteoporos Int, 2013. 24(9): 2499–2507.

[58] Namba S, Yamaoka-Tojo M, Hashikata T, Ikeda Y, kitasato L, Hashimoto T, Tojo T, Takahira N, Masuda T, Ako J. Long-term warfarin therapy and biomarkers for osteoporosis and atherosclerosis. BBA Clinical, 2015; 4: 76–80.

[59] Hall JG. *Fetal warfarin syndrome.* In: Buyse ML, ed. Birth Defects Encyclopedia. Cambridge: Blackwell, 1990: 731–732.

[60] Woo C, Chang LL, Ewing SK, Baur DC. Osteoporotic Fractures in Men Study Group. *Single-point assessment of warfarin use and risk of osteoporosis in elderly men.* J Am Geriatr Soc, 2008. 56:1171–1176.

[61] Misra D, Zhang Y, Peloquin C, Choi HK, Kiel DP, Neogi T. *Incident long- term warfarin use and risk of osteoporotic fractures: propensity-score matched cohort of elders with new on-set atrial fibrillation.* Osteoporos Int, 2014. 25(6): 1677–1684.

[62] Moe S, Drueke T, Cunningham J, Goodman W, Martin K, Olgaard K, Ott S, Spraque S, Lameire N, Eknoyan G; Kidney Disease: Improving Global Outcomes (KDIGO). *Definition, evaluation, and classification of renal osteodystrophy: a position statement from Kidney Disease: Improving Global Outcomes (KDIGO).* Kidney Int, 2006. 69(11): 1945–1953.

[63] Tanko LB, Christiansen C, Cox DA, Geiger MJ, McNabb MA, Cummings S R. *Relationship between osteoporosis and cardiovascular disease in postmenopausal women.* J Bone Miner Res, 2005. 20(11): 1912–1920.

[64] Adragao T, Herberth J, Monier Faugere MC, Branscum AJ, Ferreira A, Frazao JM, Curto JD, Malluche HH. Low bone *volume—a risk factor for coronary calcification in hemodialysis patients.* Clin J Am Nephrol, 2009. 4(2): 450-455.

[65] Toussaint ND, Lau KK, Strauss BJ, Olkinghorne KR, Kerr PG. *Association between vascular calcification, arterial stiffness and bone mineral density in chronic kidney disease.* Nephrol Dial Transplant, 2008. 23(2): 586–593.

[66] Taal MW, Roe S, Masud T, Green D, Porter C, Cassidy MJ. *Total hip bone mass predicts survival in chronic hemodialysis patients.* Kidney Int, 2003. 63(3): 1116–1120.

[67] Matuszkiewicz-Rowinska J, Skorzewska K, Radowicki S, Sokalski A, Przedlacki J, Niemczyk S, Wlodarczyk D, Puka J, Switalski M. *The benefits of hormone replacement therapy in pre-menopausal women with oestrogen deficiency on haemodialysis.* Nephrol Dial Transplant, 1999. 14(5): 1238–1243.

[68] Kohlmeier M, Sauoe J, Shearer MJ, Schaefer K, Asmus G. *Bone health of adult hemodialysis patients is related to vitamin K status.* Kidney Int, 1997. 51: 1218–1221.

[69] Fusaro M, Tripepi G, Noale M, Plebani M, Zaninotto M, Piccoli A, Naso A, Miozzo D, Giannini S, Avolio M, Foschi A, Rizzo MA, Gallieni M. *Vertebral fracture and vascular calcifications study group.* Curr Vasc Pharmacol, 2015.13(2): 248–258.

[70] McCabe KM, Adams MA, Holden RM. *Vitamin K status in chronic kidney disease.* Nutrients, 2013. 5: 4390–4398.

[71] Holden RM, Morton AR, Garland JS, Pavlov A, Day AG, Booth SL. *Vitamin K and D status in stages 3-5 chronic kidney disease.* Clin J Am Soc Nephrol, 2010; 5: 590–597.

[72] Cranenburg EC, Schurgers LJ, Uiterwijk HH, Beulens JW, Dalmeijer GW, Westerhuis R, Magdeleyns EJ, Herfs M, Vermeer C, Laverman GD. *Vitamin K intake and status are low in hemodialysis patients*. Kidney Int, 2012. 82: 605–610.

[73] Westenfeld S, Krueger T, Schlieper G, Cranenburg EC, Magdeleyns EJ, Heidenreich S, Holzmann S, Vermeer C, Jahnen-Dechent W, Ketteler M, Floege J, Schurges LJ. *Effect of vitamin K2 supplementation on functional vitamin K deficiency in hemodialysis patients: a randomized trial*. Am. J. Kidney Dis, 2012.59: 186–195.

Chapter 7

Vitamin K2 and its Impact on Tooth Epigenetics

Jan Oxholm Gordeladze, Maria A. Landin,

Gaute Floer Johnsen, Håvard Jostein Haugen and

Harald Osmundsen

Additional information is available at the end of the chapter

http://dx.doi.org/10.5772/66383

Abstract

The impact of nutritional signals plays an important role in systemic-based «models» of dental caries. Present hypotheses now focus both on the oral environment and other organs, like the nervous system and brain. The tooth is subjected to shear forces, nourishing and cleansing, and its present "support system" (the hypothalamus/parotid axis) relays endocrine signaling to the parotid gland. Sugar consumption enhances hypothalamic oxidative stress (ROS), reversing dentinal fluid flow, thus creating an enhanced vulnerability to the oral bacterial flora. The acid, produced by the oral bacterial flora, then leads to erosion of the dentine, and an irreversible loss of dental enamel layers. This attack brings about inflammatory responses, yielding metalloproteinase-based "dissolution". However, vitamin K2 (i.e. MK-4/MK-7) may come to the rescue with its antioxidant property, locally (mouth cavity) or systemically (via the brain), thus sustaining/preserving hormone-induced dentinal fluid flow (encompassing oxidative stress) and boosting/magnifying bodily inflammatory responses. However, sugars may also reduce the tooth's natural defences through endocrine signaling, thus enhancing acid-supported enamel dentine erosion. Vitamin K2 sustains and improves the salivary buffering capacity via its impact on the secretion/flow of calcium and inorganic phosphates. Interestingly, primitive cultures' diets (low-sugar and high-K2 diets) preserve dental health.

Keywords: K2, SXR, endocrine interaction, deiodinases, TH receptors

INTECH
open science | open minds

© 2017 The Author(s). Licensee InTech. This chapter is distributed under the terms of the Creative Commons Attribution License (http://creativecommons.org/licenses/by/3.0), which permits unrestricted use, distribution, and reproduction in any medium, provided the original work is properly cited. (cc) BY

1. Introduction

The "perfect" and healthy tooth is properly designed to cope with the rough environment in the oral cavity, since virtually cleanses itself in an inside-out manner. Dental caries is supposed to be a result of the phenomenon that a tooth's fluid flow has been halted or even reversed, thus compromising tooth's defense system. It is well known that the local enamel demineralization process, aided by bacterial acid, is vastly stimulated by nutritional conditions, specifically by today's refined carbohydrates (sucrose and corn syrup = free glucose + fructose). Hence, the ensuing process renders the tooth vulnerable, and part of this process begins in the hypothalamus, resulting in alterations of the fluid flow passing through the dentine layer [1]. Not to forget: in this context, nutritional phenomena play very important roles (both systemically and locally).

In the aftermath of acid-induced enamel demineralization of the breakdown of the dentin layer is accomplished by the body's own matrix metalloproteinase (MMP) enzymes [2, 3], a process which follows as a result of an untoward and galloping inflammatory response to an enhanced acid secretion. The present caries process begins as a more or less dormant, but reversible inflammation ("dentinitis"), while developing into a nonreversible dentin phase of caries after a while. This biological phenomenon is very much like reversible versus irreversible pulpitis and the terms gingivitis and periodontitis, while referring to the periodontium only [4].

However, there is a consensus that the process of dental caries recognization is multifactorial, as well as systemically based. It may not be sufficient to decrease the detrimental process rendered by the sugar intake with the ensuing enhancement of bacterial number and strains, but also boost the body's defense mechanisms with an antioxidant-rich diet that may be composed of fruits and vegetables, as well as vitamin K2.

Some research reports document that vitamin K2 can assist in significantly reducing dental caries [5, 6]. A larger body of research, however, is necessary in order to establish the mode, by which this vitamin may augment local defense mechanisms by altering saliva composition, while also systemically, via influencing the hypothalamus, as well as endocrine aspects of the parotid gland.

This concept of systemically delivered impact underscores an important shift in paradigm, from a traditional 'acid theory' explaining the development of dental caries that carries a plethora of implications for the prevention of dental decay in the future. Furthermore, it will necessary to document, via the mechanism of action of vitamin K2, how this small molecule affects gene regulation of the intimate play of osteoblast- and osteoclast-like cells in the organ layers constituting developing and full-frown teeth. The present chapter attempts to create a synthesis of current knowledge and recent research reports and ongoing research projects, with the intention of shedding new light on the impact of K2 on dental health.

2. Backing of a new concept

2.1. Oral and systemic stress responses with common denominators

The definition of the stress concept dates back some 60–70 years [7]. It was meant to feature the process of how irritants caused a bodily reaction, and how the body dealt with it. If a stressor was defined as local, e.g. acid-induced enamel demineralization or irritation of the periodontal tissues by plaques, the body might provoke a specific, but local reaction. This would be controlled inflammatory responses that will remain similar throughout the entire body. Hence, they are defined as "local adaptation syndromes" (LAS). In the dentin of a tooth, "dentinitis" represents this local inflammatory response.

A focal reaction is often mild, but resides rather quickly. However, this limited response may develop into a systemic and an exaggerated variant, thus "threatening" the entire body via the endocrine system. The present response is named "GAS" ("general adaption syndrome"), because its "attack" on the organism pretty much resembles the systemic type reaction, even though it represents a local type of stress reaction [7]. The hypothalamus/pituitary/adrenal axis serves as the mechanism in charge of the body's general, untoward reactions.

The essence of the problem resides with the ingestion of refined carbohydrates (i.e., mono and disaccharides in particular) that locally escalate the growth of microorganisms within the oral cavity and their production of acid. This inflammation causes a rapid loss of minerals, such as dental enamel, and it is named "dentinitis" and serves as a local adaptation, being a part of an LAS.

However, subsequent to this local reaction, small-molecular sugar entities (like sucrose, glucose, and fructose) exert a major impact on the body, when absorbed. Blood-sugar spikes are therefore counteracted by emerging dentinal fluid flows through the tooth by coordinated signals emanating from the hypothalamus. This adaptation (GAS) is chiefly endocrine, and affects the entire body. Hence, the present hypothalamic-parotid axis serves as the endocrine axis that is instrumental in maintaining the dental health [8]. The essence of it is the following: local irritation is magnified in the presence of a GAS response [7, 9]. Sugar molecules (mostly monosaccharides), with their marked effect on the whole organism, magnifies the local acid attack by triggering the GAS response. As a consequence, the tooth is rendered more vulnerable to the acid exposure [1], whether the acid is produced from sugar entities (via bacteria) or comes from various carboxylic acids in the diet.

2.2. How the tooth develops caries

The tooth is fed by the alimentary elements delivered by fluid flow through the dentin which may be halted or reversed, when impacted by a systemic stressor, like excessive sugar intake [10, 11]. This allows bacteria in the oral cavity to attach to the tooth, where they enhance the local concentration of acid, leading to the well-known demineralization of the enamel surface. Consequently, the flow of fluid through the tooth is enhanced by parotid hormone [8]. The secretion of parotid hormone, as well as the secretion of insulin [12, 13] is regulated by the hypothalamus, influencing GAS, and eventually affecting the adrenal glands. Finally, a normal inflammatory reaction (LAS) occurs with a corresponding enhanced metabolism, i.e., (1) an increase in reactive oxygen species (ROS) production, and (2) activation of MMP's (e.g., collagenase). Normally, tissue inhibitors of metalloproteinases (TIMP's) serve to neutralize activated MMP's, when the body has regained its control over the inflammatory process. Antioxidants aid in the control inflammation by hampering an ROS activity, thus minimizing the necessity for stimulating MMP activation. It is known that optimal nutrition exerts an important role in this process. Temporary challenges are followingly managed by the dampening inflammation known to cause reversible state, while healing may occur. This process is called dentinitis [14, 15]. Excessive irritability leads to galloping inflammatory processes that are mainly irreversible in the tooth and recognized in the dentin by the name of caries.

2.3. A systemic approach and the impact of oxidation control

The systemic "angle" of caries may be construed as a link to diabetes mellitus. When enhanced blood glucose is registered in the hypothalamus, the production of free radicals like "reactive oxygen species (ROS)" is enhanced. These molecules serve as a warning signal for the hypothalamic gland to down-regulate the secretion of parotid hormone, while simultaneously upregulating the insulin secretion. Antioxidant loading has since long been known to manage the glucose-induced free radical storm sweeping the hypothalamic gland [16].

Antioxidants counteract the detrimental effects of the free radical damage, as shown in the so-called Asian Paradox, where heavy cigarette smoking is paralleled with reduced rates of coronary disease and cancer amongst consumers of green tea [17–19] that is famous for its antioxidant properties. However, along with the fact that systemically administered antioxidant effects of green tea reduce the incident of dental caries, vitamin K2 may prove to be an even more potent antioxidant [20–24].

2.4. A summary of the features of vitamin K2

Vitamin K2 is known as menaquinones, while vitamin K1 is phylloquinone. The quinones display oxygen-containing ring structures that render them suitable for the transport of electrons [25–27]. K2 was added to the vitamin K category of molecules, since it may be produced in the body from K1 [28]. K1 is deemed essential to blood clotting, and accordingly, our body has "invented" modes to recycle K1 for repeated use. Hence, it has rendered itself less dependent on a constant dietary intake of K1.

Vitamin K2 takes several forms that are linked to the structure of their side chains (e.g., MK4 and MK7). While MK4 is the form produced by our bodies produced from K1, supplemental MK4 is entirely synthetic. MK7 is a (more) biologically active form that entertains a longer half-life. Therefore, it is often the preferred or recommended supplement [29], especially in the treatment of osteoporosis and untoward soft tissue calcifications (ref).

Vitamin K2 is bounded to the transcription factor SXR/PXR (ref) and may serve as a cofactor of vitamin K dependent carboxylases. This enzyme, when associating with vitamin K2, will change the structure of proteins by the process of gamma-carboxylation, or the SXR/PXR-vitamin K2 complex may work to enhance the expression of a set of genes that are responsive to the presence of vitamin K2 (ref). Some examples of these processes are osteocalcin, located in bones and teeth, and matrix GLA protein expressed in cardiovascular (i.e., soft) tissues. Both of these protein structures require the vitamins A and D, as well as vitamin K2 for their production [30, ref]. The carboxylation of osteocalcin by vitamin K2 allows it to attract and retain calcium that is good for bones [31, 32]. The "opposite" process is observed in cardiovascular (i.e., soft) tissues, since matrix GLA proteins allow for calcium to be deposited in arteries, when uncarboxylated, but shed or blocks the "entry" of calcium, when carboxylated with the assistance of sufficient vitamin K2 [33–35].

Dietary K2 is processed in the liver and released into the circulation via high and low density lipoproteins that make them readily available for uptake into extrahepatic tissues [36–38]. Fermented foods such as cheese have significantly higher levels of K2 than milk. The higher levels are obtained from bacterial sources. Natto (which is fermented soy) is, without any doubt the more potent source of vitamin K2 [30]. Most K2 supplements are cultured from natto; however, synthetic products with unsurpassed bioavailability and stability are now to be available in the market.

Menaquinones are taken up and stored in several tissues throughout the body. Some of the highest concentrations to be found are in the pancreas and the salivary glands [39]. Hence, it can be construed that there exists a close relationship between both of these exocrine/endocrine glands through the hypothalamus. High levels of vitamin K2 are also located in the brain, heart, and bone [40, 41] which definitely is of significance for many disease states, including dental caries, which has been shown to be associated with oxidative stress [6, 24].

2.5. How dental tissue is nourished

Saliva brings nutrients from the outside of the tooth directly to the inside [42]. The fluid holds active ingredients, like minerals and enzymes, as well as buffering agents. The free cytosolic calcium levels sustain the more important or critical role in the signaling potential of the salivary glands' contents [43–45]. Taken calcium's dependency on vitamin K2–assisted carboxylation related to osteocalcin and matrix GLA proteins for granted, findings to come may reveal that vitamin K2 exerts an impact on salivary signaling and composition and activation potency. Since long, one has known that insulin [46] and the exocrine secretions of the pancreas [47] are partially dependent on the presence of vitamin K2.

Furthermore, saliva serves as an important player in the maintenance of proper mineralization of the teeth's enamel. The saliva buffers demineralization seen with acid-induced mineral dissolution, and it delivers building blocks for remineralization "on request." The optimal pH for tooth health is variable [48], and is more related to saliva composition and flow. However, it is yet not known how saliva is connected to vitamin K2, but it has been associated with its contents of the pH-buffering inorganic phosphate, decreasing the counts of lactobacilli in the oral cavity [5].

Some types of cheese have been asserted to display anticariogenic properties [49], and this is due to its contents of fermented bacteria that produce higher amount of vitamin K2 [30, 50]. This favorable feature would serve as a source of systemically located vitamin K2, rather than locally delivered. However, there is also a possibility that vitamin K2 is absorbed across the mucous membranes of the oral cavity. Quite successfully, one has applied ubiquinone topically to subdue periodontal inflammation induced by controlled oxidative stress [51].

Vitamin K2's effect on the tooth's outer surface can be seen via its impact on saliva component distribution. This secondary prevention, i.e., the "remineralization success" is mainly relying on whether the saliva composition is altered in order to produce a so-called facilitating "remineralization microenvironment."

2.6. Interpretation of "historical" data collected by Price

The famous set of data, collected by Price, when visiting groups of primitive cultures from different parts of the world, should be well known to vitamin K2 enthusiasts. Some groups visited were still primitive as to their alimentation culture and customs, while other groups and subgroups had adapted to modern civilized diets and ways of living.

Price subsequently analyzed food samples from these groups. And, with little deviation, he registered that the "primitive diets" were high in vitamins A and D, along with factor named "Activator X". This particular ingredient could be retrieved from butter by grass fed animals [5]. He launched the idea that it was a fat soluble nutrient that has now been identified as or linked to menaquinones, vitamin K2 [6].

3. Evaluation of "the" hypothesis

3.1. The support of a systemic theory of dental caries and K2 as a critical component

The "old" theory of dental health, called the 'acid theory,' has been linked to the oral environment as an isolated and unique process, involving the bacteria-acid axis, as it is the only cause-effect relationship.

The systemic version or theory of dental caries acknowledges the effect of refined carbohydrates on oral cavity through the impact of the hypothalamus and the endocrine system. Earlier, free radicals like ROS, typically having been construed as "exhaust energy" from mitochondria. *However, they are now thought of as "critical signals," conveyed to the hypothalamus in order to influence the secretion of hormones, like parotid hormone and insulin. The major task to be undertaken now is: find which nutrients optimally affect the hypothalamus and which would maintain the centrifugal fluid flow through the teeth.* Antioxidants such as EGCG of green tea have proven effective. Vitamin K2, however, may serve as a more potent nutrient.

3.2. What we hope to learn from Dr. Price's discovery

The exocrine functions of the salivary glands, i.e., the composition of the saliva, are nutritionally related. As for the prevention of dental caries, optimum nutrition with fat soluble vitamins like K2 may serve as much more significant factor, than the role of traditional dental recommendations which goes like this: *Eat less sugar to minimize the production of bacterial acids in the oral cavity*. Dental disease will be construed and recognized as an inflammation related degenerative lifestyle disease in line with cardiovascular incidents, as well as with bone brittleness (osteoporosis) and diabetes mellitus.

4. Some recent articles "establishing" the vitamin K2 effect on teeth

It has been reported that vitamin D3 acts synergistically with vitamin K2 to prevent bone loss. A recent study evaluated the impact of vitamin K2 and vitamin D3, as an alimentary supplement in conjunction with scaling and root planning; SRP, as conventional periodontal, on gingival expression of the interleukins IL-1β and IL-10, serum bone-specific alkaline phosphatase (B-ALP), as well as tartrate-resistant acid phosphatase (TRAP, subtype 5b), and the steady state levels of alveolar bone calcium in rats subjected to experimentally provoked periodontitis. Alveolar bone mass in the periodontitis group was markedly larger than the ones in the other experimental groups. No significant differences were seen in the gingival contents of IL-1β and IL-10, blood B-ALP, TRAP-5b, and calcium, nor in alveolar bone mass between the groups receiving SRP and vitamins, and the experimental group receiving SRP alone. Furthermore, vitamin D3 and K2 alone, or combined, failed to affect gingival levels ofIL-1β and IL-10, as well as blood B-ALP and TRAP-5b, or the alveolar bone mass, as compared with traditional periodontal treatment *per se* [52].

Mesenchymal stem cells have often been used for tissue engineering in regenerative medicine. The present application focused on the features of stem cells obtained from human exfoliated deciduous teeth (SHED) in comparison with dental pulp stem cells (DPSCs) and bone marrow-derived mesenchymal stem cells (BMMSCs). Cells retrieved from various sources displayed MSC characteristics (i.e., fibroblastic morphology and MSC markers). Their growth rate markedly elevated, as compared with that of DPSCs and BMMSCs. Furthermore, it was demonstrated that some 4400 genes altered their expression by a factor of 2.0 or more. A higher gene expression in SHED was witnessed for genes participating in reaction pathways such as "cell proliferation" and "extracellular matrix" [53].

Enhancement of intracellular Ca^{2+} concentrations is a feature commonly seen during the differentiation period of stem cells. The transient receptor potential melastatin 4 (TRPM4) serves as one of many ion channels controlling the Ca^{2+} signals in both excitable and nonexcitable cells. Nelson et al. [54] characterized TRPM4 in dental follicle stem cells (DFSCs) of the rat, and defined its impact on Ca^{2+} mediated signaling in the differentiation process. ShRNA-mediated suppression of TRPM4 decreased the channel activity, resulting in cell proliferation during osteogenesis, with a concomitantly augmented mineralization. Whole genome microarray analysis revealed that a plethora of genes, being associated with both vitamin K2 and SXR = PXR = NR1/2, were affected by TRPM4 during DFSC differentiation. These observations indicate that TRPM4 inhibit osteogenesis. The information provided suggests a link between the Ca^{2+} signaling pattern and gene expression during the differentiation process, including a recognizable influence of vitamin K2.

Wnt signaling pathways are now heavily linked to bone biology [55]. In the present review, recent advances in how Wnt/Lrp5-mediated signaling modulates osteoblast and osteocyte functioning, introduce new players in the Wnt signaling pathways, proving to play important roles in bone development. Here, emerging areas of Wnt signaling in osteoclastogenesis are discussed, as well as the progress made in translating basic studies to clinical therapeutics and diagnostics centered around inhibiting Wnt pathway antagonists. These are sclerostin, Dkk1, and Sfrp1. In a recent study, (unpublished data) Osmundsen and coworkers have shown that vitamin K2 affects the Wnt system by modulating the expression of DKK1 (a Wnt inhibitor) during the development of teeth in developing molar teeth of the mouse.

Another report aims to reveal the biological and physicochemical features of MTA = mineral trioxide aggregate, related to its potency in eliciting reparative dentinogenesis. In comparison with calcium hydroxide-based materials, MTA is more efficient. It has been asserted that the action of MTA is associated with natural wound healing processes of exposed pulps, even though MTA may also stimulate matrix formation and its mineralization *in vitro*. Physicochemical analyses have shown that MTA may also interact with phosphate-containing fluids to precipitate apatite crystals. Furthermore, MTA shows better sealing ability and maintains structural stability, however [56].

Congenital diseases of tooth roots (e.g., developmental abnormalities of short and thin roots) may lead to tooth loss [57]. Recently, studies have shown that Osterix (Osx), serving as an important transcriptional factor, along with Runx1, Runx2, SP1, and SP3 [58], all participating in osteogenesis and odontogenesis, is thought to play a vital role underlying the mechanisms that determine the developmental differences between the root and the crown. During tooth development, Osx, particularly in odontoblasts and cementoblasts, promote and sustain their differentiation as well as and mineralization. Additionally, site-specific roles of Osx in the formation of tooth root have been established. Hence, Osx is construed as a promoter of odontoblast and cementoblast differentiation, as well as a factor determining root elongation. Research featuring mechanistic properties of teeth delineates a regulatory network involving Osx expression which is controllable via either BMP-signaling or Runx2-expression, pointing to a feasible way of promoting/sustaining Osx expression experimentally [59].

Calcium hydroxide $Ca(OH)_2$ has been extensively used in the dentistry; however, its mechanism of action remains unclear. Unraveling its modes of action will provide a broader understanding of the mechanisms associated with the induced dentinogenesis, as well as helping to optimize currently available treatment modalities to ensure specific regenerative processes of tooth preservation. A compilation of articles on "mechanisms of dentinogenesis involving calcium hydroxide" is featured in this paper, and recommendations related to dentinogenic mechanisms of $Ca(OH)_2$ range from direct irritating action by the material to induction of release of biologically active molecules, like fibronectin, BMPs (bone morphogenic proteins) like BMP-4 and BMP-7, TGFs (transforming growth factors) like TGFβ, IGFs (insulin-like growth factors), antiinflammatory interleukins, alkaline phosphatase (ALP), and others [60]. It is well known that a plethora of these factors are encoded by genes that are sensitive to the impact of vitamin K2 (via the transcription factor SXR = PXR = NR1/2), as well as the vitamin A (RXR) and vitamin D (VDR) receptors [61].

The extracellular matrix (ECM) provides physical support for various tissues. However, it also contributes to the development of same, their homeostasis, and prevention of disease. More than some 200–300 ECM molecules are listed as comprising the "core matrisome" in mammals, based on analyses of whole genome sequences, and during the course of tooth development and growth, the structure-function relationship of the ECM is altered dynamically. In early phases, the basement membranes (BMs) separate into two cell layers of the dental epithelium and the mesenchyme. These matrix proteins are instrumental in cell functions like adhesion, polarity, as well as differentiation and mineralization of the enamel and dentin matrices [62, 63].

Interestingly, several of the genes known to be important in tooth development, referred to in the present paragraphs, can be retrieved as SXR and/or vitamin K2-sensitive, or as shown by Osmundsen and coworkers (personal communication), and tabulated underneath in: "Summary of information obtained from the microarray analyses."

5. Effects of mandibular injection of MK-7 on gene expression in the developing molar tooth

5.1. Methods

As an initial experiment, new born (at P1) Balb C mice were given 10 μl intra-mandibular injections of MK-7 (0.2, 2, and 10 mg/kg body-wt., dissolved in corn oil). The control mice were injected with vehicle only.

At 24 hours postinjection, the pups were killed and first right-hand side molar tooth germs were removed and transferred into RNA-Later solution.

Total RNAs were isolated from individual tooth germs and used for analysis of gene expression using deoxyoligonucleotides microarrays and real-time RT-PCR using the RNeasy Mini Kit. The quality of isolated RNA was monitored using the Agilent Bioanalyzer. RNA was isolated from three separate batches of tooth germs (three tooth germs per batch).

6. Results

Results from all dosages used suggested that numerous genes exhibited significantly altered expression. At this stage, results from pups given 2mg/kg have been more extensively analyzed.

These data results suggested that 281 genes were differentially expressed ($p < 0.05$), with changes in expression ranging from about 5- (Minpp1, Pdzd2) to about 0.05-fold (Zfp485, Slc2a5). Bioinformatics analysis (using Ingenuity Pathways Analysis) suggested that a major fraction of the observed changes in gene expression was associated with "cell death," the data suggesting a highly significant association to decreased apoptosis. This is likely mediated via altered expression of genes associated with regulation of metabolic substrates being converted to polyamine, retinoic acid-dependent regulation of apoptosis regulation (three genes), and altered osteoclast and osteoblast signaling (six genes). Several genes involved in synthesis of complex carbohydrates e.g., proteoglycans, heparin sulphate of chondroitin sulphate b, were upregulated.

Interestingly, also some genes related to enamel/dentine biosynthesis exhibited differential expression (e.g., Amelx, Ambn but not Enam). Microarray results were validated by real-time RT-PCR. Transcription factor analysis (using Ingenuity Pathways Analysis) suggested significant associations to the increased transcriptional activity of Myc (measured as changes in expression of Hspd1, Ctnnb1, Dkk1, and Psmb8 being in line with the predicted change). The downregulation of Dkk1 is interesting as denosumab treatment results in reduced Dkk1 level. Further, high Dkk1 levels have been associated with increased bone loss.

The data also suggested highly significant associations to decreased apoptosis. This is likely mediated via altered significant associations to polyamine regulation (three genes), altered retinoic, acid apoptotic regulation (three genes), and altered osteoclast and osteoblast signaling (six genes).

7. Conclusions

The results are based on the measurements from independent biological triplicates and therefore suggestive of effects of MK-7 on gene expression during the tooth development. A clear effect on gene expression was apparent also at a dosage of 0.2 mg/kg body weight. The results indicate increased transcription of genes involved in development of bone (increased biosynthesis of important carbohydrates) and of enamel/dentin.

Further investigations are, however, required to elucidate these findings. Such experiments will likely entail the establishment of a clear dose-response relationship as well as of a time-course of action. Also, effects of oral administration should be studied.

Summary of information obtained from the microarray analyses

Gene name	Description of function: *General* and bone-related	References
Lmcd1	*Transcriptional cofactor restricting GATA6 and GATA4 functioning by inhibiting DNA-binding:* Gata6 and Gata4 are transcription factors shown to be involved in TGFβ- and estrogen-mediated regulation of gene expression in osteoblasts, and thus bone mineralization to hamper the development of brittleness.	*Int J Biochem Cell Biol. 2013 Mar;45(3):696–705. J Bone Miner Res. 2014 Dec;29(12):2676-87. J Cell Physiol. 2013 Jul;228(7):1594–600.*
Dmxl2	*Protein involved in many functions including participation in signal transduction pathways, such as Notch signaling:* NOTCH signaling in BMSCs (bone marrow stromal/stem cells) is required for fracture repair performed by mature osteoblastic cells.	*J Biol Chem. 2010 Nov 5;285(45):34757–64. J Clin Invest. 2016 Mar 7. pii: 80672.*
Abcb4	*The membrane-associated protein encoded by this gene is a member of the superfamily of ATP-binding cassette (ABC) transporters. ABC proteins transport various molecules across extra- and intra-cellular membranes. The gene is heavily involved in a plethora of liver functions:* In experimental animals, i.e., Abcb4 (-/-) mice with hepatic osteodystrophy, serum RANKL and TGFβ-levels were augmented, resulting in an excess bone resorption rate, probably due to a dysregulation of genes like osteoprotegerin, osteocalcin, and osteopontin, as well as vitamin D metabolism.	*Bone. 2013 Aug;55(2):501–11. doi: 10.1016/j. bone.2013.03.012.*

Gene name	Description of function: *General* and bone-related	References
Slc12a6	*This gene is a member of the K-Cl cotransporter (KCC) family. K-Cl cotransporters are integral membrane proteins that lower intracellular chloride concentrations below the electrochemical equilibrium potential:* Human osteoblasts express functional K-Cl cotransporters in their cell membrane that seems to be able to induce activation of volume-sensitive channels by KCl, necessary for normal osteoblast membrane currents, and thus secretory functions.	*Am J Physiol Cell Physiol. 2003 Jul;285(1):C22–30.*
Mta1	*Transcriptional coregulator that can act as both a transcriptional corepressor and coactivator, and stimulates the expression of WNT1 by inhibiting the expression of its transcriptional corepressor, SIX3:* Hypoxia-induced MTA1 (via HIF-1α) stimulates growth (and inhibits differentiation) of osteoblastic (MC3T3) cells which is deemed important in the process of fracture healing.	*Eur J Med Res. 2015 Feb 3;20:10.*
Pou3f3	*This gene encodes a POU-domain containing protein that functions as a transcription factor. The encoded protein recognizes an octamer sequence in the DNA of target genes. This protein may play a role in development of the nervous system:* However, it was recently shown that this gene normally upregulates the genes of the Dlx-familiy (Dlx1, 2, 5, 6), as well as downstream genes like Gbx2, is involved in the patterning of the mammalian jaw.	*Development. 2008 Sep;135(17):2905–16.*
Dio2	*This gene belongs to the iodothyronine deiodinase family. It activates thyroid hormone by converting the prohormone thyroxine (T4) by deiodination to bioactive 3,3′,5-triiodothyronine (T3). It is highly expressed in the thyroid, but is known to be expressed many other peripheral tissues:* Cold exposure (in Misty mice) compensates for BAT (brown adipose tissue) dysfunction by increasing the expression of Acadl, Pgc1a, Dio2, and other thermogenic genes, by altering the expression of osseous Runx2 and Rankl. Genes upregulated by BMP-7 showed a strong enrichment for established osteogenic marker genes, and several others (MFI2, HAS3, ADAMTS9, HEY1, DIO2 and FGFR3) in osteoblasts. Furthermore, for DIO2 seems to impact osteoblastic differentiation. Outer ring deiodination (ORD) activity was seen in bone extracts of whole skeleton, bone marrow, and MC3T3-E1 osteoblasts. [1,25(OH)2VD]-treatment induced D2 activity, while estradiol, PTH, forskolin, leptin, TNFα, TGFβ, and dexamethasone did not.	*J Bone Miner Res. 2013 Sep;28(9):1885-97. doi: 10.1002/jbmr.1943. Bone. 2009 Jul;45(1):27–41. doi: 10.1016/j.bone.2009.03.656. Epub 2009 Mar 21. Endocrinology. 2005 Jan;146(1):195–200. Epub 2004 Oct 7*
Camk4	*Camk4 is a member of he serine/threonine protein kinase (PK) family, and the Ca^{2+}-calmodulin-dependent protein kinase subfamily. It serves as a multifunctional serine-threonine protein kinase, and has been implicated in transcriptional control of lymphocytes, neurons, as well as male germ cells:* Silencing of CaMK1β obliterates the proliferation ability of osteoblasts, as well as expression of c-Fos. However, this does not influence the skeleton markersRunx2, Osterix, and/or Osteocalcin. CaMKs activate pathways mediated by CREB and NFATc1. Inhibition of CaMKs obliterates CREB phosphorylation, lowering c-Fos, and NFATc1 expression, and thus osteoclastogenesis activated by NF-κB ligand (RANKL). Finally, BMP-receptor signaling in stem cells from human exfoliated deciduous (SHED) teeth enhances the expression of genes like BMP-4, Runx2, as well as DSPP.	*Bone. 2008 Oct;43(4):700–7. Nat Med. 2006 Dec;12(12):1410–6. Epub 2006 Nov 26 J Endod. 2011 Dec;37(12):1647–52. doi: 10.1016/j.joen.2011.08.023. Epub 2011 Oct 6.*

Gene name	Description of function: *General* and bone-related	References
Ppp2r2b	*This gene encodes the family of phosphatase 2 regulatory B subunits. Protein phosphatase 2 functions as one of four main Ser/Thr phosphatases, and is involved in the inhibitory control system of cellular division:* Loss of estrogen during menopause causes changes in the female body. Hence, it is expected that HRT-associated gene expression is due to the changes in the DNA methylation profile (DMP). Of the DMP genes, ACBA1, CCL5, FASLG, PPP2R2B, and UHRF1 were differentially expressed, all of which are associated with HRT or estrogenic regulation. All genes were also associated with bone mineral content (BMC), while ABCA1, FASLG, and UHRF1 were also associated with body adiposity.	*Bahl A1, Pöllänen E, Ismail K, Sipilä S, Mikkola TM, Berglund E, Lindqvist CM, Syvänen AC, Rantanen T, Kaprio J, Kovanen V, Ollikainen M.*
Stard4	*Cholesterol homeostasis is regulated by sterol regulatory element (SRE)-binding proteins (SREBPs) and by liver X receptors (LXRAs). When sterols are depleted, LXRs are inactive and SREBPs bind promoter SREs and activate genes involved in cholesterol turnover. The protein STAR is involved in this process, and it is homologous to a family of proteins STARD4:* Estrogen-containing hormone replacement therapy (HRT) leads to a relief of typical menopausal symptoms, benefits bone and muscle health, and is associated with tissue-specific gene expression profiles. Hence, It is plausible that part of the HRT-associated gene expression is due to changes in the DNA methylation profile. The gene expression patterns of white blood cells (WBCs) and their associations with body composition, including muscle and bone measures of monozygotic (MZ) female twin pairs discordant for HRT were assessed. Of genes with differentially methylated regions (DMRs), five (ACBA1, CCL5, FASLG, PPP2R2B, and UHRF1) were also differentially expressed. These have been associated with HRT or estrogenic regulation, but were also associated with bone mineral content (BMC). Additionally, ABCA1, FASLG, and UHRF1 were also related to the body adiposity.	*Bahl A1, Pöllänen E2, Ismail K1, Sipilä S2, Mikkola TM2, Berglund E3, Lindqvist CM3, Syvänen AC3, Rantanen T2, Kaprio J1, Kovanen V2, Ollikainen M1.*
Tyk2	*This gene encodes the Janus protein kinases (JAKs). These proteins associate with and activate cytokine receptors with ensuing phosphorylation (activation) of receptor subunits. It is also a component of the interferon signaling pathways. It may therefore play a role in anti-viral immunity:* Interleukin-23 (IL-23) belonging to the IL-6/IL-12 family that plays a key role in autoimmune and inflammatory disorders. IL-23 binding to dendritic cells, macrophages and monocytes triggers the activation of Jak2 and Tyk2 which in turn phosphorylates STAT1, STAT3, STAT4, and STAT5 as well as induce formation of STAT3-STAT4 heterodimers. IL-23 is essential for the survival and/or expansion of inflammatory Th17 cells which, when activated by IL-23, sustain osteoclasto-genesis via the production of IL-17 (stimulator of the NF-kappa B) of mesenchymal cells. As a group, the IL-17 - IL-23 "axis" includes Th17 cells that play a major role in the development and maintenance of autoimmune arthritic inflammation.	*Scand J Immunol. 2010 Mar;71(3):134-45. doi: 10.1111/j.1365–3083.2009.02361.x.*
Pdcd5	*This gene encodes a protein that is upregulated during apoptosis. The encoded protein is a regulator of K(lysine) acetyltransferase-5, involved in transcription, DNA damage response and the cell cycle control, by blocking its degradation:* The programmed cell death gene (PDCD5) was overexpressed in an osteosarcoma (OS) cell line, MG-63. The results indicate that PDCD5 can induce apoptosis and G(2) phase arrest in MG-63 cells. Furthermore PDCD5 expression in established xenografted tumors was associated with a decrease in tumor size and weight. Furthermore, it was found that the Ras/Raf/MEK/ERK signaling pathway was hampered, leading to the inhibition of cyclin B and CDK1, and to the activation of caspases 3 and 9, respectively. These results are consistent with the G(2) phase arrest observed.	*Cell Signal. 2012 Aug;24(8):1713-21. doi: 10.1016/j.cellsig.2012.04.011. Epub 2012 Apr 25.*

Gene name	Description of function: *General* and bone-related	References
Psenen	*Presenilins are required for intramembranous processing of transmembrane proteins, such as the Notch proteins. Signaling by Notch receptors mediates a wide range of developmental cell fates.* *Titanium implant surfaces with modified topographies improve osteogenic properties in vivo. The activation of signaling stem cell pathways (such as TGFβ/BMP, Wnt, FGF, Hedgehog, and Notch) was characterized subsequent to incubations (24 and 72 h) with BCs to SLA and modSLA surfaces in the absence of osteogenic cell culture supplements.* *Key regulatory genes belonging to the TGFβ/BMP (TGFBRs, BMPRs, ACVRs, SMADs, Wnts, FZD1, FZDs, LRP5, NFATCs, PYGO2, LEF1) and Notch species (including PSENEN) pathways were upregulated on the modified surfaces. These data correlate with an increased expression of osteogenic markers (e.g. BSP and osteocalcin, as well as BMP2 and BMP6.* *These finding indicate that activation of proosteogenic cell signaling pathways by modSLA and SLA surfaces leads to enhanced osteogenic*	*_Clin Oral Implants Res. 2014 Apr;25(4):475–86. doi: 10.1111/clr.12178. Epub 2013 Apr 21.*
Rhob	*RHOB (Ras Homolog Family Member B) is a Protein Coding gene. Diseases associated with RHOB include oculo auricular syndrome and sertoli cell-only syndrome. Among its related pathways are Signaling by GPCR and Developmental Biology. GO annotations related to this gene include GTP binding and GDP binding. An important paralog of this gene is RHOA.* *Defects (mild affection to complete destruction) in the sealing zone were observed in the OPG-deficient animals. Resorption lacunae were not detected, indicating the loss of osteoclast-mediated bone resorption activity. Treatment with OPG resulted in a significant decrease in the expression of a cluster of instrumental genes (like for instance) Rho guanine nucleotide exchange factors (RhoGEFs), RhoGTPases, ROCK1 and ROCK2. This resulted in damage to or destruction of the sealing zone, thus inhibiting osteoclast-mediated bone resorption.*	*Int J Mol Med. 2014 Sep;34(3):856–62. doi: 10.3892/ ijmm.2014.1846. Epub 2014 Jul 10.*
Rs1	*The present gene encodes an extracellular protein serving an important organizational role in the retina. The protein encoded is assembled and secreted as a homooligomeric complex. Mutations in the present gene are lead to X-linked retinoschisis with ensuing severe loss in vision.* *G-protein-coupled receptors (GPCRs) are key regulators of skeletal homeostasis and important in fracture healing. It was earlier shown that blockade of G(i) signaling in maturing osteoblasts enhanced cortical and trabecular bone formation and prevented age-related bone loss in female mice. Furthermore, activation of G(s) signaling induced massive trabecular bone formation, but a concomitant cortical bone loss. Here, "labile" tibial fractures, where endogenous G(i) signaling are blocked by PTX, or G(s) signaling activated by Rs1, were achieved.* *Inhibition of endogenous G(i) activity gave a smaller callus, but enhanced net bone formation in both mice, irrespective of age.* **PTX treatment lowered the expression of Dkk1 and upregulated Lef1 mRNA upon fracture healing, indicating endogenous G(i) signaling in maintaining Dkk1 expression, while suppressing Wnt signaling. On the contrary, mice with activated Gs signaling demonstrated an increase in the initial callus size with enhanced callus bone production**. *These results indicate that G(i) blockade and G(s) activation are important for proper fracture healing.* *It was previously asserted that Rs1 constitutively activated Gs-coupled GPCR, under the control of the 2.3 kb Col I promoter, enhancing the steady state level mineral mass in trabecular bone of femurs. In this article, it was further concluded that Gs-signaling in OBs on enhanced intramembranous bone formation in calvariae of Col1(2.3)/Rs1 mice. Rs1 calvariae displayed a dramatic increase in bone volume with partial loss of cortical structure. Gene expression analysis of calvarial OBs showed that genes were affected by Rs1 signaling, featuring processes like: (a) differentiation, (b) synthesis of cytokine/growth factors, (c) angiogenesis, (d) coagulation, as well as (e) energy metabolism.*	*J Bone Miner Res. 2015 Oct;30(10):1896–904. doi: 10.1002/jbmr.2540. Epub 2015 May 14* *Exp Cell Res. 2015 May 1;333(2):289–302. doi: 10.1016/j. yexcr.2015.02.009. Epub 2015 Feb 20.*

Gene name	Description of function: *General* and bone-related	References
Trpm5	*This gene encodes a member of the transient receptor potential (TRP) protein family. This protein plays a role in taste transduction. It is activated by lower concentrations of intracellular Ca^{2+}, and inhibited by higher concentrations.* *Elevation of intracellular Ca^{2+} is commonly observed during stem cell differentiation (e.g., osteoblastogenesis), but cease after process completion. These findings suggest an inhibitory role for TRPM4 (Ca^{2+} ion channel) on osteogenesis while it appears to be required for adipogenesis. The data provide a link between the Ca^{2+} signaling pattern and gene expression during stem cell differentiation.*	*Stem Cells. 2013 Jan;31(1):167–77. doi: 10.1002/stem.1264.*
Amelx	*This gene encodes a member of the amelogenin family of extracellular matrix proteins. Amelogenins are involved in biomineralization during tooth enamel development.* *Research on enamel matrix proteins (EMPs) is centered on the understanding of their role in enamel biomineralization, as well as of their bioactivity for tissue engineering. It was shown that mRNA expression of AMELX and AMBN in mandibular alveolar and basal bones RNA-positive for AMELX. Furthermore, AMELX and AMBN mRNA levels varied according to: 1) ontogenic stage, and 2) tissue-type. In conclusion, it was asserted AMELX and AMBN may function as growth factor-like molecules in jaws, where they might play a role in bone physiology via autocrine/paracrine pathways, and especially during adaptation of stress-induced remodeling.* *Thymosin beta 4 is associated with RUNX2 expression through the Smad and Akt signaling pathways in mouse dental epithelial cells.* *Thymosin β4 (Tβ4) is associated with the initiation and development of the tooth germ, via enhancement of RUNX2. The transcription factor regulates the expression of genes involved in odontogenesis, like amelogenin, X-linked (Amelx), ameloblastin (Ambn), as well as enamelin (Enam).* *It appeared that the mDE6 mouse epithelial cell line expressed Runx2, Amelx, Ambn and Enam, and yielded calcified matrices upon the induction of calcification.*	*PLoS One. 2014 Jun 16;9(6):e99626. doi: 10.1371/journal.pone.0099626. eCollection 2014.* *Int J Mol Med. 2015 May;35(5):1169–78. doi: 10.3892/ijmm.2015.2118. Epub 2015 Mar 2.*
DKK1	*The present gene encodes a member of the dickkopf protein family. It is secreted, including two cysteine rich regions, and it parttakes in embrygenesis due to its inhibition of Wnt-ediated signaling. Enhanced DKK1 levels in bone marrow and blood correlates with bone osteolysis in patients suffering from multiple myeloma.* *In this article, the authors review advances and discrepancies in how Wnt/Lrp5 signaling regulates osteoblasts and osteocytes, and describe new players in Wnt signaling pathways exerting important roles in bone development, i.e., Wnt signaling in osteoclastogenesis, inhibition of Wnt pathway antagonists, such as sclerostin, Dkk 1, and Sfrp1.*	*Gene. 2012 Jan 15;492(1):1–18. doi: 10.1016/j.gene.2011.10.044. Epub 2011 Nov 3. Monroe DG1, McGee-Lawrence ME, Oursler MJ, Westendorf JJ.*

8. Emulation of the interaction between genes and microRNA species known to be instrumental in the development of the osteoblastic/ odontoblastic phenotype by Mir@nt@n

The bioinformatics program Mir@nt@n, developed by Le Bechek et al. [64], was used to arrive at high stringency interactions between microRNA species known to be instrumental in the development and stability of osteoblastic and odontoblastic cells from stem cells. The dental = osteoblastic/dentinoblastic genes (being significantly regulated by Vitamin K2 in the present study) described in the present array were fed into the program along with other genes and microRNA-species known to be instrumental in the development and stability (both positively and negatively) of the osteoblastic and/or odontoblastic phenotype:

Lmcd1, Dmxl2, Abcb4, Slc12a6, Mta1, Pou3f3, Dio2, Camk4, Ppp2r2b, Stard4, Tyk2, Pdcd5, Psenen, Rhob, Rs1, Trpm5, Amelx, DKK1, SP1, SP3, SP7, Runx2, Runx1, NR1/2, ADRB3, Foxc2, PGC1α, PPARA, PPARG, Dio2, UCP1, Adipoq, LEP, BETA3AR/ADRB3R/B3AR, hsa-mir-155, and c/EBPB.

Hsa-mir-196a, hsa-mir-16, hsa-mir-455, hsa-mir-339, hsa-mir-125b, hsa-mir-328, hsa-mir-16, hsa-mir-149, hsa-mir-125b, hsa-mir-760, hsa-mir-133, hsa-mir-29, hsa-mir-27, hsa-mir-23, hsa-mir-320, hsa-mir-26b, hsa-mir-21, hsa-mir-302, hsa-mir-132, and hsa-mir-223.

9. Major findings

The genes, significantly modulated (directly or indirectly) by vitamin K2, are presented in **Table 1**.

Of major interest here, from a regulatory point of view, and as a minimal "cluster" of necessary and sufficient genes, are probably the following species: RUNX1, RUNX2, SP1, SP3, and DIO2, along with the microRNA-species 149, 328, 339, and 760 (see **Figure 1**). It is well known that the osteoblast and odontoblast phenotypes are "determined" and "stabilized" by the RUNX- and SP-families of transcription factors (upregulated), as well as the

Dental genes affected (directly or indirectly) by Vitamin K2

Lmcd1	=	LIM and cysteine rich domains 1
Dmxl2	=	Dmx2 like protein
Abcb4	=	ATP-binding cassette subfamily B member 4
Slc12a6	=	Solute carrier family 12 member 6
Mta1	=	Metastasis associated 1
Pou3f3	=	POU doman, calss 3 transcription factor
Dio2	=	Deiodinase 2
Camk4	=	Calcium/calmodulin-dependent protein kinase type I
Ppp2r2b	=	Serine/threonine-protein phosphatase 2A, subunit B
Stard4	=	StAR-related lipid transfer protein 4
Tyk2	=	Tyk2 tyrosine kinase 2
Pdcd5	=	Programmed celle death 5 (acetyltransferase 5)
Psenen	=	Presenilin - part of a secretase complex
Rhob	=	Ras homolog Family Member B
Rs1	=	Retinochisin 1
Trpm5	=	Transient receptor potential cation channel subfamily M member 5
Amelx	=	Amelogenin, important for tooth enamel development
DKK1	=	DKK1 dickkopf WNT signaling pathway inhibitor 1

For relations to tooth (enamel and dentine) development, see tabulation of specific gene/protein effect related to bone homeostasis

Table 1. "Dental" genes affected directly or indirectly by exposure to vitamin K2 (MK-7).

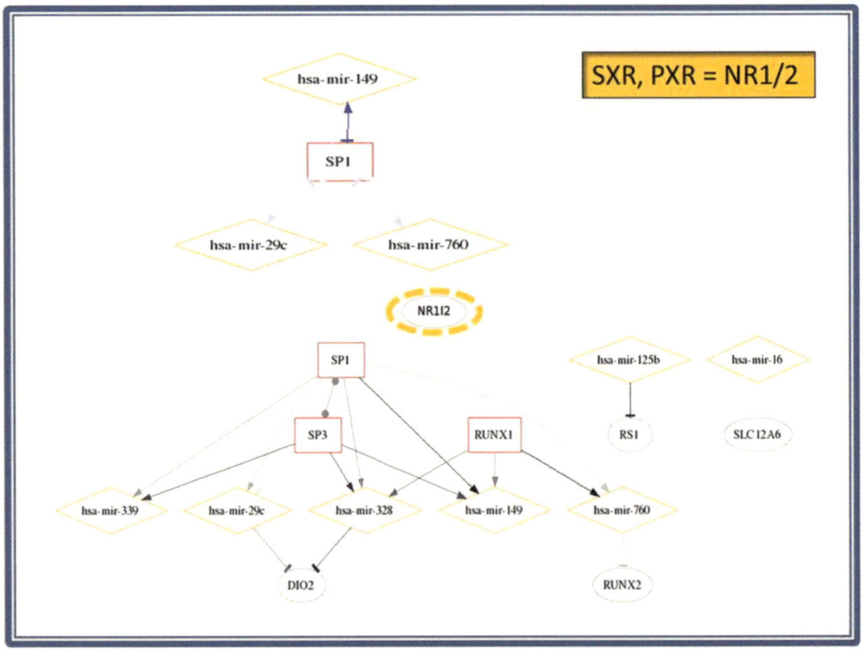

Figure 1. Interactions between transcription factors, "functional" genes, and microRNA species as emulated in the bioinformatics program Mir@nt@n.

microRNA-species 149, 328, and 339 (downregulated). Recently, it was shown [58] that mir-760 is involved in the effect of vitamin K2, since it associates with the transcription factor NR1/2 = SXR = PXR [65].

Using high/maximal stringency emulations rendered by the bioinformatics program Mir@ nt@n [64], it was quite interesting to find that the gene DIO2 (deiodinase2) that encodes the enzyme transforming T4 to T3 in peripheral tissues, was associated with has-mir-760, also found to exert an impact on the levels of Runx2, as well as being involved in the steady state of SP1 and SP3, transcription factors upstream of the Runx species deemed to be markers of the osteoblast/odontoblast phenotype (see **Figure 2**). It therefore does not come as a surprise that bone tissue is heavily dependent on DIO2 activation to function properly, i.e., replenishing "lost" osteoblasts from precursor cell, as well as proper functioning of differentiated osteoblasts/dentinoblasts to maintain bone/dentine mass at a stable level [66]. It may though come as a surprise to many that, in fact, vitamin K2 serves a rather prominent role in this process.

Finally, when applying low-stringency criteria to the Mir@nt@n-emulation process, a larger and less rigid network of mutual interactions was obtained (see **Figure 3**). From the interactions predicted, one may hypothesize the following: It is not trivial to ingest a dose that is too small to see the broad spectrum of beneficial effects of vitamin K2 on osteoblasts/odontoblasts. Furthermore, the dose should be titrated to ensure proper levels and characteristics and amounts of bodily beige versus white adipocytes (confer the postulated

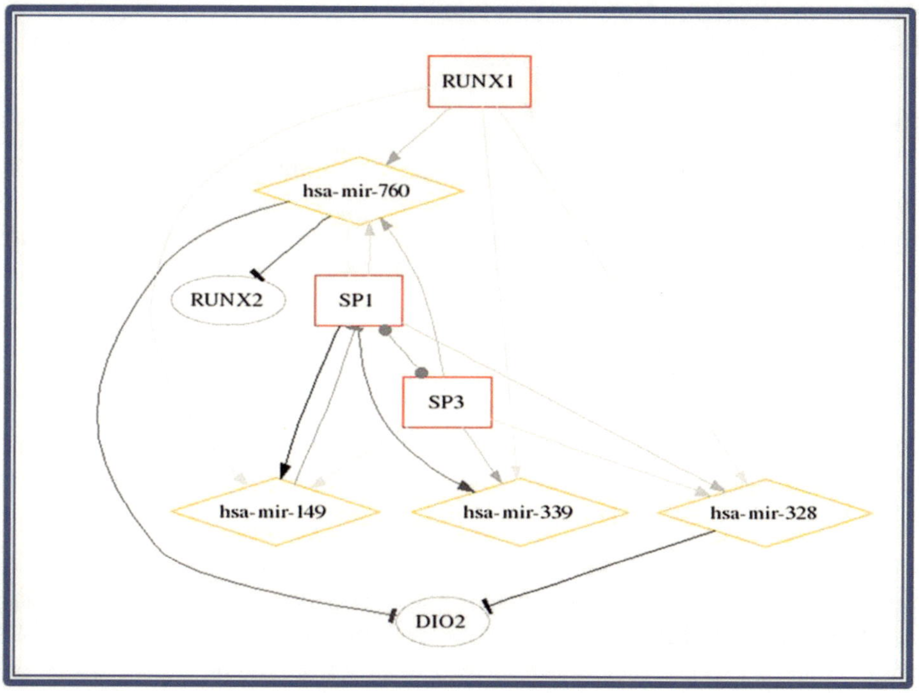

Figure 2. Interactions (high/maximal stringency emulation) of transcription factors, microRNA species, and differentiation / function-related genes in tooth germs from the rat.

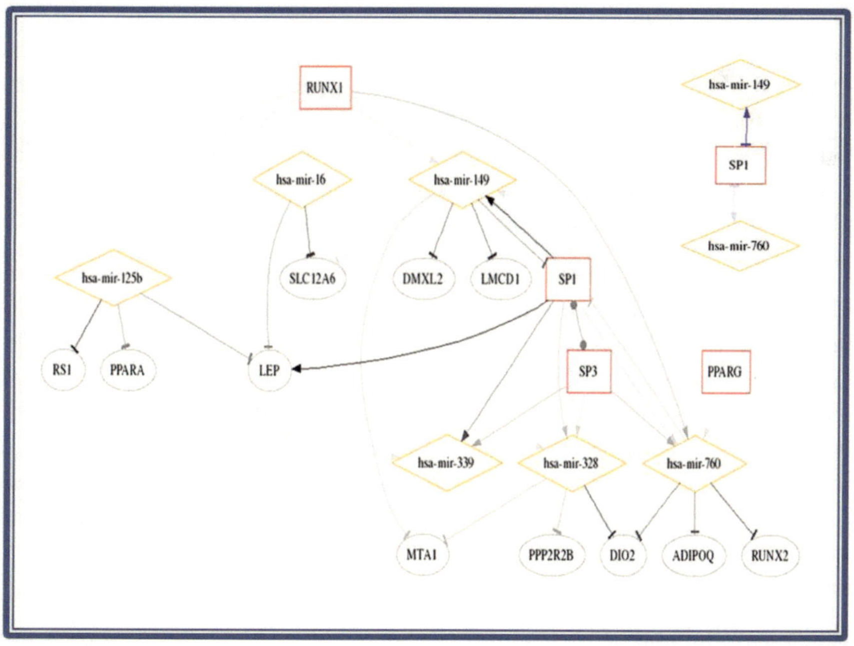

Figure 3. Extended interaction scheme (low stringency emulation) of transcription factors, microRNA species, and differentiation-related genes in tooth germs from the rat.

impact of vitamin K on mir-760 on SP1 and mir-149 with reciprocal regulatory loops), and the mir-760 "junction" between RUNX1, SP1, SP3, and PPARG versus DIO2, ADIPOQ, and RUNX2, which are all part of a mutually interacting network regulated by vitamin K2 odontoblasts/osteoblasts.

Finally, it should be emphasized that vitamin K2 (MK-7) upregulates Amelx and DKK1 in tooth germs, the former is instrumental in the building and maintenance of tooth enamel, and thus their resilience toward enamel erosion; the latter, DKK1 (dickopf1 WNT signaling pathway inhibitor 1), takes part in the modulation of osteoclast-related bone degradation, and in this context, the healthy transition between osteoclast-induced resorption and renewal of bone tissue with microcracks [67].

10. Pertinent question: is the dental filling material toxic to the living tooth? Contemplations on the making of live and artificial teeth

Monomers from methacrylate based dental materials both prior to and post polymerization have demonstrated adverse effects both *in vitro* and *in vivo* in terms of cytotoxicity [68], mutagenicity/genotoxicity [69–72], negative effects on fertility [73], xenoestrogenicity [74–76], and allergy induction [77]. The degree of cytoxicity will vary from the type assay used, materials tested, time intervals for testing, and cell types tested [78].

It is most pertinent to perform *in vitro* cytotoxicity testing on cells from cell types and tissues relevant to the area of *in vivo* placement of dental materials [79]. Recent studies on elution of monomers available in both dental composites and methacrylate- and epoxy-based root canal sealers looked at reactions in submandibular salivary gland acinar cells for the evaluation of cytotoxicity, cell proliferation, and apoptosis [69, 80]. The findings in such studies are of great interest and importance, but the authors of one of these studies [80] stated that their model would have been more realistic had they utilized human primary cells from direct target tissue. Tissues that often share the closest proximity to dental fillings are certainly not salivary glands but rather gingiva, mucosa, and, in particular, pulp tissues [79].

The pulp is a loose connective tissue within a nonresilient capsule of dentin and enamel. Pulpal inflammation is considered a protective mechanism and can either be of an acute or chronic nature. Acute and chronic responses are related to the "magnitude and duration of the insult [81]." Inflammation will inevitably cause vasodilation, increased vessel permeability which in turn will result in relatively large changes in tissue pressure [82]. Bacterial infection is the most common reason for pulpal inflammation, but any insult or stimuli will most probably result in a response. It is an established fact that many of the constituents in dental adhesive resin are cytotoxic [81], and the difference in cytoxicity varies among commercial materials commonly used by public dentists in Norway [unpublished results in a report to the Norwegian National Directorate of Health].

This project aims at elucidating the cellular effects of "leachables" (residual monomers) from dental filling materials exerted on dental pulp stem cells (DPSCs) *ex vivo*. It is important to ensure that the cells used in the study are, indeed, stem cells. The International Society for Cellular Therapy has released a position statement wherein they list three criteria to define human stem cells: (1) adherence to plastic, (2) specific surface antigen expression, and (3) multipotent differentiation potential [83]. The cells to be used in this project fulfil all three criteria [84], and were isolated in accordance with a published procedure described by Sorrentino et al. [85].

11. Characterization of the DPSCs

Dental specimens were obtained from extraction after signed informed consent, and the pulp was exposed by cutting the tooth, while maintaining sterile conditions: The enamel of the tooth crown is partially cut, following the sagittal plane, applying a diamond bur. Thereafter, the cut is completed using a piezoelectric ultrasound scalpel to avoid overheating of the tissue. The pulp is then treated collagenase and dispase for 1 hour at 37°C, and then incubated in a bioSpherix chamber under normoxic conditions [85].

Phenotyping of the DPSCs yielded a CD-profile very much like the one seen for bone marrow mesenchymal stem cells (BM-MSCs) with an approximately identical percentage of cells expressing CD10 (CALLA), CD 13 (Aminopeptidase N), CD29 (β1-integrin), CD44 (H-CAM, Pgp-1), CD49acd (VLA-1,3,4 = α1,3,4-integrin), CD54 (I-CAM-1), CDw90 (THY-1), CD105 (Endoglin, TGFβ-R), CD140b (PDGF-Rb), CD146 (M-CAM), CD147 (Neurothelin/basigin), CD166 (Alcam, CD6-ligand), and also comparable amounts of GD2 (Neural ganglioside).

12. Tissue engineering using stem cells: can it be avoided?

It has been asserted that tissue engineering might be the future of endodontics [86]. It is stated in the abstract that pulpal regeneration after tooth injury is not easily accomplished, since the infected pulp is required for tooth extraction or root canal therapy. It is further asserted that an ideal form of therapy might consist of regenerative approaches where diseased or necrotic pulp tissues are removed and replaced with healthy pulp tissue to revitalize the affected tooth. The authors list different techniques, ranging from stem cell therapy, the use of growth factors, pulp implants, implant of 3D cell printed in hydrogels, injectable scaffolds, bioactive materials, the use of co-enzymes, and root canal revascularization. However, despite alleged advantages of the subject approaches, they also suffer major disadvantages like low cell survival, lack of de novo production of pulp, necrosis of reinfected pulp, and lack of vascularity, and requirement for precise root canal fitting.

By determining the cut point of toxicity (i.e., cell death/enhanced apoptosis and lack of proper differentiation induced by the leakage of monomers of endodontic filling materials), it is possible to develop new filling materials without an acute and long term detrimental effect on DPSCs. Hence, the development of a test battery to check the monomers that may diffuse into the root canal, for cytotoxicity and ability to attain proper and functional cell phenotypes (i.e., odontoblasts, neural lattice, and endothelial cells constituting blood vessels) seems mandatory. The present project description aims to define such a test battery using highly sophisticated techniques like proteomics (including phosphoproteomics) and mass cytometry.

The advantage of using such techniques resides with the fact that extremely few cells may be used in complex arrays of incubation conditions, while still yielding reliable results. The technology described in the present project outline, may also enable the definition of a minimal and sufficient array of variables, which precisely describes a robust test battery to be implemented as a gold standard to be adopted in the development of endodontal biomaterial fillings in the future.

Author details

Jan Oxholm Gordeladze[1*], Maria A. Landin[2], Gaute Floer Johnsen[3], Håvard Jostein Haugen[3] and Harald Osmundsen[2]

*Address all correspondence to: j.o.gordeladze@medisin.uio.no

1 Department of Molecular Medicine, Section for Biochemistry, Institute of Basic Medical Science, University of Oslo, Oslo, Norway

2 Department of Oral Biology, Faculty of Dentistry, University of Oslo, Oslo, Norway

3 Institute of Clinical Odontology, Faculty of Odontology, University of Oslo, Norway

References

[1] Southward K. The systemic theory of dental caries. Gen Dent 2011:59(5):367–73.

[2] Sorsa T, Tjäderhane L, Salo T. Matrix metalloproteinases (MMPs) in oral diseasesOral Dis 2004;10(6):311–8.

[3] Tjäderhane L, Larjava H, Sorsa T, et al. The activation and function of host matrix metalloproteinases in dentin matrix breakdown in caries lesions.J Dent Res 1998;77(8):1622–9.

[4] Listgarten MA. Serum IgA and IgG antibodies to Treponema vincentii and Treponema denticola in adult periodontitis, juvenile periodontitis and periodontally healthy subjects. J Clin Periodontol 1986;13(5):418–30.

[5] Nutrition and Physical Degeneration, 8th ed. Weston Price DDS. Price-Pottenger Nutrition Foundation Inc.; pp. 362–369, 398.

[6] On the Trail of the Elusive X-Factor: A Sixty-Two-Year-Old Mystery Finally Solved http://www.westonaprice.org/fat-soluble-activators/x-factor-is-vitamin-k2: 2013-09-14.

[7] The Stress of Life. Hans Selye MD. New York: McGraw-Hill Book Co.; 1978. pp.3–54, 57–168, 259–264.

[8] Leonora J, Tieche JM, Steinman RR. Further evidence for a hypothalamus-parotid gland endocrine axis in the rat. Arch Oral Biol 1993;38(10):911–6.

[9] Kesim S, Unalan D, Esen C, Ozturk A. The relationship between periodontal disease severity and state-trait anxiety level. J Pak Med Assoc 2012;62(12):1304–8.

[10] Leonora J, Tieche JM, Steinman RR. Stimulation of intradentinal dye penetration by feeding in the rat. Arch Oral Biol 1993;38(9):763–7.

[11] Zhang Q, Szalay AA, Tieche JM, et al. Cloning and functional study of porcine parotid hormone, a novel proline-rich protein. J Biol Chem 2005;280(23):22233–44.

[12] Leloup C, Tourrel-Cuzin C, Magnan C, et al. Mitochondrial reactive oxygen species are obligatory signals for glucose-induced insulin secretion. Diabetes 2009;58(3):673–81. http://dx.doi.org/10.2337/db07-1056.

[13] Steinman RR, Leonora J. Effect of selected dietary additives on the incidence of dental caries in the rat. J Dent Res 1975;54(3):570–7.

[14] Liu Y, Chen M, Yao X, et al. Enhancement in dentin collagen's biological stability after proanthocyanidins treatment in clinically relevant time periods. Dent Mater 2013;29(4):485–92. http://dx.doi.org/10.1016/ j.dental.2013.01.013.

[15] Liu Y, Dusevich V, Wang Y. Proanthocyanidins rapidly stabilize the demineralized dentin layer. J Dent Res 2013;92(8):746–52. http://dx.doi.org/10.1177/0022034513492769.

[16] Leloup C, Casteilla L, Carrière A, et al. Balancing mitochondrial redox signaling: a key point in metabolic regulation. Antioxid Redox Signal 2011;14(3):519–30. http://dx.doi.org/10.1089/ars.2010.3.

[17] Barayuga SM, Pang X, et al. Methamphetamine decreases levels of glutathione peroxidases 1 and 4 in SH-SY5Y neuronal cells: protective effects of selenium. Neurotoxicology 2013;37:240–6. http://dx.doi.org/10.1016/j.neuro.2013.05.009.

[18] Ravenel MC, Salinas CF, et al. Methamphetamine abuse and oral health: a pilot study of "meth mouth". Quintessence Int 2012;43(3):229–37.

[19] Sumpio BE, Cordova AC, et al. Green tea, the "Asian paradox", and cardiovascular disease. J Am Coll Surg 2006;202(5):813–25.

[20] Nugala B, Namasi A, et al. Role of green tea as an antioxidant in periodontal disease: the Asian paradox. J Indian Soc Periodontol 2012;16(3):313–6.

[21] Yu H, Oho T, Tagomori S, Morioka T. Anticariogenic effects of green tea. Fukuoka Igaku Zasshi 1992;83(4):174–80.

[22] Ferland G. Vitamin K, an emerging nutrient in brain function. BioFactors 2012;38(2):151–7. http://dx.doi.org/10.1002/biof.1004. Epub 2012 Mar 15.

[23] Vervoort LM, Ronden JE, Thijssen HH. The potent antioxidant activity of the vitamin K cycle in microsomal lipid peroxidation. Biochem Pharmacol 1997;54(8):871–6.

[24] Oldenburg J, Marinova M, Müller-Reible C, Watzka M. The vitamin K cycle. Vitam Horm 2008;78:35–62. http://dx.doi.org/10.1016/S0083-6729(07)00003-9.

[25] Sharma C, Suhalka P, Sukhwal P, et al. Curcumin attenuates neurotoxicity induced by fluoride: an in vivo evidence. Pharmacogn Mag 2014;10(37):61–5. http://dx.doi.org/10.4103/0973-1296.126663.

[26] Atmaca N, Atmaca HT, Kanici A, Anteplioglu T. Protective effect of resveratrol on sodium fluoride-induced oxidative stress, hepatotoxicity and neurotoxicity in rats. Food Chem Toxicol 2014;70(5):191–7. http://dx.doi.org/10.1016/j.fct.2014.05.011. Epub 2014 May 22.

[27] Davidson VL. Electron transfer in quinoproteins. Arch Biochem Biophys 2004;428(1):32–40.

[28] Thijssen HH, Drittij MJ, Vermeer C, et al. Menaquinone-4 in breast milk is derived from dietary phylloquinone. Br J Nutr 2002;87(3):219–26.

[29] Sato T, Schurgers LJ, Uenishi K. Comparison of menaquinone-4 and menaquinone-7 bioavailability in healthy women. Nutr J 2012;12(11):93. http://dx.doi.org/10.1186/1475-2891-11-93.

[30] Kate R-B. Vitamin K2 and the Calcium Paradox: How a Little-Known Vitamin Could Save Your Life. Mississauga, Canada: Wiley; pp. 182–185.

[31] Knapen MHJ, Schurgers LJ, Vermeer C. Vitamin K2 supplementation improves hip bone geometry and bone strength indices in postmenopausal women. Osteoporos Int 2007;18(7):963–72.

[32] Hosoi T. Clinical implications of under carboxylated osteocalcin. Clin Calcium 2009;19(12):1815–21.

[33] Kyla Shea M, Holden Rachel M. Vitamin K status and vascular calcification: evidence from observational and clinical studies. Adv Nutr 2012;3(2):158–65.

[34] Kyla Shea M, O'Donnell Christopher J, et al. Vitamin K supplementation and progression of coronary artery calcium in older men and women. Am J Clin Nutr 2009;89(6):1799–807.

[35] Dalmeijer GW, van der Schouw YT, et al. The effect of menaquinone-7 supplementation on circulating species of matrix Gla protein. Atherosclerosis 2012;225(2):397–402.

[36] Patergnani S, Suski JM, Agnoletto C, et al. Calcium signaling around mitochondria associated membranes (MAMs). Cell Commun Signal 2011;22(9):19. http://dx.doi.org/10.1186/1478-811X-9-19.

[37] Ferland G. Vitamin K and the nervous system: an overview of its actions. Adv Nutr 2012;3(2):204–12. http://dx.doi.org/10.3945/an.111.001784.

[38] Schurgers LJ, Vermeer C. Determination of phylloquinone and menaquinones in food, effect of food matrix on circulating vitamin K concentrations. Department of Biochemistry and Cardiovascular Research Institute. Maastricht, The Netherlands: Maastricht University. Haemostasis 2000;30:298–307.

[39] Komai M, Shirakawa H. Vitamin K metabolism. Menaquinone-4 (MK-4) formation from ingested VK analogues and its potent relation to bone function. Clin Calcium 2007;17(11):1663–72.

[40] Li J, Lin JC, Wang H, et al. Novel role of vitamin K in preventing oxidative injury to developing oligodendrocytes and neurons. J Neurosci 2003;23(13):5816–26.

[41] Ferland G. Vitamin K and brain function. Semin Thromb Hemost 2013;39(8):849–55. http://dx.doi.org/10.1055/s-0033-1357481. Epub 2013 Oct 9.

[42] Dowd FJ. Saliva and dental caries. Dent Clin North Am 1999;43(4):579–97.

[43] Ambudkar IS. Regulation of calcium in salivary gland secretion. Crit Rev Oral Biol Med 2000;11(1):4–25.

[44] Ambudkar IS. Dissection of calcium signaling events in exocrine secretion. Neurochem Res 2011;36(7):1212–21.

[45] Ambudkar IS. Polarization of calcium signaling and fluid secretion in salivary gland cells. Curr Med Chem 2012;19(34):5774–81.

[46] Choi HJ, Yu J, et al. Vitamin K2 supplementation improves insulin sensitivity via osteocalcin metabolism: a placebo-controlled trial. Diabetes Care 2011;34(9):e147.

[47] Thomas DD, Krzykowski KJ, Engelke JA, et al. Exocrine pancreatic secretion ofphospholipid, menaquinone-4, and caveolin-1 in vivo. Biochem Biophys Res Commun 2004;319(3):974–9.

[48] Dawes C. What is the critical pH and why does a tooth dissolve in acid? J Can Dent Assoc 2003;69(11):722–4.

[49] Telgi RL, Yadav V, et al. In vivo dental plaque pH after consumption of dairy products. Gen Dent 2013;61(3):56–9.

[50] Elder SJ, Haytowitz DB, et al. Vitamin K contents of meat, dairy, and fast food in the U.S. Diet. J Agric Food Chem 2006;54(2):463–7.

[51] Yoneda T, Tomofuji T, et al. Anti-aging effects of co-enzyme Q10 on periodontal tissues. J Dent Res 2013;92(8):735–9. http://dx.doi.org/10.1177/ 0022034513490959.

[52] Aral K, Alkan BA et al. Therapeutic effects of systemic vitamin K2 and vitamin D3 on gingival inflammation and alveolar bone in rats with experimentally induced periodontitis. J Periodontol 2015;85(5):666–73. DOI: 10.1902/jop.2015.140467

[53] Nakamura S, Yamada, Y et al. Stem cell proliferation pathways comparison between human exfoliated deciduous teeth and dental pulp stem cells by gene expression profile from promising dental pulp. J Endodont 2009;35(11):1536–42. DOI: 10.1016/j.joen.2009.07.024

[54] Nelson P, Tran TDN et al. Transient receptor potential melastatin 4 channel controls calcium signals and dental follicle stem cell differentiation. Stem Cells 2013;31(1):167–77. DOI: 10.1002/stem.1264

[55] Monroe DG, McGee-Lawrence ME et al. Update on Wnt signaling in bone cell biology and bone disease. Gene 2012;492:1–18.

[56] Okiji K, Yoshiba K. Reparative dentinogenesis induced by mineral trioxide aggregate:a review from the biological and physicochemical points of view. Int J Dentist. DOI:10.1155/2009/464280 Volume 2009 (2009), Article ID 464280, pp 1-12.

[57] He YD, Sui BD et al. Site-specific function and regulation of Osterix in tooth root formation. Int Endodont J. DOI: 10.111/irj.12585. 49(12): 1124–1131.

[58] Gordeladze JO, Stakkestad Ø et al. Regulatory loops consisting of transcription factors and microRNA species determine the mineralizing characteristics of cell phenotypes –implications for bone engineering and prevention of soft tissue mineralization. In: "Cells and Biomaterial in Regenerative Medicine", 2014, ed. Daniel Eberli, http://dx.doi.org/10.5772/59149, pp 71–102.

[59] Sinha KM, Zhou X. Genetic and molecular control of Osterix in skeletal formation. J Cell Biochem 2013; 114(5):975–984.

[60] Sangwan P, Sangwan A et al. Tertiary dentinogenesis with calcium hydroxide: a review of proposed mechanisms. Int Endodont J 2013;46:3–19.

[61] Example of transcription factor (TF) regulation of the ALP gene: http://www.sabiosciences.com/chipqpcrsearch.php?gene=ALPL&factor=Over+200+TF&species_id=0&ninfo=n&ngene=n&nfactor=y

[62] Hynes RO, Naba A. Overview of the matrisome – an inventory of extracellular matrix constituents and functions. Cold Spring Harb Perspect Biol 2012;4:a004903.

[63] Yoshizaki K and Yamada Y. Gene evolution and functions of the extracellular matrix proteins in teeth. Orthod Waves 2013 March 1;72(1):1–10.

[64] Le Bechek A, Portales-Casmar E et al. Mir@nt@n: a framework integrating transcription factors, microRNAs and their targets to identify sub-network motifs in a meta-regulatory network model. Bioinformatics 2011;12:67.

[65] Horie-Inoue K, Inoue S. Steroid and xenobiotic receptor mediates a novel vitamin K2 signaling pathway in osteoblastic cells. J Bone Miner Metab 2008;26:9–12.

[66] Lecka-Czernik B, Stechschultea LA. High bone mass in adult mice with diet-induced obesity results from a combination of initial increase in bone mass followed by attenuation in bone formation; implications for high bone mass and decreased bone quality in obesity, http://dx.doi.org/10.1016/j.mce.2015.01.001, pp 35–41.

[67] Tai N, Inoue D. Anti-Dickkopf1 (Dkk1) antibody as a bone anabolic agent for the treatment of osteoporosis. Clin Calcium 2014 Jan;24(1):75–83. DOI: CliCa14017583.

[68] Huang FM, et al. Cytotoxicity of dentine bonding agents on human pulp cells is related to intracellular glutathione levels. Int Endodont J 2010;43:1091–1097.

[69] Samuelsen JT, et al. HEMA reduces cell proliferation and induces apoptosis in vitro. Dental Mater 2008;24:134–140.

[70] Schweikl H, et al. Cytotoxic and mutagenic effects of dental composite materials. Biomaterials 2005;26:1713–1719.

[71] Tadin A, et al. Genotoxicity in gingival cells of patients undergoing tooth restoration with two different dental composite materials. Clin Oral invest 2014;18:87–96.

[72] Walther UI, et al. Cytotoxicity of ingredients of various dental materials and related compounds in L2- and A549 cells. J Biomed Mater Res 2002;63:643–649.

[73] Al-Hiyasat AS, Darmani H, Elbetieha AM. Leached components from dental composites and their effects on fertility of female mice. Eur J Oral Sci 2004;112:267–272.

[74] Ruse ND. Xenoestrogenicity and dental materials. J Can Dent Assoc 1997;63:833–836.

[75] Tarumi H, et al. Estrogenicity of fissure sealants and adhesive resins determined by reporter gene assay. J Dent Res 2000;79:1838–1843.

[76] Wada H, et al. In vitro estrogenicity of resin composites. J Dent Res 2004;83:222–226.

[77] Raap U, Stiesch M, and Kapp A. Contact allergy to dental materials. J German Soc Dermatol 2012;10:391–396; quiz 397.

[78] Goldberg M. In vitro and in vivo studies on the toxicity of dental resin components: a review. Clin Oral Invest 2008;12:1–8.

[79] Huang FM, Chang YC. Cytotoxicity of dentine-bonding agents on human pulp cells in vitro. Int Endodont J 2002;35:905–909.

[80] Lodiene G, et al. Detection of leachables and cytotoxicity after exposure to methacrylate- and epoxy-based root canal sealers in vitro. Eur J Oral Sci 2013;121:488–496.

[81] Dahl JE, ØRstavik DAG. Responses of the pulp–dentin organ to dental restorative biomaterials. Endodont Top 2007;17:65–73.

[82] Tonder KJ. Vascular reactions in the dental pulp during inflammation. Acta Odontol Scand 1983;41:247–256.

[83] Dominici M, et al. Minimal criteria for defining multipotent mesenchymal stromal cells. The International Society for Cellular Therapy position statement. Cytotherapy 2006;8:315–317.

[84] Signore M, et al. Identity and ranking of colonic mesenchymal stromal cells. J Cell Physiol 2012;227:3291–3300.

[85] Sorrentino A, et al. Isolation and characterization of CD146+ multipotent mesenchymal cells. Exp Hem 2008;36:1035–1046.

[86] Deepak BS, et al. Tissue engineering: is it the future of endodontics? People's J Sci Res 2011; 4(1):76–82.

Section 4

Vitamin K2 and its Effect on Various Organ Systems

Chapter 8

Anti-Inflammatory Actions of Vitamin K

Stephen J. Hodges, Andrew A. Pitsillides,
Lars M. Ytrebø and Robin Soper

Additional information is available at the end of the chapter

http://dx.doi.org/10.5772/63891

Abstract

Naphthoquinone compounds have received attention for their ability to regulate diseases from bacterial and parasite infections through to chronic human diseases. Inflammation is widely considered to be at the root of many chronic diseases. The reports of anti-inflammatory activity of naphthoquinones, including vitamin K1 (phylloquinone) and vitamin K2s (menaquinones), are of interest due to their very low toxicity. Most of the evidence for the anti-inflammatory mechanisms of vitamin K suggests a role in the inhibition of the cell signalling complex nuclear factor kappa-B (NF-κB).

Keywords: vitamin K1, vitamin K2, menaquinone, cell signalling, NF-κB

1. Introduction

From the 1960s through to the late 1990s, the F. Hoffman-La Roche laboratories in Basal, Switzerland, were the centre of much of the innovative investigation into the role of vitamins in physiology and pathophysiology. At that time, their laboratories were the leaders in exploring newer concepts of biological actions of vitamins. For vitamin K, an exciting potential development was in the analgesic and anti-inflammatory properties of naphthoquinone compounds; an area that had not previously been considered [1, 2].

The explosion in the understanding of several areas since then, such as cytokine and chemokine biology and physiology, molecular biology of signalling pathways and the genetic translation of these signals, has facilitated the ability to explore molecular events in the cell. With these innovations have come newer understandings of the role of vitamin K in physiology that go

INTECH
open science | open minds

© 2017 The Author(s). Licensee InTech. This chapter is distributed under the terms of the Creative Commons Attribution License (http://creativecommons.org/licenses/by/3.0), which permits unrestricted use, distribution, and reproduction in any medium, provided the original work is properly cited.

beyond the gamma-carboxylation of specific protein glutamyl residues in the vitamin K–dependent proteins [3, 4].

2. Inflammation

The answer to the fundamental question 'What is Inflammation?' is, of course, complex. For many years, discussion on inflammation revolved around the immune system and the ability to distinguish 'self' from 'non-self'. This required an understanding of the ancient 'innate' immune system and the 'adaptive' immune system. There has been a sea change in thinking over the past three decades towards an appreciation of a primary role in the initiation of the inflammatory response residing in the tissues [5].

The cells of a tissue, or tissues, in an organism are primed to respond when they are exposed to an unusual stimulus. The response leads to a set of consequences, such as eicosanoid, cytokine, chemokine release, alterations in metabolic activity and genes becoming activated or switched off. These events can cause further cell activation either intrinsically within the tissue or extrinsically such as drawing leukocytes into the tissue. The whole process would be described as an inflammatory response.

The sequelae to the initiating events can lead to an expansion of the response, or the challenge can be nullified and the tissues return to an original state. If the inflammatory state continues, it can either persist as a long-term process, encountered as chronic diseases, or in severe cases, the explosive changes overwhelm the organism leading to tissue necrosis and organ failure [6, 7].

There are reasonable arguments to suggest that normal ageing shares some similarities with such chronic inflammatory disease [8, 9]. Indeed, the term 'inflammaging' was coined to describe this possibility [9]. This case has also been made for specific diseases of ageing, such as osteoporosis [10, 11].

There is also an innate anti-inflammatory process involving leukocytes and a number of biological mediators, such as cytokines; however, with respect to vitamin K, this area of research has not been the subject of focused research.

Irrespective of how we currently define inflammation, it is a biological process that serves a purpose in the preservation of the organism, sometimes at the cost of some part of the whole, but the government of the system can all too easily slip free of controls [12].

3. Vitamin K and inflammation

The pharmaceutical development of the anti-inflammatory role of activated protein C, a vitamin K–dependent protein, is being established at the boundary of inflammation and coagulation [13]. We will not consider this important subject here as, while vitamin K is required for generation of the mature functional protein, it is the role of protein C as a serine protease that is central to the mechanism of limiting microthrombus formation in organ tissue beds in sepsis [14].

Several important chronic diseases with an inflammatory background have been associated with vitamin K deficiency. These include cystic fibrosis, inflammatory bowel disease, pancreatitis, chronic kidney disease and osteoporosis [15–19]. This review will not address the relationship between vitamin K and these diseases, despite our ongoing interests. We have instead focused our attention on the proposed cellular and molecular aspects of vitamin K in regulating inflammation. With respect to this position, a subject that receives continuing interest is the potential role of vitamin K in the amelioration of signal transduction pathways; specifically, how vitamin K may be able to modulate the stimulus received at the surface of a cell and the message transmission to the cell nucleus for interpretation and response.

3.1. Introduction to cell-to-nucleus signalling

Within cells, there are several primary pathways that are important in communicating the exposure to an insult or injury. Early work on avian retroviruses, and in particular reticuloendotheliosis virus strain T (Rev-T), identified an oncogene (v-rel) capable of transforming avian lymphoid cells [20–22]. Once this oncogene had been characterized in avian systems, work progressed quickly to look at mammalian homologues. This research revealed a mouse homologue, c-rel, which was also found to have homologous character to the fruit fly Drosophila gene '*dorsal*' [23–25]. The Drosophila *dorsal* gene is known to play a fundamental role in dorsal-ventral development of the fly larvae, acting as a translocation factor in the cell nucleus to regulate gene expression. Around the same time, another nuclear translocation factor, with sequencing homology to these signalling factors, was discovered, which became known as nuclear factor kappa-B (NF-κB) [26].

3.1.1. NF-κB

We present a greatly simplified overview of NF-κB signalling, and we recommend that interested readers find some of the excellent review literature that has been published [e.g. 27–34] in a complex and engrossing story.

NF-κB is present in nearly all cells and participates in a diverse range of biological functions including inflammation, immunity, differentiation, cell growth, tumourigenesis and apoptosis, that is, from birth to death of a cell. This pathway is a central conductor of the molecular orchestra that is important for normal cell function, but it can become over-activated on a more general level [35] or to a greater degree leading to tumourigenesis [36, 37].

We now know that NF-κB is a family of proteins that exist as hetero- or homodimers that have been conserved from primitive organisms through to man [30]. The dimers are normally quiescent in the cytosol of cells through the close association with a regulatory inhibitory protein (IκB), first identified through an inspirational series of denaturation-renaturation experiments [38]. NF-κB becomes activated by the removal of IκB from the complex, which requires IκB phosphorylation by cytosolic kinase enzymes (IKK) [39]. The IκB protein is then tagged, by ubiquitination, for degradation in the proteasome [34, 40]. A family of IκBs have now been identified and together with the IKK enzymes that phosphorylate the IκB proteins and a scaffold IKK modulator (NEMO; IKKγ) [41], function at the heart of the system of NF-κB activation and regulation.

Once dissociated from its inhibitory chaperone, NF-κB moves to the cell nucleus where it recognizes specific nucleotide promoter sequences that activates a number of genes. This is known for over 200 genes.

The complexity of the activation of NF-κB creates multiple tiers of regulatory control and, therefore, also potential focal points for the development of therapeutic agents. Vitamin K and its derivatives have been investigated for their ability to intervene in the activation of NF-κB.

3.1.2. NF-κB and vitamin K

There is evidence that vitamin K can regulate the activation of the NF-κB pathway. Other signalling pathways also control cell functions and are, of course, important. Furthermore, in the complex nature of multiple signalling pathways there is also considerable crosstalk between these pathways [42]. There are also some reports that vitamin K has regulatory functions on signalling pathways, other than NF-κB, such as the mitogen-activated protein (MAP) kinases, but in this discussion, we focus on vitamin K and NF-κB.

Most of the research on vitamin K regulation of the NF-κB pathway has been done on cultured cells, with a large element coming from researchers in Japan with a core interest in the molecular mechanisms regulating cancer cells.

Hepatocellular carcinoma cells (HCC) are known to overexpress several oncogenes and also down-regulate tumour suppressor genes [43, 44]. Vitamin K2, in the form of menaquinone-4, has been investigated for its growth regulating effects in three hepatocellular cancer cell lines [45]. Focussing on the proto-oncogenic expressed protein cyclin D1, which regulates the cell cycle at the G_1-S transition, and which is itself regulated by NF-κB [46]. These researchers found menaquinione-4 inhibited both cyclin D1 mRNA expression and protein synthesis. In HCC activated with phorbol ester or cytokines agonists in order to stimulate NF-κB signalling, vitamin K was found to ameliorate cyclin D1 activation and NF-κB promoter binding was inhibited. It was also noted that there was inhibition of the phosphorylation of IκB, with suppression of IKK kinase activity.

The same research group has considered the effects of vitamin K2 on the expression of matrix metalloproteinase (MMP) enzymes [47]. These enzymes are implicated in metastatic tumour invasion [48]. HCC in culture were treated with phorbol ester and the expression of three MMP proteins and their corresponding mRNA examined. The study found that menaquinone-4 suppressed basal and phorbol ester stimulated NF-κB activation, which translated into the suppression of both MMP mRNA and protein expression levels. In these experiments, vitamin K2 was also found to suppress the activation of the mitogen-activated protein (MAP) kinase signalling pathway.

Two other publications from these researchers, again using HCC, explored the effects of vitamin K2 to regulate cancer cell biology [49, 50]. In the first they examined potential modulation of HCC growth by the anti-cancer therapeutic 5-fluorouracil, through inhibition of cyclin D and innate activated NF-κB signalling. Their other study investigated the effects of vitamin K2 on protein kinase C (PKC), a known mediator in multiple stimulated cell responses in HCC [51]. Working in an adjunctive role with 5-fluorouracil, menaquinone-4 was found to augment the growth inhibition induced by 5-fluorouracil alone, this being most effective when the cells were pretreated with vitamin K2. From the discussion above on the NF-κB pathway, any finding of an inhibition of kinase activities may be expected to have an effect on NF-κB activation. In their PKC study, these researchers found delineation in the response to vitamin K2 between the PKC isoforms. All the PKC isozymes were found to be involved in NF-κB activation, but vitamin K2 inhibited the NF-κB activation through its actions on only two of these, PKCα and PKCε. This selectivity was not investigated further.

The above discussion focuses on directed inflammatory stimuli using laboratory reagents; however, there are numerous agents that can promote an inflammatory response, including infective agents like viruses and bacteria. We are all exposed to bacterial endotoxins or lipopolysaccharides (LPS), every day and without much effect, but when these endotoxins are shed in an injured tissue, the results are registered quickly through cell surface Toll-like receptors that trigger, among other signalling pathways, NF-κB [52].

A potent, and key, cytokine in the response to endotoxin insult is interleukin-6 (IL-6) [53]. Up-regulation of IL-6 synthesis is under NF-κB control [54], and this cytokine can feed back into the inflammatory response. This feedback is achieved via interaction with its own specific cell surface receptor that activates the Janus kinase (JAK) signal transducer and activator of transcription 3 (STAT3) pathway [55].

We have reported a vitamin K-mediated suppression of IL-6 release in LPS-challenged primary human fibroblast cell cultures [56]. This investigation found that vitamin K2, in the form of menaquinone-4, was a more potent anti-inflammatory compound than vitamin K1 in this primary cell system.

Other cell and animal experiments have looked at the possible regulation of inflammation by vitamin K inhibition of IL-6 release following endotoxin challenge. These include studies using germ-free rats, THP-1 human monocyte-like cells and murine RAW264.7 macrophage-like cells [57, 58] and have noted a significant reduction in the LPS-stimulated cellular IL-6 mRNA levels. In the vitamin K supplemented animal experiments, there was a suppression of the hepatic mRNA levels of the inflammation response protein macrophage migration inhibitory factor (MIF), compared to animals fed a vitamin K-deficient diet. This study also found that supplementation with vitamin K1 also caused a rise in hepatic tissue menaquinone-4 levels [57]. Extension of these *in vivo* findings in the two cell lines, THP-1 and RAW264.7, demonstrated that the suppression of LPS-stimulated IL-6 mRNA by vitamin K2 was consistent with inhibition of NF-κB activity via the restricted phosphorylation of IKK kinases [58]. These researchers also questioned whether these events involved a classical vitamin K gamma-carboxylation-mediated process; co-culture experiments with warfarin found that these inhibitory effects of vitamin K on NF-κB were independent of vitamin K–dependent protein gamma-carboxylation [58]. The widely appreciated role of IL-6 in inflammatory diseases makes the findings of an inhibitory activity for vitamin K in IL-6 release an increasingly attractive prospect.

One of the dominant areas of NF-κB understanding is in osteoclastogenesis. Activation of this pathway in hematopoietic stem cells by receptor activator of nuclear factor kappa-B ligand (RANKL), a member of the tumour necrosis ligand superfamily, has been found to be a prominent initiator of osteoclast formation [59, 60]. Once precursor cells have been positioned to develop into osteoclasts by RANKL, NF-κB further facilitates development through inhibition of apoptosis [61].

Menaquinone-7, a more hydrophobic homologue of vitamin K2 than menaquinone-4, has been found to limit rat osteoclast formation under various challenges [62]. Later studies determined that the anti-osteoclastogenic actions of menaquinone-7 were mediated by the inhibition of NF-κB [63].

A central feature of all research on the role of vitamin K in the inhibition of NF-κB is the high concentrations that were required to demonstrate an effect of the order of 10^{-5} to 10^{-4} M in cell culture and pharmacological dosing in animal experiments. The relatively high concentrations required suggests that the anti-inflammatory activities of vitamin K are not working through specific receptors and it is, therefore, more likely that the anti-inflammatory activities are due to other characteristics of vitamin K.

3.2. Other anti-inflammatory vitamin K–like compounds

Some supporting data on the potential for vitamin K to serve a physiological anti-inflammatory role can be found in studies using other vitamin K–like compounds. Below, we focus on two compounds that have a common 2-methyl-1,4-naphthoquinone functional group as vitamin

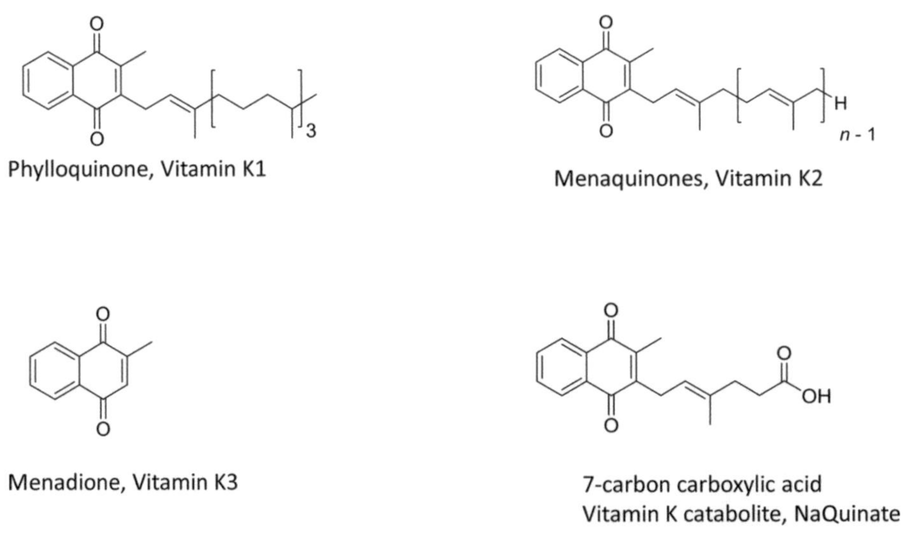

Figure 1. The molecular structures for phylloquinone, vitamin K1; menaquinone vitamins, vitamins K2; menadione, vitamin K3; NaQuinate, the 7-carbon carboxylic acid catabolite derived from all forms of vitamin K1 and K2.

http://dx.doi.org/10.5772/63891

K1 and K2, namely menadione, or vitamin K3, and a 7-carbon carboxylic acid catabolite of vitamin K, NaQuinate (**Figure 1**).

3.2.1. Menadione and inhibition of inflammation

Menadione, or vitamin K3, was for a long time considered to be a synthetic compound. Research that started in the 1990s demonstrated that menadione was an important derivative from vitamin K1 that was converted in the rat to make the vitamin K2 congener menaquinone-4 [64–67]. The enzyme responsible for the conversion of menadione to menaquinone-4, UBIAD1 [68] has a wide distribution across species and likely plays a significant role in the vitamin K status of many animals (see chapter by O'Neil et al., this volume).

Seminal work on the amelioration of chronic inflammatory arthritis in a rabbit antigen-induced model found a pronounced anti-inflammatory effect in animals receiving oral menadione. This was related to a significant decrease in synoviocyte glucose-6-phosphate dehydrogenase activity in the treated rabbits [69] and reduced macrophage numbers in the treated joint (personal communication). It is unclear if this was due to a direct effect on the joint tissue as a latter study using menadione-epoxide in the same model was also found to have a significant anti-inflammatory activity [70]. Alternative explanation for these effects includes the possible conversion of these menadione and menadione-epoxide molecules into vitamin K2, as menaquinone-4, through the action of UBIAD1 [68]. This would be consistent with lapine caecal bacterial conversion due to known coprophagy in this species (see chapter by O'Neil et al., in this volume).

Menadione has also been shown to inhibit NF-κB translocation to the nucleus in tumour necrosis factor (TNF)-α stimulated human embryonic kidney (HEK)293 cells [71]. Intriguingly, such effects were not observed in similar experiments using vitamin K1 or K2 in these cells. There are several possible reasons for this divergence in activity between the naphthoquinone homologues. However, menadione is known to be inert as a co-factor for the vitamin K–dependent gamma-carboxylation reaction [72], suggesting that alternative mechanisms, related to their redox properties, rather than protein modification are likely to be more dominant.

These researchers also found that menadione was capable of suppressing LPS-induced NF-κB nuclear translocation and TNF-α release from murine macrophage-like RAW 264 cells [71]. Additionally, in a murine model of acute lung injury/acute respiratory distress syndrome (ARDS), which occurs in the setting of acute severe illness complicated by systemic inflammation, menadione also attenuated the LPS-induced severity of lung injury and suppressed the increase in serum TNF-α level. This occurred concomitantly with inhibition the LPS-evoked nuclear translocation of NF-κB in lung tissue.

Together, these data indicate that menadione has a capacity to ameliorate inflammation. The *in vivo* examples identified above can be described as rigorous tests of the anti-inflammatory potential of this molecule. The mechanism of action is likely to centre on NF-κB, but the strong suppression of glucose-6-phosphate dehydrogenase in the rabbit arthritis model and, thereby, the regulation of intracellular NADPH, may indicate another important feature of the anti-inflammatory character of menadione.

3.2.2. Inhibition of inflammation by a vitamin K catabolite, NaQuinate

Vitamin K1 and K2s are broken down in the liver, and this is the only organ that has been reported to be responsible for the catabolism of these vitamins [73]. Early metabolism studies in men using radioactive vitamin K1 has revealed the catabolites to be side-chain shortened carboxylic acid products, the more abundant being a 5-carbon aliphatic acid, while the less abundant is a 7-carbon aliphatic acid containing a single double bond (**Figure 2**) [74]. These compounds are glucuronidated in the liver for excretion in the bile or through the kidneys. In human studies, the level of the catabolites varies with the amount of vitamin K ingested at pharmacological doses, as either phylloquinone or menaquinone-4 [75] or with dietary phylloquinone intake [76].

We have found that both of these acid catabolites cannot participate in the gamma-carboxylation reaction and are, in fact, inhibitors or the vitamin K gamma-carboxylase enzyme (Soper and Hodges, unpublished data).

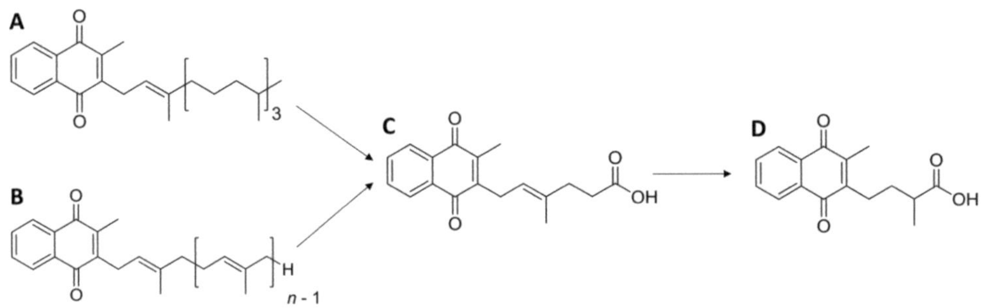

Figure 2. The metabolism of vitamin K1 (A) and vitamins K2 (B) with the common generation is the 7-carbon carboxylic acid catabolite, NaQuinate (C), which is further broken down to the 5-carbon carboxylic acid catabolite (D).

Our recent studies with the 7-carbon carboxylic acid catabolite of vitamin K, NaQuinate (**Figure 1**), found significant attenuation of LPS-challenged osteoblast-like MG63 cell IL-6 release [77]. This has likely *in vivo* implications, as in humans NaQuinate is usually present at high levels following pharmacological vitamin K dosing, irrespective of the source of vitamin K as either vitamin K1 or any of the vitamin K2 vitamers (**Figure 2**) [75]. Parallel experiments in MG63 cells using 1,25-$(OH)_2$-vitamin D3 or interleukin-1β as agonists showed similar suppression of IL-6 release by NaQuinate (Soper and Hodges, unpublished data). Interestingly, LPS-challenged MG63 cells were substantially less affected by the 5-carbon carboxylic acid vitamin K catabolite (**Figure 2**), which only differs from NaQuinate by 2 carbon atoms and a carbon-carbon double bond [77].

These results are in agreement with our earlier findings with LPS-challenged primary human fibroblast cultures [56]. Furthermore, if the carboxylic acid function on the catabolites is blocked with a methyl group, the LPS-induced release of IL-6 is greatly reduced in MG63 cells

http://dx.doi.org/10.5772/63891

[77]. The reason for the disparity in inhibitory activities between the two closely related vitamin K catabolites is not known, and a free acid function is a common structural element of many widely used anti-inflammatory agents.

It should be noted that, in comparison to the descriptions above of the effects of vitamin K on NF-κB and other anti-inflammatory markers, the concentration of NaQuinate required to achieve an effect was substantially lower, with an IC_{50} of 2–5 × 10^{-7} M.

More recently, *in vivo* investigations have been performed using NaQuinate in two murine models of bone loss, induced through either ovariectomy (OVx), or as a result of limb disuse following sciatic neurectomy (NTx). In the OVx model, we found, using several micro-computed tomography measurement parameters, that tibial trabecular bone loss was significantly ameliorated in NaQuinate treated animals (0.75 mg/kg); the tibiae in the sham-operated control group being indistinguishable from the NaQuinate-treated oviarectomised animals [77]. We found that the more aggressive bone loss following NTx was also greatly reduced in the NaQuinate-treated animals. Moreover, we found substantial reduction in the inhibition of bone loss in both of the mouse models when a NaQuinate derivative containing a methyl ester group was used to block the carboxylic acid function, suggesting that a free carboxylic acid function is required for effective inhibition of bone loss in these models.

4. Concluding remarks

Vitamin K has been found to have anti-inflammatory activity in an increasing number of studies. This inhibitory activity would appear to be directed through inhibition of NF-κB signalling. In cell culture experiments, it is often mentioned that the cells needed to be primed with vitamin K before being challenged with agonists. Furthermore, experimental results are immerging that show inhibition of cytokine release in agonist-challenged cell culture and animal models, which in the light of vitamin K deficiency in chronic diseases, may reflect on a vitamin K role in organ homeostasis. Therefore, the defining vitamin K sufficiency on the basis solely of adequate functional blood clotting factors is likely to be an over-simplification and needs to be considered more fully.

The question is open as to the value of vitamin K in a therapeutic or a prophylactic role. For example, a vitamin K deficient, seriously ill patient is unlikely to benefit greatly from a vitamin K intervention, but a patient with a chronic inflammatory disease may. This can only be highlighted in large community studies, which need to be run over many years in large numbers of volunteers.

However, there are already some indications of prophylactic benefit of increased vitamin K intake in large population studies in Japan. Exploring consumption of 'natto', a live menaquinone-7 producing bacteria (*Bacillus subtilis* subsp) culture food product from fermented soybean that is culturally more favoured in the east than the west of Japan found that ingestion substantially increases human circulating menaquinone-7 levels and that bacterial gut colonization was evident several days after a single 80 g natto portion [78]. Intriguingly, prefectural sales of natto also linked consumption with a notable decrease in age-related hip fractures in eastern Japanese population [78].

Given the safety of vitamin K, even in high pharmaceutical doses, supplementation may have a benefit, beyond meeting coagulation demands than is generally perceived, particularly in early chronic inflammatory diseases and inflammaging.

Acknowledgements

This work was facilitated, in part, by the award of a senior fellowship to SJH by The Northern Norway Regional Health Authority.

Author details

Stephen J. Hodges[1,2*], Andrew A. Pitsillides[1], Lars M. Ytrebø[2] and Robin Soper[3]

*Address all correspondence to: shodges@rvc.ac.uk

1 Department of Comparative Biomedical Sciences, The Royal Veterinary College, London, UK

2 Anesthesia and Critical Care Research Group, The University Hospital of North Norway and UIT-The Arctic University of Norway, Tromsø, Norway

3 Cancer Research UK Centre for Drug Development, Strategy and Research Funding, London, UK

References

[1] Hanck A, Weiser H. Physiological and pharmacological effects of vitamin K. Int J Vitam Nutr Res. 1983;24 (Suppl):155–170.

[2] Weiser H, Kormann A. Pharmacological properties of vitamin K1 and analogues: antidote functions, analgesic and antiphlogistic properties. Unpublished results presented as a poster at the 17th Steenbock Symposium June 1987, Current Advances in Vitamin K Research, University of Wisconsin, Madison, USA, 1988.

[3] Suttie JW. Synthesis of vitamin K-dependent proteins. FASEB J. 1993;7:445–452.

[4] Vermeer C. γ-Carboxyglutamate-containing proteins and the vitamin K-dependent carboxylase. Biochem J. 1990;266:625–636.

[5] Matzinger P. The danger model: a renewed sense of self. Science. 2002;296(5566):301–5.

[6] Medzhitov R. Origin and physiological roles of inflammation. Nature. 2008;454:428–435.

[7] Ashley NT, Weil ZM, Nelson RJ. Inflammation: mechanisms, costs, and natural variation. Annu Rev Ecol Evol Syst. 2012;43:385–406.

[8] Franceschi C, Bonafè M, Valensin S, Olivieri F, De Luca M, Ottaviani E, De Benedictis G. Inflamm-aging. An evolutionary perspective on immunosenescence. Ann N Y Acad Sci. 2000;908:244–54.

[9] Franceschi C, Capri M, Monti D, Giunta S, Olivieri F, Sevini F, Panourgia MP, Invidia L, Celani L, Scurti M, Cevenini E, Castellani GC, Salvioli S. Inflammaging and anti-inflammaging: a systemic perspective on aging and longevity emerged from studies in humans. Mech Ageing Dev. 2007;128(1):92–105.

[10] Lencel P, Magne D. Inflammaging: the driving force in osteoporosis? Med Hypotheses. 2011;76(3):317–21.

[11] Sanguineti R, Puddu A, Mach F, Montecucco F, Viviani GL. Advanced glycation end products play adverse proinflammatory activities in osteoporosis. Mediat Inflamm. 2014;2014:975872.

[12] Angus DC, van der Poll T. Severe sepsis and septic shock. N Engl J Med. 2013;369(9): 840–851.

[13] Esmon CT. The anticoagulant and anti-inflammatory roles of the protein C anticoagulant pathway. J Autoimmun. 2000;15(2):113–6.

[14] Allen KS, Sawheny E, Kinasewitz GT. Anticoagulant modulation of inflammation in severe sepsis. World J Crit Care Med. 2015;4(2):105–15.

[15] Kleinman RE, Fracchia MS. Vitamin K and cystic fibrosis: give me a double, please. Am J Clin Nutr. 2010;92(3):469–70.

[16] Nakajima S, Iijima H, Egawa S, Shinzaki S, Kondo J, Inoue T, Hayashi Y, Ying J, Mukai A, Akasaka T, Nishida T, Kanto T, Tsujii M, Hayashi N. Association of vitamin K deficiency with bone metabolism and clinical disease activity in inflammatory bowel disease. Nutrition. 2011;27(10):1023–8.

[17] Sikkens EC, Cahen DL, Koch AD, Braat H, Poley JW, Kuipers EJ, Bruno MJ. The prevalence of fat-soluble vitamin deficiencies and a decreased bone mass in patients with chronic pancreatitis. Pancreatology. 2013;13(3):238–42.

[18] McCabe KM, Adams MA, Holden RM. Vitamin K status in chronic kidney disease. Nutrients. 2013;5(11):4390–8.

[19] Iwamoto J. Vitamin K_2 therapy for postmenopausal osteoporosis. Nutrients. 2014;6(5): 1971–80.

[20] Breitman ML, Lai MMC, Vogt PK. The genomic RNA of avian reticuloendotheliosis virus REV. Virology. 1980;100:450–461.

[21] Chen ISY, Mak TW, O'Rear JJ, Temin HM. Characteristics of reticuloendotheliosis virus strain T DNA and isolation of a novel variant of reticuloendotheliosis virus strain T by molecular cloning. J Virol. 1981;40:800–811.

[22] Wong TC, Lai MMC. Avian reticuloendotheliosis virus contains a new class of oncogene of turkey origin. Virology. 1981;111:289–293.

[23] Steward R. Dorsal, an embryonic polarity gene in Drosophila, is homologous to the vertebrate proto-oncogene, c-rel. Science. 1987;238(4827):692–4.

[24] Rushlow CA, Han K, Manley JL, Levine M. The graded distribution of the dorsal morphogen is initiated by selective nuclear transport in Drosophila. Cell. 1989;59(6): 1165–77.

[25] Grumont RJ, Gerondakis S. Structure of a mammalian c-rel protein deduced from the nucleotide sequence of murine cDNA clones. Oncogene Res. 1989;4(1):1–8.

[26] Sen R, Baltimore D. Multiple nuclear factors interact with the immunoglobulin enhancer sequences. Cell. 1986;46(5):705–16.

[27] Barnes PJ. Nuclear factor-kappa B. Int J Biochem Cell Biol. 1997;29(6):867–70.

[28] Ghosh S, May MJ, Kopp EB. NF-kappa B and Rel proteins: evolutionarily conserved mediators of immune responses. Annu Rev Immunol. 1998;16:225–60.

[29] Moynagh PN. The NF-κB pathway. J Cell Sci. 2005;118(Pt 20):4589–92.

[30] Gilmore TD. Introduction to NF-κB: players, pathways, perspectives. Oncogene. 2006;25(51):6680–4.

[31] Liu F, Xia Y, Parker AS, Verma IM. IKK biology. Immunol Rev. 2012;246(1):239–53.

[32] Kanarek N, Ben-Neriah Y. Regulation of NF-κB by ubiquitination and degradation of the IκBs. Immunol Rev. 2012;246(1):77–94.

[33] Hinz M, Scheidereit C. The IκB kinase complex in NF-κB regulation and beyond. EMBO Rep. 2014;15(1):46–61.

[34] Wertz IE. TNFR1-activated NF-κB signal transduction: regulation by the ubiquitin/proteasome system. Curr Opin Chem Biol. 2014;23:71–7.

[35] Nasto LA, Seo HY, Robinson AR, Tilstra JS, Clauson CL, Sowa GA, Ngo K, Dong Q, Pola E, Lee JY, Niedernhofer LJ, Kang JD, Robbins PD, Vo NV. ISSLS prize winner: inhibition of NF-κB activity ameliorates age-associated disc degeneration in a mouse model of accelerated aging. Spine (Phila Pa 1976). 2012;37(21):1819–25.

[36] Del Prete A, Allavena P, Santoro G, Fumarulo R, Corsi MM, Mantovani A. Molecular pathways in cancer-related inflammation. Biochem Med (Zagreb). 2011;21(3):264–75.

[37] DiDonato JA, Mercurio F, Karin M. NF-κB and the link between inflammation and cancer. Immunol Rev. 2012;246(1):379–400.

[38] Baeuerle PA, Baltimore D. Activation of DNA-binding activity in an apparently cytoplasmic precursor of the NF-κB transcription factor. Cell. 1988;53(2):211–7.

[39] Ghosh S, Baltimore D. Activation in vitro of NF-kappa B by phosphorylation of its inhibitor I kappa B. Nature. 1990;344(6267):678–82.

[40] Glickman MH, Ciechanover A. The ubiquitin-proteasome proteolytic pathway: destruction for the sake of construction. Physiol Rev. 2002;82(2):373–428.

[41] Yamaoka S, Courtois G, Bessia C, Whiteside ST, Weil R, Agou F, Kirk HE, Kay RJ, Israël A. Complementation cloning of NEMO, a component of the IkappaB kinase complex essential for NF-κB activation. Cell. 1998;93(7):1231–40.

[42] Ghosh S, Dass JF. Study of pathway cross-talk interactions with NF-κB leading to its activation via ubiquitination or phosphorylation: a brief review. Gene. 2016 pii: S0378-1119(16)30152-4.

[43] Thorgeirsson SS, Grisham JW. Molecular pathogenesis of human hepatocellular carcinoma. Nat Genet. 2002;31(4):339–46.

[44] Ozturk M, Arslan-Ergul A, Bagislar S, Senturk S, Yuzugullu H. Senescence and immortality in hepatocellular carcinoma. Cancer Lett. 2009;286(1):103–13.

[45] Ozaki I, Zhang H, Mizuta T, Ide Y, Eguchi Y, Yasutake T, Sakamaki T, Pestell RG, Yamamoto K. Menatetrenone, a vitamin K2 analogue, inhibits hepatocellular carcinoma cell growth by suppressing cyclin D1 expression through inhibition of nuclear factor κB activation. Clin Cancer Res. 2007;13(7):2236–45.

[46] Hinz M, Krappmann D, Eichten A, Heder A, Scheidereit C, Strauss M. NF-κB function in growth control: regulation of cyclin D1 expression and G0/G1-to-S-phase transition. Mol Cell Biol. 1999;19(4):2690–8.

[47] Ide Y, Zhang H, Hamajima H, Kawaguchi Y, Eguchi Y, Mizuta T, Yamamoto K, Fujimoto K, Ozaki I. Inhibition of matrix metalloproteinase expression by menatetrenone, a vitamin K2 analogue. Oncol Rep. 2009;22(3):599–604.

[48] Westermarck J, Kähäri VM. Regulation of matrix metalloproteinase expression in tumor invasion. FASEB J. 1999;13(8):781–92.

[49] Zhang H, Ozaki I, Hamajima H, Iwane S, Takahashi H, Kawaguchi Y, Eguchi Y, Yamamoto K, Mizuta T. Vitamin K2 augments 5-fluorouracil-induced growth inhibition of human hepatocellular carcinoma cells by inhibiting NF-κB activation. Oncol Rep. 2011;25(1):159–66.

[50] Xia J, Matsuhashi S, Hamajima H, Iwane S, Takahashi H, Eguchi Y, Mizuta T, Fujimoto K, Kuroda S, Ozaki I. The role of PKC isoforms in the inhibition of NF-κB activation by vitamin K2 in human hepatocellular carcinoma cells. J Nutr Biochem. 2012;23(12):1668–75.

[51] Diaz-Meco MT, Moscat J. The atypical PKCs in inflammation: NF-κB and beyond. Immunol Rev. 2012;246(1):154–67.

[52] Zhang G, Ghosh S. Molecular mechanisms of NF-κB activation induced by bacterial lipopolysaccharide through Toll-like receptors. J Endotoxin Res. 2000;6(6):453–7.

[53] Rossol M, Heine H, Meusch U, Quandt D, Klein C, Sweet MJ, Hauschildt S. LPS-induced cytokine production in human monocytes and macrophages. Crit Rev Immunol. 2011;31(5):379–446.

[54] Libermann TA, Baltimore D. Activation of interleukin-6 gene expression through the NF-kappa B transcription factor. Mol Cell Biol. 1990;10(5):2327–34.

[55] Heinrich PC, Behrmann I, Haan S, Hermanns HM, Müller-Newen G, Schaper F. Principles of interleukin (IL)-6-type cytokine signalling and its regulation. Biochem J. 2003;374(Pt 1):1–20.

[56] Reddi K, Henderson B, Meghji S, Wilson M, Poole S, Hopper C, Harris M, Hodges SJ. Interleukin 6 production by lipopolysaccharide-stimulated human fibroblasts is potently inhibited by naphthoquinone (vitamin K) compounds. Cytokine. 1995;7(3): 287–90.

[57] Ohsaki Y, Shirakawa H, Hiwatashi K, Furukawa Y, Mizutani T, Komai M. Vitamin K suppresses lipopolysaccharide-induced inflammation in the rat. Biosci Biotechnol Biochem. 2006;70(4):926–32.

[58] Ohsaki Y, Shirakawa H, Miura A, Giriwono PE, Sato S, Ohashi A, Iribe M, Goto T, Komai M. Vitamin K suppresses the lipopolysaccharide-induced expression of inflammatory cytokines in cultured macrophage-like cells via the inhibition of the activation of nuclear factor κB through the repression of IKKα/β phosphorylation. J Nutr Biochem. 2010;21(11):1120–6.

[59] Anderson DM, Maraskovsky E, Billingsley WL, Dougall WC, Tometsko ME, Roux ER, Teepe MC, DuBose RF, Cosman D, Galibert L. A homologue of the TNF receptor and its ligand enhance T-cell growth and dendritic-cell function. Nature. 1997;390(6656): 175–9.

[60] Yasuda H, Shima N, Nakagawa N, Yamaguchi K, Kinosaki M, Mochizuki S, Tomoyasu A, Yano K, Goto M, Murakami A, Tsuda E, Morinaga T, Higashio K, Udagawa N, Takahashi N, Suda T. Osteoclast differentiation factor is a ligand for osteoprotegerin/osteoclastogenesis-inhibitory factor and is identical to TRANCE/RANKL. Proc Natl Acad Sci U S A. 1998;95(7):3597–602.

[61] Vaira S, Alhawagri M, Anwisye I, Kitaura H, Faccio R, Novack DV. RelA/p65 promotes osteoclast differentiation by blocking a RANKL-induced apoptotic JNK pathway in mice. J Clin Invest. 2008;118(6):2088–97.

[62] Yamaguchi M, Ma ZJ. Inhibitory effect of menaquinone-7 (vitamin K2) on osteoclast-like cell formation and osteoclastic bone resorption in rat bone tissues in vitro. Mol Cell Biochem. 2001;228(1–2):39–47.

[63] Yamaguchi M, Weitzmann MN. Vitamin K2 stimulates osteoblastogenesis and suppresses osteoclastogenesis by suppressing NF-κB activation. Int J Mol Med. 2011;27(1):3–14.

[64] Thijssen HH, Drittij-Reijnders MJ. Vitamin K distribution in rat tissues: dietary phylloquinone is a source of tissue menaquinone-4. Br J Nutr. 1994;72(3):415–25.

[65] Davidson RT, Foley AL, Engelke JA, Suttie JW. Conversion of dietary phylloquinone to tissue menaquinone-4 in rats is not dependent on gut bacteria. J Nutr. 1998;128(2): 220–3.

[66] Ronden JE, Drittij-Reijnders MJ, Vermeer C, Thijssen HH. Intestinal flora is not an intermediate in the phylloquinone-menaquinone-4 conversion in the rat. Biochim Biophys Acta. 1998;1379(1):69–75.

[67] Thijssen HH, Vervoort LM, Schurgers LJ, Shearer MJ. Menadione is a metabolite of oral vitamin K. Br J Nutr. 2006;95(2):260–6.

[68] Nakagawa K, Hirota Y, Sawada N, Yuge N, Watanabe M, Uchino Y, Okuda N, Shimomura Y, Suhara Y, Okano T. Identification of UBIAD1 as a novel human menaquinone-4 biosynthetic enzyme. Nature. 2010;468:117–121.

[69] Pitsillides AA, Blake SM, Glynn LE, Bitensky L, Chayen J. Amelioration by menadione of the experimental chronic immune arthritis in the rabbit. Cell Biochem Funct. 1990;8(4):221–6.

[70] Pitsillides AA, Blake SM, Glynn LE, Frost GT, Bitensky L, Chayen J. The effect of menadione epoxide on the experimental immune arthritis in the rabbit. Int J Exp Pathol. 1991;72(3):301–9.

[71] Tanaka S, Nishiumi S, Nishida M, Mizushina Y, Kobayashi K, Masuda A, Fujita T, Morita Y, Mizuno S, Kutsumi H, Azuma T, Yoshida M. Vitamin K3 attenuates lipopolysaccharide-induced acute lung injury through inhibition of nuclear factor-kappaB activation. Clin Exp Immunol. 2010;160(2):283–92.

[72] Buitenhuis HC, Soute BA, Vermeer C. Comparison of the vitamins K1, K2 and K3 as cofactors for the hepatic vitamin K-dependent carboxylase. Biochim Biophys Acta. 1990;1034(2):170–5.

[73] Losito R, Owen CA Jr, Flock EV. Metabolic studies of vitamin K1-14C and menadione-14C in the normal and hepatectomized rats. Thromb Diath Haemorrh. 1968;19(3): 383–8.

[74] Shearer MJ, Barkhan P. Studies on the metabolites of phylloquinone (vitamin K1) in the urine of man. Biochim Biophys Acta. 1973;297(2):300–12.

[75] Harrington DJ, Soper R, Edwards C, Savidge GF, Hodges SJ, Shearer MJ. Determination of the urinary aglycone metabolites of vitamin K by HPLC with redox-mode electrochemical detection. J Lipid Res. 2005;46(5):1053–60.

[76] Harrington DJ, Booth SL, Card DJ, Shearer MJ. Excretion of the urinary 5C- and 7C-aglycone metabolites of vitamin K by young adults responds to changes in dietary phylloquinone and dihydrophylloquinone intakes. J Nutr. 2007;137(7):1763–8.

[77] Soper RJ, Oguz C, Emery R, Pitsillides AA, Hodges SJ. Vitamin K catabolite inhibition of ovariectomy-induced bone loss: structure–activity relationship considerations. Mol Nutr Food Res. 2014;58(8):1658–66.

[78] Kaneki M, Hodges SJ, Hosoi T, Fujiwara S, Lyons A, Crean SJ, Ishida N, Nakagawa M, Takechi M, Sano Y, Mizuno Y, Hoshino S, Miyao M, Inoue S, Horiki K, Shiraki M, Ouchi Y, Orimo H. Japanese fermented soybean food as the major determinant of the large geographic difference in circulating levels of vitamin K2: possible implications for hip-fracture risk. Nutrition. 2001;17(4):315–21.

Chapter 9

Vitamin K2: Implications for Cardiovascular Health in the Context of Plant-Based Diets, with Applications for Prostate Health

Michael S. Donaldson

Additional information is available at the end of the chapter

http://dx.doi.org/10.5772/63413

Abstract

Vitamin K was originally discovered as a blood coagulation factor. But observations regarding intakes in populations and health outcomes lead to a deeper understanding of the differences between vitamins K1 and K2. Studies of warfarin-treated rats and MGP - deficient mice led to understanding the central role of MGP in controlling calcification of arteries. A sensitive biomarker assay was then developed, based on a particular species of matrix γ–carboxylation protein OR matrix GLA protein (MGP). Warfarin therapy in people, especially those suffering from chronic kidney disease, was found to cause the highest level of this biomarker desphospho-uncarboxylated MGP (dp-ucMGP). Intervention studies with vitamin K2 brought down levels of dp-ucMGP and also led to relief of some disease endpoints. The process of varicose vein formation includes a role for vitamin K, implicating a lack of vitamin K in the development of varicoceles, which leads to benign prostate hyperplasia. It is likely that much good will be accomplished using vitamin K2 in interventions. Complex, multifaceted diseases will not be treated by single-nutrient solutions. The best interventions will be those which combine vitamin K2 treatment with a healthy diet rich in fruits and vegetables, combined with a healthy lifestyle.

Keywords: menaquinone, menaquinone-4 (MK-4), menaquinone-7 (MK-7), cardiovascular, calcification, prostate, plant-based diet

INTECH
open science | open minds

© 2017 The Author(s). Licensee InTech. This chapter is distributed under the terms of the Creative Commons Attribution License (http://creativecommons.org/licenses/by/3.0), which permits unrestricted use, distribution, and reproduction in any medium, provided the original work is properly cited.

1. Introduction

The story of vitamin K goes back to the 1930s, when Henrik Dam at the University of Copenhagen isolated an antihemorrhagic vitamin that was fat soluble but different from previously isolated vitamins A, D, and E [1]. He found high concentrations of the vitamin in hog liver fat and hemp seed, but found it to be virtually absent in cod liver oil while testing a variety of animal organs, hen eggs, cereals, seeds, vegetables, and various fats and oils including butter fat. The initial quantification was all based on the time required to coagulate blood from a chicken. The term "vitamin K" was used as short for the German term "koagulation." Because of the type of assay used, all vitamin K factors, whether K1 or K2, were thought to be useful only for coagulation of the blood.

At the same time as Henrik Dam was working out the details of the antihemorrhagic vitamin, there was another investigator working in the USA with a vitamin factor not found in cod liver oil, but which worked synergistically with them to promote proper mineralization, bone growth, and to prevent dental caries [2]. Dr. Weston Price found that the amount of the factor in mammalian milk varied with the "nutrition of the animal," with highest levels from milk of cows that were consuming rapidly growing green grass. Because Dr. Price's assay was based on the release of iodine from hydroiodic acid, a test for peroxides, no connection was made between his discovery and any other vitamin activity. What Price found was the activity of vitamin K2, formed in mammary glands from phylloquinone, vitamin K1, found in abundance in rapidly growing green grass [3]. No connection was made between the anticoagulation activity of K1 and the mineral-directing activity of vitamin K2 until the modern research era.

Because of the way vitamin K was discovered for its coagulation function, it was assumed that this was its only function for many years. The Recommended Daily Allowance (RDA) for vitamin K is based on its coagulation function. Though both phylloquinone (phyllo—from plants) and menaquinones support coagulation, as we explore in this chapter, there are clear differences between the functioning of phylloquinone and the menaquinones, and they are not fully interchangeable.

This chapter is a review of vitamin K2 research in the area of cardiovascular health, especially dealing with arterial calcification. It is organized in a loose chronological order, following the themes of the research as the field matured. First, there were observations about dietary intake of K1 and K2 and health outcomes, forming hypotheses to be tested further. Associations were discovered at this stage. Harmful observations with warfarin-type drugs, which are vitamin K antagonists, were seen early on in the research cycle as well. Hypotheses were then explored with animal studies, looking for mechanisms and biomarkers—to help get endpoints that were quicker to develop than mortality and disease. Then came the rise of biomarker studies, using markers for vitamin K status (dp-ucMGP), as well as markers for risk factors, such as arterial calcification, pointing to disease outcomes. As the science matured even further, intervention studies related to K2 and biomarker changes emerged. The final stage is now emerging with intervention studies looking not just at biomarkers but disease outcomes, thus tying all of the research together.

http://dx.doi.org/10.5772/63413

2. Observational studies of dietary vitamin K intake and health outcomes

Observational studies help form hypotheses by finding associations between factors that may or may not be related to the health outcome of interest. One of the early population-based studies in the 1990s examined the vitamin K status of 113 postmenopausal women [4]. Dietary intake of vitamin K was assessed along with examinations for the presence or absence of aortic calcified lesions. Blood samples were assayed for osteocalcin, a vitamin K-dependent protein that is responsible for proper deposition of minerals in bone tissue. It was thought that some forms of osteocalcin might be a marker for vitamin K status. The first type had low affinity for hydroxyapatite, while the second had high affinity. Women with calcified lesions ($n = 34$) had a lower intake of total vitamin K as well as a higher amount of low-affinity osteocalcin. Together these results indicated that these women had impaired vitamin K status that might be related to their atherosclerotic lesions.

A groundbreaking study took place in the Netherlands which changed the way vitamin K2 was thought of afterward. First, a very careful survey of foods eaten in the Netherlands was carried out by interview-based dietary intakes of over 5400 people, guided by a validated food frequency questionnaire (FFQ). For many animal-based foods, the quantitation of vitamin K2 had not been done, so this was carried out and reported as well. The mean intake of K1 and K2, respectively, varied from 124 and 9.3 μg/day in the lowest quartile to 375 and 45 μg/day in the upper quartile [5]. A report of health outcomes from the Rotterdam Study in 2004 from 4800 of these subjects revealed that even though vitamin K2 was a minor part of the total vitamin K intake, only K2 and not K1 intake was associated with a lower risk of disease [6]. For people in the upper third of intakes of K2 compared to the lowest third, there were decreases in relative risks of coronary heart disease (CHD) mortality (57%), all-cause mortality (26%), and severe aortic calcification (52%). Even though intakes of K2 were only about 10% as much as the amounts of K1, their effect on cardiovascular disease was greater. Cheese was the primary source of menaquinones in this cohort in the Netherlands, not exactly highly regarded as a heart healthy food. This fact made confounding by other "healthy" nutrients less likely and made the results more robust.

Similar results have been seen in a second study, from the Prospect-EPIC cohort, also from the Netherlands [7]. About 16,000 women aged 49–70 were followed for 8 years. Vitamin K1 and K2 intakes were estimated from a FFQ. Vitamin K2 intake varied from <20 to >36 μg/day across quartiles. For every increase of 10 μg of K2, there was a 9% reduction in hazard ratio of risk of CHD, with the effects coming mainly from menaquinone subtypes MK-7, MK-8, and MK-9. There was no association between intake of K1 and CHD, as seen in the Rotterdam Study.

Two large cohort studies have been analyzed for associations between vitamin K intake and CHD. When the Nurses' Health Study (NHS) cohort was analyzed for an association between dietary vitamin K intake and cardiovascular outcomes, it was found that K1 intake was associated with a 21% decrease in multivariate relative risk of total CHD [8]. The association was attenuated by adjustments for other dietary factors and lifestyle patterns, so that it was not apparent to the authors whether the results were due to vitamin K1 or that K1 was just a marker for a lifestyle pattern associated with a high intake of K1. Median intakes of K1 for the lowest and highest quintiles were 87 and 300 μg/day for the NHS cohort. In another large cohort, the Health Professionals' Follow-up Study, there was a decrease in the relative risk of total CHD across increasing quintiles of vitamin K1 intake. However, when the results were adjusted for lifestyle and other dietary factors, the trend was no longer significant [9]. Neither of these large cohorts reported dietary intakes of K2, perhaps because the database for menaquinone concentrations in the USA was not complete at that time, nor is it fully available at the time of writing this chapter (only partial data are available on MK-4 but none on higher menaquinones), 17 years after such data were obtained for the Rotterdam Study [5]. The negative results from these cohorts for vitamin K1 and CHD only reinforce just how striking the results were from the Dutch studies for vitamin K2. The findings from the Netherlands were not expected or anticipated by many.

The question of whether vitamin K intake is related to arterial calcifications has been probed in two observational studies. In a cross-sectional study of 1689 women, dietary intakes of vitamin K1 and K2 were estimated with an FFQ, and standard screening mammograms were assessed for the presence of breast arterial calcifications [10]. Unadjusted results showed an inverse association between intake of vitamin K2 and breast arterial calcifications, but adjustments for aging, smoking, diabetes, and dietary factors made the association no longer significant. Adjustment for diabetes may have been unwise, as vitamin K2 intake has also been shown to reduce the risk of diabetes among 38,000 Dutch men and women in the Prospect-EPIC cohort mentioned previously [11]. So, this adjustment may have attenuated the results enough to make the association no longer statistically significant.

In another cross-sectional study, 564 postmenopausal women were examined for an association between coronary calcifications and intake of vitamins K1 and K2 [12]. Women were chosen from the Prospect-EPIC cohort study. In this cohort, cheese contributed 54%, milk products contributed 22%, and meat contributed 15% of the K2 intake. The mean intake of K2 ranged from 18.0 ± 4.5 to 48.5 ± 9.0 in the lowest and highest quartiles, respectively. Examinations found that 62% of the women had coronary calcifications. In the model adjusted for age and cardiovascular risk factors, increased menaquinone intakes were associated with a decreased calcification prevalence ratio of 0.80 (95% CI: 0.65–0.98), comparing highest to lowest quartile.

A more recent observational study reported findings contrary to those found in the Prospect-EPIC cohort and the Rotterdam Study. The PREDIMED cohort is a Spanish study to examine the effect of adoption of the Mediterranean Diet on cardiovascular, cancer, and all-cause mortality. The intakes of vitamins K1 and K2 were estimated by FFQ, and endpoints of cardiovascular, cancer, or all-cause mortality were tracked for a median follow-up of 4.8 years. Energy-adjusted intakes of vitamins K1 and K2, respectively, ranged from 170 and 18.4 μg/day in the lowest quartiles for each vitamin to 626 and 57.5 μg/day in the upper quartiles [13]. People in the upper quartile consumed about twice as many vegetables, especially leafy greens, as those in the lowest quartile. The upper quartile of vitamin K intake in this cohort adopting the Mediterranean Diet was substantially higher than seen in the other observational studies of other European or American cohorts. No protective effects for higher intakes of menaquinones were seen in this cohort for cardiovascular mortality, cancer mortality, or all-cause mortality. However, high intakes of phylloquinone lead to a reduced hazard ratio of 0.54 and

http://dx.doi.org/10.5772/63413

0.64 for cancer mortality and all-cause mortality. It is possible that enough vitamin K1 and other plant-based protective nutrients such as folate, vitamin C, fiber, potassium, and magnesium were supplied by the diet that even those participants who had low intakes of menaquinones were still not at an elevated risk compared to other subjects in the PREDIMED cohort. Participants who increased their intake of either vitamins K1 or K2 or both during the course of the study experienced decreased risk of cardiovascular mortality (K1 only), cancer mortality, and all-cause mortality. The main conclusion from this observational study perhaps is that eating plants is good for you, and the Mediterranean Diet is generally beneficial, as seen in another report from this cohort [14].

3. Vitamin K antagonist studies

Another line of evidence that led to the discovery of the role of menaquinones was the effect of blood thinning drugs such as warfarin and coumarin. While this class of drugs has been very helpful in preventing strokes in the short term, it has also caused damage long term, as people are often prescribed blood thinners for many years. In 1998, it was reported by Price et al. [15] that warfarin caused calcification of the elastic lamellae in rat arteries and heart valves within a period of 2 weeks, with increasing intensity each week. Vitamin K1 was given concurrently to maintain normal blood coagulation. At the time, menaquinone was not even mentioned in the article, not even in the discussion. The discovery of importance was that warfarin had negative side effects for arterial calcification that were not counteracted by vitamin K1.

This discovery nearly coincided in time with work on an MGP-deficient mouse model. Matrix γ-carboxylation protein, or matrix Gla protein (MGP), was originally discovered in bone tissue but is actually expressed in many tissues of the body, including vascular smooth muscle cells and chondrocytes in cartilage. MGP requires the activation by vitamin K in order to bind calcium ions and prevent crystallization of calcium. These MGP-deficient mice developed normally to term but died within 2 months as a result of extensive arterial calcification, which led to blood vessel rupture [16]. Also seen was the inappropriate calcification of cartilage, including the growth plate of bones. This MGP-deficient mouse model clearly showed that MGP has a central, active role in preventing calcifications of arterial walls and also of cartilage. This research coupled together with the warfarin-caused calcification pointed to a central role for MGP in controlling arterial calcification.

Further work on the interrelationship between vitamin K1 and K2 was spurred on by feeding studies in rats. When rats were made vitamin K deficient, then fed only K2 as MK-4, they accumulated MK-4 especially in the pancreas, aorta, fatty tissues, and brain. Liver and serum levels of MK-4 were low. When vitamin K–deficient rats were fed only K1, they accumulated K1 in the liver, heart, and fatty tissues, and they also accumulated MK-4 in the same way as the rats that were fed MK-4, indicating that there was conversion from phylloquinone to MK-4 [17]. So, with a warfarin-rat model that was able to induce arterial calcification and knowing that there were differences between K1 and K2 distributions in the rats and that K2 prevented heart disease in the Rotterdam Study, Spronk and coworkers set out to see how to prevent arterial calcification [18]. When warfarin-treated rats were fed K1, the rats got arterial calcification, as shown before [15], even at the highest tested dose of K1. But when the warfarin-treated rats were fed K2 as MK-4 or K1 together with MK-4 simultaneously, the arterial calcification was prevented. The picture was becoming clearer. Further studies have shown that in this rodent model warfarin treatment not only causes arterial calcification but functionally augments aortic peak velocity, aortic valve-peak gradient, and carotid pulse-wave velocity [19].

The work in rats spurred investigators to look at the effect of anticoagulants in people. In one study, aortic heart valves were examined that had been replaced during routine surgery. Some patients received preoperative marcoumar treatment, for between 16 and 35 months, with a mean of 25 months. When compared with patients who did not have any blood thinner treatment, there was about twice as much calcification on the valves from patients who had received the marcoumar [20]. The mean calcified area on the valve went from 16% in the untreated group to 37% in the anticoagulant group. In a cross-sectional study, coronary artery calcium scores and valvular calcium scores were compared between patients on long-term use of anticoagulants and patients without any anticoagulant therapy [21]. The Agaston calcium scores were about double in the anticoagulant treatment group, indicating that the effects of anticoagulants seen in mice and rats are also present in people, even when the treatment was only for a couple of years.

These initial results have been confirmed by further studies. Rennenberg et al. [22] examined 19 patients younger than 55 years of age who had used coumarins for more than 10 years but did not have other cardiovascular risk factors. These patients were compared with 18 matched healthy controls. When they examined femoral arteries, they found the coumarin users had 8.5 times the chance of having arterial calcification compared to the healthy controls. Fourteen of 19 coumarin users, but only 4 of 18 controls had femoral arterial calcifications. Another cross-section examination of low-risk atrial fibrillation patients found that both age and use of oral anticoagulants were related to increased coronary calcium score [23]. And as length of time using the anticoagulants increased, the coronary calcium score also increased, going from 53 ± 115 for no use to 90 ± 167 for 6–60 months, and to 236 ± 278 for >60 months of use. These findings were also confirmed in a series of 133 oral anticoagulant users matched by age, gender, and Framingham cardiovascular risk score [24]. Agaston calcium scores increased from 79.6 ± 159.8 for use of 2.5 ± 1.5 months, to 142.4 ± 306.0 for 18.7 ± 8.8 months, to 252.5 ± 399.3 for 86.4 ± 47.1 months of use.

In women undergoing screening mammography who took warfarin, breast arterial calcifications were also more common with increasing length of warfarin treatment [25]. Prevalence of breast arterial calcifications increased from 25.0% for <1 year of therapy to 74.4% for >5 years of therapy. So, these calcifications can appear in peripheral tissues as well, not just in the aorta. To show this peripheral effect further, and in men, after completing the breast arterial calcification study, Han and O'Neill examined radiographs of ankles and feet, retrospectively, and checked records for warfarin use prior to the x-ray [26]. They found a significant increase, from 19% to 38% prevalence, in peripheral arterial calcifications in people who had been using warfarin for at least 5 years prior to their x-ray. While these drugs could be termed "anticoagulants," the preferred term for many of these authors is vitamin K antagonists, for this is their mode of action.

Calcification of arteries had originally been thought of as a one-way process, without reversibility, similar to the thinking about coronary plaque. However, just as regression can be seen of atherosclerotic plaques [27], so calcification of arteries, too, is a dynamic process. Using the warfarin-treated rat as a model for arterial calcification, Schurgers et al. [28] first fed rats for 6 weeks on the diet to induce calcification. Then the warfarin treatment was stopped and rats were fed normal levels of K1, or high levels of K1 or K2 (as MK-4). Normal levels of vitamin K1 continued to progressively increase calcification, but both forms of vitamin K at high doses reversed arterial calcification by about 50%. Vitamin K1 does not work as long as warfarin is present, as it inhibits the conversion of K1 into MK-4. But when the warfarin treatment is stopped, this research clearly showed that this calcification process could be reversed by high doses of vitamin K, especially K2.

4. Biomarker research studies

One difficulty in this field of research is determining the functional vitamin K status of an individual. A blood test of vitamin K levels is not sufficient. The amount of vitamin K in the blood is very small and generally only reflects the vitamin K1 that was consumed within the last 4 hours or so, as K2 levels are too low to assay in blood, and K1 clears from the blood with triglycerides. As research progressed, it became increasingly apparent that there were more functions for vitamin K than originally discovered. Coagulation was only the most immediately obvious function of vitamin K in the liver. But the observation studies and vitamin K antagonist research indicated more functions beyond coagulation, dealing with regulation of calcification throughout the body. McCann and Ames [29] elaborated on this multifunction vitamin, indicating that triage theory helps us understand the distribution of vitamin K to various organs. Triage theory states that the most critical functional needs are met first in the body (coagulation) when there is a shortage of a micronutrient. Then when there is an abundance of the micronutrient, all of the secondary functions important to long-term health are also met.

For these reasons, and possibly others, functional tests for vitamin K status for these secondary functions beyond coagulation were sought. Osteocalcin, a vitamin K–dependent protein found in bone, can be measured in the circulation as well. The ratio of carboxylated to undercarboxylated or uncarboxylated osteocalcin is one biomarker for functional vitamin K status. However, this applies more to the status of vitamin K as it applies to bones. Since MGP is involved in arterial calcification, assays for determining the concentrations of various forms of MGP were developed [30, 31]. Of the various forms of MGP, the dephosphorylated, uncarboxylated form has been most closely related to arterial calcification. Among coumarin users an elevated dp-ucMGP level was found compared to controls (1439 ± 481 pM vs. 299 ± 163 pM, respectively) [22]. In a cohort of 101 chronic kidney disease patients, the level of dp-ucMGP increased with increased severity of the disease [32]. Plasma dp-ucMGP was also independently associated with aortic calcification, and a concentration greater than 921 pM was a predictor of all-cause mortality in a crude analysis.

What about when people on dialysis are also taking oral anticoagulants? Among 160 hemodialysis patients in Belgium, the 23 who were treated with anticoagulants had much higher circulating concentrations of dp-ucMGP, 5604 pM (interquartile range: 3758, 7836 pM) and 1939 pM (interquartile range: 1419, 2841pM) for the anticoagulant treated and non-treated groups [33].

In a study of 147 patients with symptomatic severe calcific valvular aortic stenosis, the levels of dp-ucMGP were associated with cardiac function and long-term mortality in multivariate analysis [34]. Increasing severity of disease was related to dp-ucMGP concentrations in a study of 179 patients with chronic heart failure [35].

The dp-ucMGP assay was checked for correlation with vitamin K status and coronary artery calcification in a study of older adults without cardiovascular disease [36]. While the assay did correlate well with plasma phylloquinone, uncarboxylated prothrombin, and serum uncarboxylated osteocalcin, there was no association between dp-ucMGP levels and coronary artery calcification. Shea and coworkers [36] presented data that are not consistent with the other reports on this assay. Perhaps the assay works better for much higher levels of dp-ucMGP, such as found in disease states. This study looked at older adults without clinical cardiovascular disease, whose levels of dp-ucMGP were much lower than subjects with cardiovascular disease (CVD). As suggested by the authors, the coronary artery calcification analyzed in this report may have been more in the intimal layer, rather than in the medial layer, where MGP has a greater role [28]. A more recent study involving 200 health women found a borderline statistically significant relationship between dp-ucMGP and coronary artery calcification, as well as a strong relationship between dp-ucMGP and vitamin K status (ratio of carboxylated osteocalcin) [37]. The results in [36] appear to the exception, as there is a consistent relationship between dp-ucMGP, vitamin K status, and health outcomes involving arterial calcification in all of the other studies examined.

Other studies have generally found that the biomarker dp-ucMGP does correlate with vitamin K status and disease outcomes related to arterial calcification. In the EPIC-NL cohort, 518 participants were identified as diabetic at baseline [38]. After 11.2 years of follow-up, incidence of CVD was significantly associated with baseline concentrations of dp-ucMGP, but not other species of MGP. The hazard ratio per standard deviation (HRSD) of dp-ucMGP for all CVD was 1.21 (95% CI 1.06–1.38), for peripheral artery disease HRSD = 1.32 (95% CI 1.07–1.65), and for heart failure HRSD = 1.85 (95% CI 1.42–2.17). The prospective Longitudinal Aging Study, Amsterdam (LASA) examined 577 people aged >55 years who were free of CVD at the baseline [39]. There were 40 incident cases of CVD during the 5.6 years of follow-up. For the highest tertile compared to the lowest tertile of dp-ucMGP, there was a hazard ratio of 2.69 (95% CI 1.09–6.62) for being diagnosed with CVD. The carboxylated form of MGP was not related to risk of CVD.

Two Czech Republic prospective studies have examined the usefulness of the dp-ucMGP as a biomarker to predict cardiovascular mortality. From the EUROASPIRE III and EUROASPIRE-stroke surveys, 799 patients were examined who had already experienced a myocardial infarction, coronary revascularization, or first ischemic stroke. After a median follow-up of 5.6 years, 159 patients died. In the fully adjusted model, the patients in the highest quartile of dp-ucMGP (≥977 pM) had higher risk of all-cause and cardiovascular mortality, HRR 1.89 (95% CI 1.32–2.72) and 1.88 (95% CI 1.18–2.61), respectively [40]. For those subjects in the upper quartile of dp-ucMGP who also had heart failure, indicated by an elevated circulating brain natriuretic peptide level >100 ng/L, mortality risk was further increased, HRR 4.86 (95% CI 3.15–7.49) [41]. In a random sample from the general population from the Czech post-MONICA study, Mayer et al. [42] found that aortic stiffness, as measured by pulse wave velocity, was related to vitamin K status. Compared to the lowest quartile, the upper quartile of dp-ucMGP (≥671 pM) has an increased odds ratio of 1.73 (95% CI 1.17–2.5).

5. Intervention studies

One of the first intervention types was to confirm the utility of various species of MGP as biomarkers for vitamin K status. If you improve someone's status by oral supplementation, the biomarker should reflect this improvement in a dose-dependent manner. So, in 2012, Dalmeijer and coworkers [43] reported a randomized, double-blind placebo-controlled trial (RCT) of 60 people taking 0, 180, or 360 μg/day of vitamin K2 as menaquinone 7 (MK-7) for 12 weeks. Assays were performed for three different species of MGP: desphospho-uncarboxylated MGP (dp-ucMGP), desphospho-carboxylated MGP (dp-cMGP), and total uncarboxylated MGP (t-ucMGP). Vitamin K status was also measured using the ratio of uncarboxylated to carboxylated osteocalcin. (Note that the research field on the role of vitamin K2 in bones matured earlier than the field of cardiovascular effects of K2, so osteocalcin was well established as a vitamin K2 marker by this time.) After 12 weeks of the supplements, the osteocalcin ratio decreased significantly, with a 60% drop at 180 μg dose and a 74% drop at the 360 μg dose. The amount of dp-ucMGP decreased significantly and dose-dependently as well, by 31% and 46% at 180 and 360 μg, respectively. There were no changes in the placebo group, as expected. Changes in other species of MGP (dp-cMGP and t-ucMGP) were not different between placebo and the supplement groups. This study was one of the first intervention trials to validate the usefulness of dp-ucMGP as a biomarker for vitamin K status. Observational studies had been carried out, but the intervention studies took the research one more step toward maturity.

In the same year, another RCT was reported of 42 Dutch men and women randomized to receive 0, 10, 20, 45, 90, 180, or 360 μg/day of vitamin K2 as MK-7. The ratio of uncarboxylated to carboxylated osteocalcin was determined along with the concentration of dp-ucMGP. The upper three doses (90, 180, and 360 μg/day) increased the carboxylation of osteocalcin and decreased the amount of dp-ucMGP. In these healthy adults aged 18–45, no adverse effects were seen on the generation of thrombin, indicating that coagulation factors were not perturbed by the additional supply of vitamin K2. This is reasonable, for the coagulation factors are generally all carboxylated. Only the extrahepatic vitamin K–dependent proteins seem to suffer when there is a shortage of vitamin K, as explained by the triage theory [29].

Patients with chronic kidney disease and those undergoing dialysis have been shown repeatedly to suffer with high levels of uncarboxylated vitamin K–dependent proteins and have high levels of arterial calcification as well. (The most deficient group is the same patients taking vitamin K antagonist drugs concurrently.) In order to prepare for a RCT with a disease endpoint of stabilizing or reversing arterial calcification in this patient group, a supplement trial was conducted in the Netherlands with 50 hemodialysis patients [44]. An age-matched healthy control group was selected also for comparisons. The hemodialysis patients were randomized into groups taking 45, 135, or 360 μg/day of vitamin K2 as MK-7 for 6 weeks. Measurements were taken for the levels of uncarboxylated osteocalcin (ucOC), dp-ucMGP, and PIVKA-II. PIVKA-II is a prothrombin liver protein that is only seen in the circulation under situations of severe vitamin K deficiency, such as found in hemodialysis patients. At baseline hemodialysis, patients, compared to healthy controls, had 4.5-fold higher dp-ucMGP levels and 8.4-fold higher uncarboxylated osteocalcin levels. PIVKA-II levels were detectable in 49 of the 50 hemodialysis patients. There was a dose-dependent response to the MK-7 treatment, with the 45 μg dose being little different than a placebo, the 135 μg dose giving 37%, 11%, and 34% changes in dp-ucMGP, ucOC, and PIVKA-II, respectively, which was almost significant, and the 360 μg dose yielding statistically significant changes of 61%, 34%, and 42% decreases in dp-ucMGP, ucOC, and PIVKA-II. This short trial showed both the severity of the vitamin K deficiency in this hemodialysis patient group as well as the effectiveness of a relatively high dose (360 μg/day) of MK-7 in bringing down functional markers of vitamin K deficiency.

As a follow-up to the Westenfeld Study [44] just reviewed, another larger study with hemodialysis patients was carried out with slightly higher doses, but administered three times a week by a nurse after dialysis [45]. This method was used to increase patient compliance. Doses for the 200 patients were 360, 720, and 1080 μg of MK-7 three times as week for 8 weeks. This works out to equivalent daily doses of 154, 309, and 463 μg/day. After 8 weeks, levels of dp-ucMGP decreased by 17, 33, and 46% in the three dosage groups, respectively. Results here were similar to those in [44], but with less decrease in relative change in dp-ucMGP at the highest dose (46% vs. 61%). However, the absolute differences before and after intervention were greater in [45]. Absolute changes in dp-upMGP in the Westenfeld Study were −404, −730, and −978 pM at 45, 135, and 360 μg/day [44]. In a study by Caluwé and coworkers [45], absolute changes in dp-ucMGP were −566, −962, and −1487 pM for the equivalent daily doses of 154, 309, and 463 μg/day. So, the outcomes were very similar, especially given that the second trial was 8 weeks long rather than just 6 weeks.

How widespread is extrahepatic vitamin K insufficiency? A cross-sectional sample of 896 healthy individuals showed that dp-ucMGP levels increased with age, staying around 200 pM until about age 40 and then increasing up to over 600 pM for those >70 years of age [46]. Levels of dp-ucMGP decreased in children and adults when given supplemental MK-7, again showing that dp-ucMGP was a biomarker that responded to extrahepatic vitamin K status, and that many adults had less than optimal levels of vitamin K. These levels are not optimal, neither are they as severe as hemodialysis patients, who averaged around 3000 pM but ranged to over 7000 pM in some individuals [44, 45]. Hemodialysis patients who took vitamin K antagonists had a mean dp-ucMGP of about 5600 pM [33]. But the values >600 pM in the >70 age group are close to the range at which excess heart disease mortality occurred in other studies, >977 pM in [40] and >921 pM in [32]. Calcification of arteries takes time and finally takes a toll on the elderly if not protected against through the years.

In one of the first RCT studies to report a disease endpoint, Knapen and coworkers examined arterial stiffness in 244 women after supplementation for 3 years with 180 μg/day of MK-7 [47]. Previous work by this research group had linked increased levels of dp-ucMGP with arterial stiffness [48]. Compared to the placebo group, dp-ucMGP levels dropped by about 50%. The K2 treatment resulted in improvements in the whole group, but the best results for improving stiffness were seen in those women who started with the worse condition, with a stiffness index β above the median of 10.8. For these women, there were improvements in distention, compliance, distensibility, Young's Modulus, and the local carotid pulse wave velocity. Not all became normal, as acute phase markers interleukin-6, C-reactive protein, and tumor necrosis factor-α remained abnormal as well as the markers of endothelial dysfunction vascular cell adhesion molecule and E-selectin.

A Polish RCT study examined disease endpoints for vascular calcification and progression of atherosclerosis in 42 women who had chronic kidney disease but were not undergoing dialysis [49]. The women were split into three groups followed for 270 days: one taking 90 μg/day of MK-7 (K), one group taking 90 μg/day of MK-7 plus 10 μg/day of vitamin D as cholecalciferol (K+D), and one taking only 10 μg/day of vitamin D (D). The results were that the intervention slowed the progression of atherosclerosis but did not significantly slow the increase in coronary artery calcium score. The reason for this lack of success is in the dose of MK-7 used. As noted in the studies by Westenfeld et al. [44] and Caluwé et al. [45], a daily dose of around 360 μg is needed to significantly reduce dp-ucMGP. Changes in dp-ucMGP were reported to be significant in this Polish trial, but actual change was from 1077.1 ± 507.7 to 961.5 ± 506.7, or a change of only 115.6 pM. While this change may have been statistically significant, it was clinically irrelevant. This study also reveals why it is important to develop research stepwise to ensure that interventions will be successful. If the treatment is not expected to yield a large change in biomarker studies, how would it yield a successful clinical result?

A rationale and study protocol has been published for the VitaK-CAC trial [50]. For this 2-year trial, the intervention will be 360 μg/day of MK-7. Patients with coronary artery disease will be monitored for coronary artery calcification as the primary endpoint. Secondary endpoints will be arterial structure and function. This study appears to be well designed to yield good results, as long as the arterial calcification seen in coronary arteries is susceptible to the MGP mechanism of action. Nutrients have a way of working in complimentary fashion, if not synergistically. The best chance for successful intervention is to use all of the known tools available, which is discussed in the section on plant-based diets below.

6. Prostate health and vitamin K2

While the focus of this chapter has been on arterial calcification and cardiovascular health, it has come to the attention of this author that there is evidence that supports a hypothesis that poor prostate health is not a hormone issue, but a cardiovascular issue [51]. The first piece of evidence comes from interventional radiologists Dr. Gat and coworkers, who were initially working on reversing infertility in men by relieving varicoceles, or varicose veins in the pampiniform venous plexus. They discovered that varicose veins in the internal spermatic vein, which normally returns blood from the testes to the kidneys, prevented normal blood flow. The one-way valves had failed in this vein, possibly causing the varicose vein, or as part of the process of forming varicose veins—the exact mechanism is debated. So instead of normal blood flow, the blood flowed retrograde through the prostatic veins. When testosterone levels were measured near the prostate gland in 12 infertile men with varicocele, the mean concentration was 3632 pmol/l compared to 27.33 pmol/l in the serum, or about 130 times higher [52]. By occluding the internal spermatic vein, Dr. Gat was able to relieve the physical pressure due to the elevated blood pressure caused by the height of the column of blood sitting in the internal spermatic vein and also made a pathway through normal venous pathways for blood to drain away from the testes without retrograde flow past the prostate. This venous occlusion surgery led to relief of benign prostatic hyperplasia (BPH), and possibly prevention of prostate cancer as well [52].

Varicose veins and destruction of the one-way valves in the internal spermatic vein were the direct cause of BPH, but what causes varicose veins? Work by Cario-Toumaniantz and coworkers [53] on the differentially expressed genes and gene products in varicose vein tissue showed an overexpression of genes involved in extracellular remodeling, including matrix Gla protein. Smooth muscle cells were seen proliferating in varicose vein tissue with high expression of MGP, particularly the uncarboxylated form of MGP. Overexpression of MGP and proliferation of smooth muscle cells have been seen before, reported by Price and coworkers investigating the effects of warfarin on arterial calcification in rats [15], and by Schurgers and coworkers reporting the reversal of warfarin-induced arterial calcification in rats [28]. In areas of calcification of arteries, there is proliferation of smooth muscle cells and increased expression of uncarboxylated MGP, similar to what was seen in varicose vein tissue [53]. It is likely then that vitamin K is involved in the mechanism by which varicose veins form, just as it has been implicated in the formation of arterial calcifications. When varicose vein tissue culture was treated with warfarin, mineralization increased, which could be inhibited by the inclusion of vitamin K in the culture media [53], indicating a direct role for vitamin K in the prevention of varicose veins.

In addition to this evidence from cellular biology, evidence from a prospective observational study showed an association between intake of vitamin K2 and poor prostate health manifested as prostate cancer. There was a significant association between menaquinone intake and advanced prostate cancer in the EPIC-Heidelberg cohort [54]. A nested case-control follow-up study also found an association between the ratio of undercarboxylated osteocalcin to carboxylated osteocalcin and high-grade prostate cancer and advanced prostate cancer [55]. Neither of these reports found a connection with vitamin K1 intake. Further evidence comes from a retrospective study of warfarin use and clinical stage of prostate cancer at diagnosis [56]. While some of the evidence for intermediate or short-term use of warfarin is conflicting, the comparison between those men who had used warfarin for at least 4 years in the 5 years before prostate cancer diagnosis and nonusers of warfarin showed an odds ratio of 2.2 (95% CI 1.03–4.81) of poor prognosis disease. This would seem to agree with the hypothesis that vitamin K antagonists would increase the possibility of forming varicoceles, leading to poor prostate health. But there could be other mechanisms of action as well.

At this point, the role of vitamin K2 in prostate health is a good hypothesis but needs further confirmation by interventional radiologists and other researchers before the link between prostate health and cardiovascular function, especially the role of vitamin K2 plays, is certain.

7. Vitamin K2 and a plant-based diet?

Reversal of atherosclerosis by a plant-based diet has already been mentioned in this chapter [27]. Dr. Esselstyn has also shown that a plant-based diet very low in fat can reverse coronary artery disease [57]. Of 198 subjects in the lifestyle intervention study, 177 were adherent to the program while 21 formed a control group of non-adherent comparison subjects. There was reversal of angiographic-verified blockages in 39 adherent subjects. Disease progression occurred in 4 (2.3%) adherent subjects, but in 11 (52.4%) of the non-adherent subjects. Coronary artery calcification was not measured.

There is abundant evidence that increased consumption of fruits and vegetables in conjunction with exercise and a healthy lifestyle is beneficial for cardiovascular health and lower cardiovascular mortality. Healthy diets have been reported to lower cardiovascular mortality risk by about 30–40% [14, 58, 59]. When combined with other lifestyle factors, the risk plummets to about 20% or less compared to the least healthy fraction of the population [60–63].

The point here is that all available resources should be used to counter the disease process and to promote healthy aging. While vitamin K2 is a valuable nutrient that is generally in short supply in the global diet, the context of the entire diet must be kept in mind. The best results will be obtained by a full complement of healthy foods. Population studies have shown that even 40–50 μg/day of menaquinones from the diet is associated with cardioprotection [6, 7] and lower risk of advanced prostate cancer and lung cancer [54, 55, 64].

Is vitamin K2 compatible with a plant-based diet? While most of the common sources of menaquinones are cheese, fermented dairy products, eggs, and meats, there are plant sources as well. While these are animal products, judicial selection could be used to maximize K2 intake without consuming a large amount of any animal-based foods. Natto is a well-known Japanese food that is very rich in MK-7, though not very popular outside of Japan. The menaquinone in natto is made by fermentation with *Bacillus subtilis var. natto*. Fermented vegetables such as sauerkraut contain a small amount of menaquinones as well. By selecting probiotic bacteria based on their production of menaquinones, the amount of vitamin K2 from a serving of fermented vegetables could be significant. More product development is needed in this area.

8. Research directions and priorities

The research on vitamin K2 and cardiovascular health has come a long way, going from observational studies of populations and warfarin studies, to developing biomarkers, finding effective dosage schedules, and begin carrying out RCT studies examining disease endpoints. Very few disease endpoints have been reported at this time, but several should be completed in the next few years. Diseases are multifaceted, so the solutions should likewise be multifaceted. Trying to reverse a complex disease with a single nutrient has generally been unsuccessful. It is likely that the best success will be found when vitamin K2 is used in the context of a whole food plant-based diet that contains some dietary source of vitamin K2, along with supplements to reverse disease damage when appropriate.

One of the major benefits seen by Dr. Weston Price in the 1930s was the prevention and reversal of dental caries using vitamin K2 [2]. This could be a very fruitful area of research that is untapped at this point. Mental health effects of vitamin K2 are likely as well, as the brain contains a significant concentration of K2. Varicose veins in general and prostate health in particular should be studied in light of the K2 research presented here in this chapter. Mechanisms of the function of MGP should be worked out along the way as well. Identification of the function of other vitamin K–dependent proteins is still lacking. There is a gamma-carboxylation–rich protein (GRP) that is quite small yet has 16 carboxylation sites and is highly expressed in cartilage. Its exact function is still unknown. So, there is work that can be done at the molecular level as well as at the public health level in furthering our understanding of vitamin K2.

Conflict of Interest

Michael Donaldson is a research scientist at the Hallelujah Acres Foundation for investigations pertaining to the Hallelujah Diet. Funding for this research has been provided by Hallelujah Acres, Inc.

Author details

Michael S. Donaldson*

Address all correspondence to: mdonaldson@myhdiet.com

Hallelujah Acres, Zillah, WA, USA

References

[1] Dam H. The antihaemorrhagic vitamin of the chick. Biochem J. 1935 Jun;29(6):1273–85.

[2] Price WA. Nutrition and physical degeneration. 6th edition. New Canaan, Conn: Keats Pub; 2003. 524 p.

[3] Masterjohn C. On the trail of the elusive X-factor: Vitamin K2 revealed. Wise traditions in food, farming and the healing arts [Internet]. 2007 [cited 2016 May 17]. Available from: http://www.westonaprice.org/health-topics/abcs-of-nutrition/on-the-trail-of-the-elusive-x-factor-a-sixty-two-year-old-mystery-finally-solved/

[4] Jie K-SG, Bots ML, Vermeer C, Witteman JCM, Grobbee DE. Vitamin K intake and osteocalcin levels in women with and without aortic atherosclerosis: a population-based study. Atherosclerosis. 1995 Jul 1;116(1):117–23.

[5] Schurgers LJ, Geleijnse JM, Grobbee DE, Pols HAP, Hofman A, Witteman JCM, et al. Nutritional intake of vitamins K1 (Phylloquinone) and K2 (Menaquinone) in the Netherlands. J Nutr Environ Med. 1999 Jan;9(2):115–22.

[6] Geleijnse JM, Vermeer C, Grobbee DE, Schurgers LJ, Knapen MHJ, van der Meer IM, et al. Dietary intake of menaquinone is associated with a reduced risk of coronary heart disease: the Rotterdam study. J Nutr. 2004 Nov;134(11):3100–5.

[7] Gast GCM, de Roos NM, Sluijs I, Bots ML, Beulens JWJ, Geleijnse JM, et al. A high menaquinone intake reduces the incidence of coronary heart disease. Nutr Metab Cardiovasc Dis NMCD. 2009 Sep;19(7):504–10.

[8] Erkkilä AT, Booth SL, Hu FB, Jacques PF, Manson JE, Rexrode KM, et al. Phylloquinone intake as a marker for coronary heart disease risk but not stroke in women. Eur J Clin Nutr. 2004 Sep 29;59(2):196–204.

[9] Erkkilä AT, Booth SL, Hu FB, Jacques PF, Lichtenstein AH. Phylloquinone intake and risk of cardiovascular diseases in men. Nutr Metab Cardiovasc Dis NMCD. 2007 Jan; 17(1):58–62.

[10] Maas AHEM, van der Schouw YT, Beijerinck D, Deurenberg JJM, Mali WPTM, Grobbee DE, et al. Vitamin K intake and calcifications in breast arteries. Maturitas. 2007 Mar 20;56(3):273–9.

[11] Beulens JWJ, van der A DL, Grobbee DE, Sluijs I, Spijkerman AMW, van der Schouw YT. Dietary phylloquinone and menaquinones intakes and risk of type 2 diabetes. Diabetes Care. 2010 Aug 1;33(8):1699–705.

[12] Beulens JWJ, Bots ML, Atsma F, Bartelink M-LEL, Prokop M, Geleijnse JM, et al. High dietary menaquinone intake is associated with reduced coronary calcification. Atherosclerosis. 2009 Apr;203(2):489–93.

[13] Juanola-Falgarona M, Salas-Salvadó J, Martínez-González MÁ, Corella D, Estruch R, Ros E, et al. Dietary intake of vitamin K is inversely associated with mortality risk. J Nutr. 2014 May 1;144(5):743–50.

[14] Martínez-González MA, Sánchez-Tainta A, Corella D, Salas-Salvadó J, Ros E, Arós F, et al. A provegetarian food pattern and reduction in total mortality in the Prevención con Dieta Mediterránea (PREDIMED) study. Am J Clin Nutr. 2014 Jul 1;100(Supplement 1):320S–8S.

[15] Price PA, Faus SA, Williamson MK. Warfarin causes rapid calcification of the elastic lamellae in rat arteries and heart valves. Arterioscler Thromb Vasc Biol. 1998 Sep 1;18(9): 1400–7.

[16] Luo G, Ducy P, McKee MD, Pinero GJ, Loyer E, Behringer RR, et al. Spontaneous calcification of arteries and cartilage in mice lacking matrix GLA protein. Nature. 1997 Mar 6;386(6620):78–81.

[17] Ronden JE, Drittij-Reijnders MJ, Vermeer C, Thijssen HH. Intestinal flora is not an intermediate in the phylloquinone-menaquinone-4 conversion in the rat. Biochim Biophys Acta. 1998 Jan 8;1379(1):69–75.

[18] Spronk HMH, Soute BAM, Schurgers LJ, Thijssen HHW, De Mey JGR, Vermeer C. Tissue-specific utilization of menaquinone-4 results in the prevention of arterial calcification in warfarin-treated rats. J Vasc Res. 2003;40(6):531–7.

[19] Krüger T, Oelenberg S, Kaesler N, Schurgers LJ, van de Sandt AM, Boor P, et al. Warfarin induces cardiovascular damage in mice. Arterioscler Thromb Vasc Biol. 2013 Nov 1;33(11):2618–24.

[20] Schurgers LJ, Aebert H, Vermeer C, Bültmann B, Janzen J. Oral anticoagulant treatment: friend or foe in cardiovascular disease? Blood. 2004 Nov 15;104(10):3231–2.

[21] Koos R, Mahnken AH, Mühlenbruch G, Brandenburg V, Pflueger B, Wildberger JE, et al. Relation of oral anticoagulation to cardiac valvular and coronary calcium assessed by multislice spiral computed tomography. Am J Cardiol. 2005 Sep 15;96(6):747–9.

[22] Rennenberg RJMW, van Varik BJ, Schurgers LJ, Hamulyak K, ten Cate H, Leiner T, et al. Chronic coumarin treatment is associated with increased extracoronary arterial calcification in humans. Blood. 2010 Jun 17;115(24):5121–3.

[23] Weijs B, Blaauw Y, Rennenberg RJMW, Schurgers LJ, Timmermans CCMM, Pison L, et al. Patients using vitamin K antagonists show increased levels of coronary calcification: an observational study in low-risk atrial fibrillation patients. Eur Heart J. 2011 Oct 1;32(20):2555–62.

[24] Schurgers LJ, Joosen IA, Laufer EM, Chatrou MLL, Herfs M, Winkens MHM, et al. Vitamin K-antagonists accelerate atherosclerotic calcification and induce a vulnerable plaque phenotype. PLoS One. 2012 Aug 29;7(8):e43229.

[25] Tantisattamo E, Han KH, O'Neill WC. Increased vascular calcification in patients receiving warfarin. Arterioscler Thromb Vasc Biol. 2015 Jan 1;35(1):237–42.

[26] Han KH, O'Neill WC. Increased peripheral arterial calcification in patients receiving warfarin. J Am Heart Assoc. 2016 Jan 26;5(1):e002665.

[27] Ornish D, Scherwitz LW, Billings JH, Gould KL, Merritt TA, Sparler S, et al. Intensive lifestyle changes for reversal of coronary heart disease. JAMA. 1998;280(23):2001–7.

[28] Schurgers LJ, Spronk HMH, Soute BAM, Schiffers PM, DeMey JGR, Vermeer C. Regression of warfarin-induced medial elastocalcinosis by high intake of vitamin K in rats. Blood. 2007 Apr 1;109(7):2823–31.

[29] McCann JC, Ames BN. Vitamin K, an example of triage theory: is micronutrient inadequacy linked to diseases of aging? Am J Clin Nutr. 2009 Oct 1;90(4):889–907.

[30] Cranenburg ECM, Koos R, Schurgers LJ, Magdeleyns EJ, Schoonbrood THM, Landewé RB, et al. Characterisation and potential diagnostic value of circulating matrix Gla protein (MGP) species. Thromb Haemost. 2010 Oct;104(4):811–22.

[31] Cranenburg ECM, Vermeer C, Koos R, Boumans M-L, Hackeng TM, Bouwman FG, et al. The circulating inactive form of matrix Gla protein (ucMGP) as a biomarker for cardiovascular calcification. J Vasc Res. 2008 Apr 10;45(5):427–36.

[32] Schurgers LJ, Barreto DV, Barreto FC, Liabeuf S, Renard C, Magdeleyns EJ, et al. The circulating inactive form of matrix Gla protein is a surrogate marker for vascular calcification in chronic kidney disease: a preliminary report. Clin J Am Soc Nephrol. 2010 Apr 1;5(4):568–75.

[33] Delanaye P, Krzesinski J-M, Warling X, Moonen M, Smelten N, Médart L, et al. Dephosphorylated-uncarboxylated matrix Gla protein concentration is predictive of vitamin K status and is correlated with vascular calcification in a cohort of hemodialysis patients. BMC Nephrol. 2014;15:145.

[34] Ueland T, Gullestad L, Dahl CP, Aukrust P, Aakhus S, Solberg OG, et al. Undercarboxylated matrix Gla protein is associated with indices of heart failure and mortality in symptomatic aortic stenosis. J Intern Med. 2010 Nov 1;268(5):483–92.

[35] Ueland T, Dahl CP, Gullestad L, Aakhus S, Broch K, Skårdal R, et al. Circulating levels of non-phosphorylated undercarboxylated matrix Gla protein are associated with disease severity in patients with chronic heart failure. Clin Sci. 2011 Aug 1;121(3):119–27.

[36] Shea MK, O'Donnell CJ, Vermeer C, Magdeleyns EJP, Crosier MD, Gundberg CM, et al. Circulating uncarboxylated matrix Gla protein is associated with vitamin K nutritional status, but not coronary artery calcium, in older adults. J Nutr. 2011;141(8):1529–34.

[37] Dalmeijer GW, van der Schouw YT, Vermeer C, Magdeleyns EJ, Schurgers LJ, Beulens JWJ. Circulating matrix Gla protein is associated with coronary artery calcification and vitamin K status in healthy women. J Nutr Biochem. 2013 Apr;24(4):624–8.

[38] Dalmeijer GW, van der Schouw YT, Magdeleyns EJ, Vermeer C, Verschuren WMM, Boer JMA, et al. Matrix Gla protein species and risk of cardiovascular events in type 2 diabetic patients. Diabetes Care. 2013 Nov 1;36(11):3766–71.

[39] van den Heuvel EGHM, van Schoor NM, Lips P, Magdeleyns EJP, Deeg DJH, Vermeer C, et al. Circulating uncarboxylated matrix Gla protein, a marker of vitamin K status, as a risk factor of cardiovascular disease. Maturitas. 2014 Feb;77(2):137–41.

[40] Mayer O, Seidlerová J, Bruthans J, Filipovský J, Timoracká K, Vaněk J, et al. Desphospho-uncarboxylated matrix Gla-protein is associated with mortality risk in patients with chronic stable vascular disease. Atherosclerosis. 2014 Jul;235(1):162–8.

[41] Mayer O, Seidlerová J, Vaněk J, Karnosová P, Bruthans J, Filipovský J, et al. The abnormal status of uncarboxylated matrix Gla protein species represents an additional mortality risk in heart failure patients with vascular disease. Int J Cardiol. 2016 Jan 15;203:916–22.

[42] Mayer O, Seidlerová J, Wohlfahrt P, Filipovský J, Vaněk J, Cífková R, et al. Desphospho-uncarboxylated matrix Gla protein is associated with increased aortic stiffness in a general population. J Hum Hypertens. 2015 May 28. DOI: 10.1038/jhh.2015.55.

[43] Dalmeijer GW, van der Schouw YT, Magdeleyns E, Ahmed N, Vermeer C, Beulens JWJ. The effect of menaquinone-7 supplementation on circulating species of matrix Gla protein. Atherosclerosis. 2012 Dec;225(2):397–402.

[44] Westenfeld R, Krueger T, Schlieper G, Cranenburg ECM, Magdeleyns EJ, Heidenreich S, et al. Effect of vitamin K2 supplementation on functional vitamin K deficiency in hemodialysis patients: a randomized trial. Am J Kidney Dis. 2012 Feb 1;59(2):186–95.

[45] Caluwé R, Vandecasteele S, Vlem BV, Vermeer C, Vriese ASD. Vitamin K2 supplementation in haemodialysis patients: a randomized dose-finding study. Nephrol Dial Transplant. 2014 Jul 1;29(7):1385–90.

[46] Theuwissen E, Magdeleyns EJ, Braam LAJLM, Teunissen KJ, Knapen MH, Binnekamp IAG, et al. Vitamin K status in healthy volunteers. Food Funct. 2014 Feb;5(2):229–34.

[47] Knapen MHJ, Braam LAJLM, Drummen NE, Bekers O, Hoeks APG, Vermeer C. Menaquinone-7 supplementation improves arterial stiffness in healthy postmenopausal women: double-blind randomised clinical trial. Thromb Haemost. 2015 Feb 19;113(5): 1135–1144.

[48] Pivin E, Ponte B, Pruijm M, Ackermann D, Guessous I, Ehret G, et al. Inactive matrix Gla-protein is associated with arterial stiffness in an adult population-based study. Hypertension. 2015 Jul 1;66(1):85–92.

[49] Kurnatowska I, Grzelak P, Masajtis-Zagajewska A, Kaczmarska M, Stefańczyk L, Vermeer C, et al. Effect of vitamin K2 on progression of atherosclerosis and vascular calcification in nondialyzed patients with chronic kidney disease stages 3-5. Pol Arch Med Wewnętrznej. 2015 Sep 25;125(9):631–40.

[50] Vossen LM, Schurgers LJ, van Varik BJ, Kietselaer BLJH, Vermeer C, Meeder JG, et al. Menaquinone-7 supplementation to reduce vascular calcification in patients with coronary artery disease: rationale and study protocol (VitaK-CAC trial). Nutrients. 2015 Oct 28;7(11):8905–15.

[51] Donaldson MS. Vitamin K: the missing link to prostate health. Med Hypotheses. 2015;84(3):219–22.

[52] Gat Y, Gornish M, Heiblum M, Joshua S. Reversal of benign prostate hyperplasia by selective occlusion of impaired venous drainage in the male reproductive system: novel mechanism, new treatment. Andrologia. 2008;40(5):273–81.

[53] Cario-Toumaniantz C, Boularan C, Schurgers LJ, Heymann M-F, Le Cunff M, Leger J, et al. Identification of differentially expressed genes in human varicose veins: involvement of matrix Gla protein in extracellular matrix remodeling. J Vasc Res. 2007;44(6): 444–59.

[54] Nimptsch K, Rohrmann S, Linseisen J. Dietary intake of vitamin K and risk of prostate cancer in the Heidelberg cohort of the European prospective investigation into cancer and nutrition (EPIC-Heidelberg). Am J Clin Nutr. 2008 Apr;87(4):985–92.

[55] Nimptsch K, Rohrmann S, Nieters A, Linseisen J. Serum undercarboxylated osteocalcin as biomarker of vitamin K intake and risk of prostate cancer: a nested case-control study in the Heidelberg cohort of the European prospective investigation into cancer and nutrition. Cancer Epidemiol Biomarkers Prev. 2009 Jan;18(1):49–56.

[56] Tagalakis V, Tamim H. The effect of warfarin use on clinical stage and histological grade of prostate cancer. Pharmacoepidemiol Drug Saf. 2010 May 1;19(5):436–9.

[57] Esselstyn CB, Gendy G, Doyle J, Golubic M, Roizen MF. A way to reverse CAD? J Fam Pract. 2014 Jul;63(7):356–64b.

[58] Reedy J, Krebs-Smith SM, Miller PE, Liese AD, Kahle LL, Park Y, et al. Higher diet quality is associated with decreased risk of all-cause, cardiovascular disease, and cancer mortality among older adults. J Nutr. 2014 Jun 1;144(6):881–9.

[59] Sijtsma FP, Soedamah-Muthu SS, de Goede J, Griep LMO, Geleijnse JM, Giltay EJ, et al. Healthy eating and lower mortality risk in a large cohort of cardiac patients who received state-of-the-art drug treatment. Am J Clin Nutr. 2015 Dec 1;102(6):1527–33.

[60] Chiuve SE, Fung TT, Rexrode KM, Spiegelman D, Manson JE, Stampfer MJ, et al. Adherence to a low-risk, healthy lifestyle and risk of sudden cardiac death among women. J Am Med Assoc. 2011 Jul 6;306(1):62–9.

[61] Chiuve SE, McCullough ML, Sacks FM, Rimm EB. Healthy lifestyle factors in the primary prevention of coronary heart disease among men: benefits among users and nonusers of lipid-lowering and antihypertensive medications. Circulation. 2006 Jul 11;114(2):160–7.

[62] Ford ES, Bergmann MM, Kröger J, Schienkiewitz A, Weikert C, Boeing H. Healthy living is the best revenge: findings from the European prospective investigation into cancer and nutrition-potsdam study. Arch Intern Med. 2009 Aug 10;169(15):1355–62.

[63] Zhang Y, Tuomilehto J, Jousilahti P, Wang Y, Antikainen R, Hu G. Lifestyle factors on the risks of ischemic and hemorrhagic stroke. Arch Intern Med. 2011 Nov 14;171(20): 1811–8.

[64] Nimptsch K, Rohrmann S, Kaaks R, Linseisen J. Dietary vitamin K intake in relation to cancer incidence and mortality: results from the Heidelberg cohort of the European prospective investigation into cancer and nutrition (EPIC-Heidelberg). Am J Clin Nutr. 2010 Mar 24;91(5):1348–1358.

Chapter 10

Menaquinone-4 Enhances Steroidogenesis in Testis Derived Tumor Cells Via the Elevation of cAMP Level

Hsin-Jung Ho, Hitoshi Shirakawa and Michio Komai

Additional information is available at the end of the chapter

http://dx.doi.org/10.5772/63982

Abstract

Naturally existing vitamin K consists of vitamins K1 and K2. Menaquinone-4 (MK-4), an analog of vitamin K2 and a product of vitamin K1 metabolism, can be detected in several organs, including the testis; however, the function of MK-4 in these tissues has not been well characterized. Recent studies have suggested that vitamin K is involved in enhancing protein kinase A (PKA) activity in several cell types, thus regulating numerous PKA-dependent biological processes. To highlight the effect of vitamin K, we focused on its role in the steroidogenic pathway. Experiments on vitamin K–deficient rats revealed a reduced expression of genes involved in the biosynthesis of cholesterol and steroid hormones in the testis. Moreover, compared with control animals, rats fed on MK-4 diet presented significantly higher testosterone levels in the plasma and testis. These results suggest that vitamin K is involved in the steroidogenic pathway in the testis. Testosterone levels were found to increase in a dose-dependent manner also in cell-based experiments upon addition of MK-4, but such an effect was not observed in vitamin K1 levels. Furthermore, the effect of MK-4 on testosterone production was abolished by the specific PKA inhibitor H89, thus confirming the regulatory role of MK-4 on PKA activation. Here, we describe how MK-4 modulates PKA activation by enhancing intracellular 3′,5′-cyclic adenosine monophosphate (cAMP) levels in testis-derived I-10 cells. The presented evidence supports the role of MK-4 in cAMP/PKA signaling and steroidogenesis.

Keywords: menaquinone-4, steroidogenesis, cAMP, protein kinase A, Leydig cells

1. Introduction

Naturally existing vitamin K comprises vitamin K1 (phylloquinone) and vitamin K2 (menaquinone). Menaquinone-4 (MK-4), an analog of vitamin K2, contains a 2-methyl-1,4-naphtho-

INTECH
open science | open minds

© 2017 The Author(s). Licensee InTech. This chapter is distributed under the terms of the Creative Commons Attribution License (http://creativecommons.org/licenses/by/3.0), which permits unrestricted use, distribution, and reproduction in any medium, provided the original work is properly cited.

quinone ring and a geranylgeranyl group (four isoprene units) as a side chain (**Figure 1**). MK-4 is not commonly synthesized by bacteria and is instead alkylated from menadione, which is supplemented in animal feeds to increase vitamin K levels. In most organs, MK-4 is converted from dietary vitamin K1 and other menaquinones [1, 2] via a process catalyzed by UbiA prenyltransferase domain containing protein 1 [3]. Furthermore, MK-4 is prescribed as a therapeutic agent for osteoporosis and to prevent fractures in Japan [4].

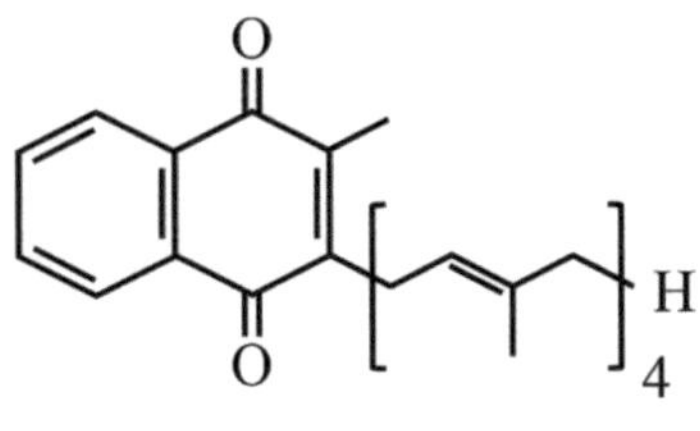

Figure 1. Chemical structure of MK-4. MK-4 has a 2-methyl-1,4-naphthoquinone ring and a geranylgeranyl group (four isoprene units) as a side chain.

Vitamin K is a well-known nutrient required for blood coagulation and bone metabolism. In recent years, novel functions of vitamin K against inflammation [5, 6], tumors [7–9], and ligand of the nuclear receptor PXR (also known as SXR) [10, 11] have been reported. These findings suggest the beneficial role of vitamin K, including MK-4, in several biological processes. In rodents, MK-4 is distributed throughout the body and is observed in high quantity in the liver, bone, brain, pancreas, and reproductive organs, even when animals are fed a low MK-4 diet (**Figure 2**) [12–15]. However, the role of MK-4 in these organs has not been well characterized. This chapter focuses on the functional effects of MK-4 on steroidogenesis in testicular Leydig cells.

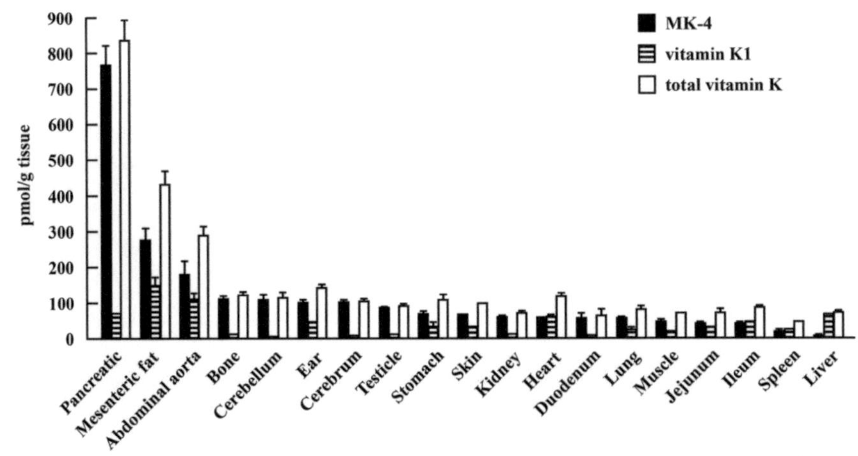

Figure 2. Vitamin K contents in rat tissues. Male Wistar rats were fed an AIN-93G diet for three weeks. Tissue levels of vitamin K and MK-4 were determined by fluorescent HPLC (reproduced with permission from Shirakawa et al. [14]).

http://dx.doi.org/10.5772/63982

2. Steroidogenesis in testicular Leydig cells

The major function of testicular Leydig cells is to produce testosterone in response to the pituitary luteinizing hormone (LH) as shown in **Figure 3**. The LH receptor (LHR) affects the 3′,5′-cyclic adenosine monophosphate (cAMP)/protein kinase A (PKA) signaling pathway by associating with G proteins containing the cytoplasmic Gαs subunit. Production and secretion of testosterone in Leydig cells are tightly regulated by intracellular cAMP, a common secondary messenger. The formation and degradation of intracellular cAMP are under the control of adenylate cyclase (AC) and cyclic nucleotide phosphodiesterase (PDE), respectively. A rise in intracellular cAMP levels activates PKA, which stimulates downstream steroidogenic proteins. Steroidogenic acute regulatory (StAR) protein transports cholesterol to the inner mitochondrial membrane of these cells to initiate steroidogenesis. Cytochrome P450scc (also known as CYP11A), a cholesterol side-chain cleavage enzyme, catalyzes a cascade of reactions that convert cholesterol to the steroid hormone precursor pregnenolone, which is then transformed into testosterone [16]. Both StAR and CYP11A constitute rate-limiting steps in the overall steroidogenesis of testosterone from cholesterol.

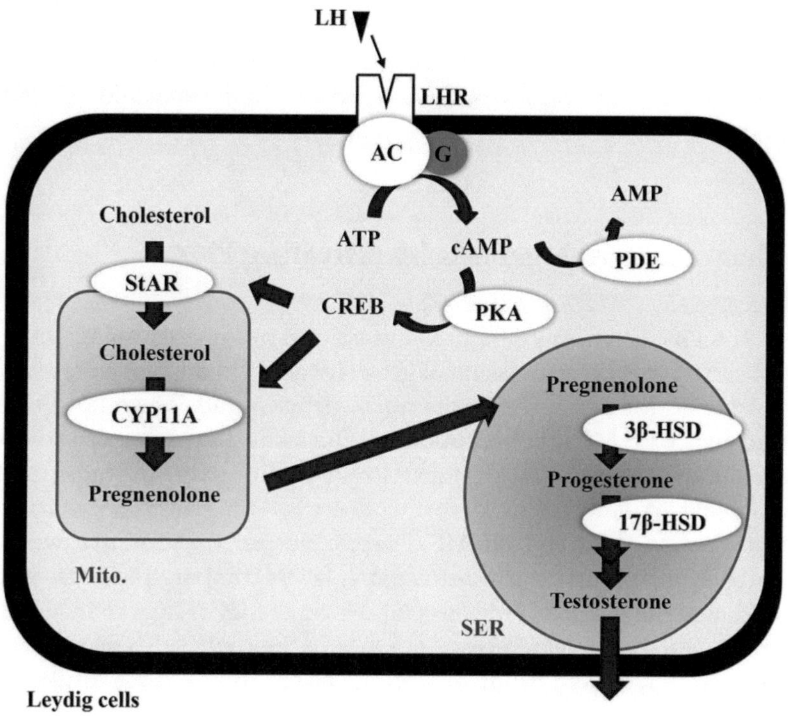

Figure 3. Steroidogenic pathway in testicular Leydig cells. AC, adenylate cyclase; CREB, cAMP response element-binding protein; CYP11A, cholesterol side-chain cleavage enzyme; G, G protein; HSD, hydroxysteroid dehydrogenase; LH, luteinizing hormone; LHR, LH receptor; Mito, mitochondria; PDE, cyclic nucleotide phosphodiesterase; PKA, protein kinase A; SER, smooth endoplasmic reticulum; StAR, steroidogenic acute regulatory protein.

Testosterone production in men is crucial for fetal development, sperm production, and the development of male secondary sex characteristics. Furthermore, increasing evidence shows that lowering testosterone production causes infertility and sexual dysfunction in men, including cases of late-onset hypogonadism (LOH) [17, 18]. The current progressive aging of the population and the stress of modern life may also contribute to LOH, resulting in sexual dysfunction, muscle weakness, and even depression. Moreover, low testosterone levels can also predict the development of several diseases, such as type 2 diabetes and cardiovascular disease [19–21]. This issue could be addressed by identifying nutrients that may boost testosterone levels.

3. Vitamin K modulates the activation of PKA in various cell types

Vitamin K has been shown to affect several biological processes by modulating the activity of PKA in different cell lines. Thus, vitamin K has been reported to enhance nerve growth factor-mediated neurite outgrowth via PKA activation in PC12D pheochromocytoma cells [22]. Vitamin K2 has also been reported to inhibit growth and invasion of hepatocellular carcinoma cells through the activation of PKA, which modulates the activities of several transcriptional factors and inhibits the small GTPase Rho [23]. Furthermore, vitamin K2 has been shown to modulate gene expression in osteoblasts upon activation of PKA [24]. Taken together, these findings indicate that vitamin K plays an important role in the activation of PKA in various cell types.

4. MK-4 enhances steroidogenesis by activating PKA

We confirmed the link between vitamin K and steroid production using DNA microarray analysis. We observed that the expression of genes involved in the biosynthesis of cholesterol and steroid hormones was decreased in vitamin K–deficient rats. The mRNA level of Cyp11a positively correlated with MK-4 concentration in the testis. Moreover, testosterone levels in the plasma and testis of vitamin K–deficient rats were significantly reduced, in spite of normal levels of plasma LH [25]. Another study further described the effects of dietary vitamin K on testosterone production. Rats fed on MK-4-supplemented diet for five weeks presented significantly higher plasma and testis testosterone levels compared to those of control rats, irrespective of changes in plasma LH levels [26]. Moreover, Cyp11a protein levels in the testis were higher in the MK-4-supplemented group than in the control. These results suggest that vitamin K is involved in steroid production in the testis. To link these results with the anti-inflammatory properties of vitamin K observed in the lipopolysaccharide (LPS)-induced models [5, 6], we examined the effects of dietary vitamin K on steroidogenesis in LPS-induced rats [27]. We found that dietary vitamin K intake affected testicular vitamin K concentration and offset the LPS-induced lowering of testosterone synthesis in the testis. In summary, testicular vitamin K plays an important role in steroidogenesis in Leydig cells.

http://dx.doi.org/10.5772/63982

We also found a relationship between vitamin K and steroidogenesis in testis-derived tumor cells. After incubation of two testis-derived cell lines, mouse I-10, and rat R2C, in the presence of MK-4, we detected a dose-dependent rise in secreted testosterone in culture medium [26]. In I-10 cells, MK-4, but not vitamin K1, led to increased levels of testosterone. We also found that menaquinone-3 and menaquinone-7 stimulated testosterone production in I-10 cells (unpublished data). These results suggest that menaquinones, which have an unsaturated isoprenyl side chain, enhance testosterone production. In I-10 cells, the production of progesterone, a testosterone precursor, increased in a dose-dependent manner following treatment with MK-4. This indicates that stimulation of steroidogenesis by MK-4 occurs upstream of progesterone synthesis (**Figure 2**). To assess the effects of MK-4 on the activation of PKA in I-10 cells, a reporter gene assay was employed (**Figure 4**). The cAMP response element-driven luciferase reporter plasmid was transfected into I-10 cells. Accordingly, MK-4 enhanced reporter activity relative to the control. Moreover, Western blot analysis revealed that MK-4 increased the expressions of Cyp11a, as well as phosphorylation levels of PKA and the cAMP response element-binding protein (CREB) in I-10 cells. In contrast, the increase in testosterone level induced by MK-4 was completely abolished by treatment with the PKA inhibitor H89 [26]. These results indicate that, unlike vitamin K1, MK-4 significantly stimulates testosterone production and may play an important role in steroidogenesis.

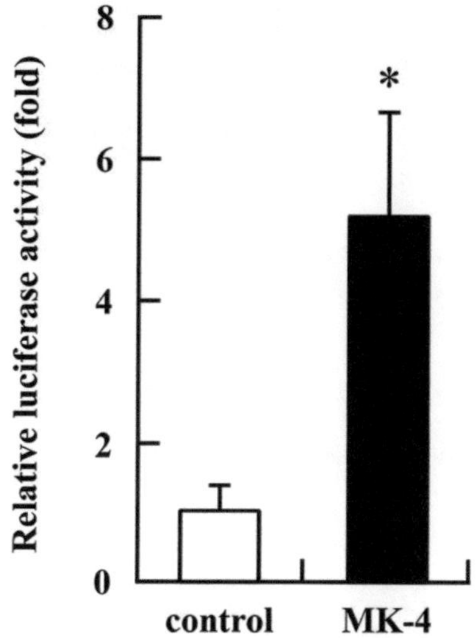

Figure 4. PKA activation following treatment with MK-4 in I-10 cells. I-10 cells were transfected with a plasmid bearing a cAMP response element fused to a luciferase reporter gene and then treated with MK-4.

Our findings were in accordance with previous studies by Ichikawa et al. [24], who showed that MK-4, but not vitamin K1 or other vitamin K2 isoform, specifically induced mRNA levels of GDF15 and STC2, whose protein levels are regulated by PKA in human and mouse osteoblasts. However, Tsang and Kamei have reported that both vitamin K1 and MK-4 promote nerve growth factor-dependent outgrowth of neuronal cells, which were blocked in the presence of a PKA inhibitor [22]. The inconsistencies between these studies may be explained by differences in the uptake, metabolism, and solubility of each vitamin K analog used. In one example, MK-4 was taken up faster than vitamin K1 by MG-63 osteosarcoma cells and HepG2 hepatoma cells by using stable isotope-labeled vitamin K1 and MK-4 [28].

In contrast to a report that MK-4 activated PKA without increasing intracellular levels of cAMP in hepatocellular carcinoma cells [23], we showed that intracellular cAMP levels increased in a dose-dependent manner by treatment with MK-4 for 1.5 h in I-10 cells (**Figure 5**). These results indicated that MK-4 enhances testosterone and progesterone production via activation of PKA as well as modulation of cAMP levels in testis cells.

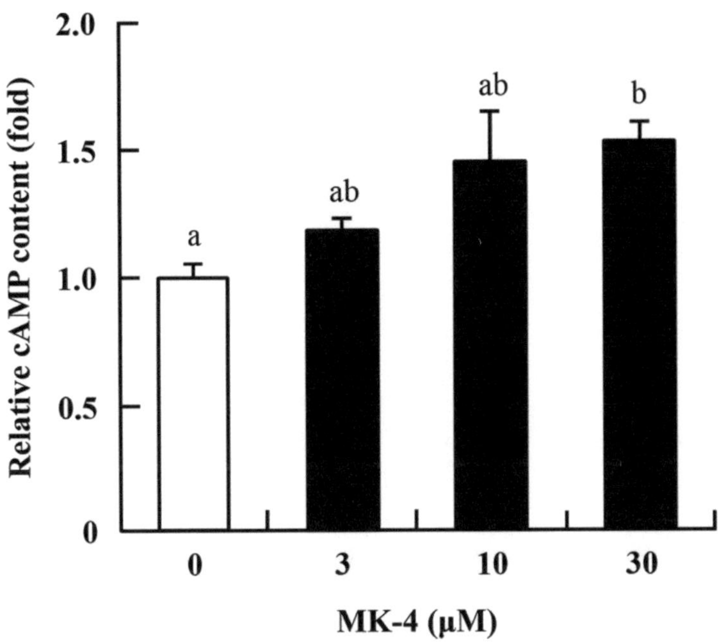

Figure 5. Intracellular cAMP levels in I-10 cells upon MK-4 treatment.

5. The MK-4 structurally related compound, geranylgeraniol

The C20 isoprenoid compound, geranylgeraniol (GGOH, **Figure 6**), is a functional side-chain component (the geranylgeranyl group) of MK-4 and has been shown to have a similar biological function as MK-4. In addition, GGOH may also have anti-tumorigenic effects against prostate cancer [29], colon cancer [30], leukemia [31], as well as anti-inflammatory activity in rats [32]. Our latest study revealed a time- and dose-dependent increase in testosterone and progesterone production in GGOH-treated I-10 cells. As expected, addition of GGOH stimulated also the PKA signaling pathway and augmented intracellular cAMP levels in I-10 cells [33].

http://dx.doi.org/10.5772/63982

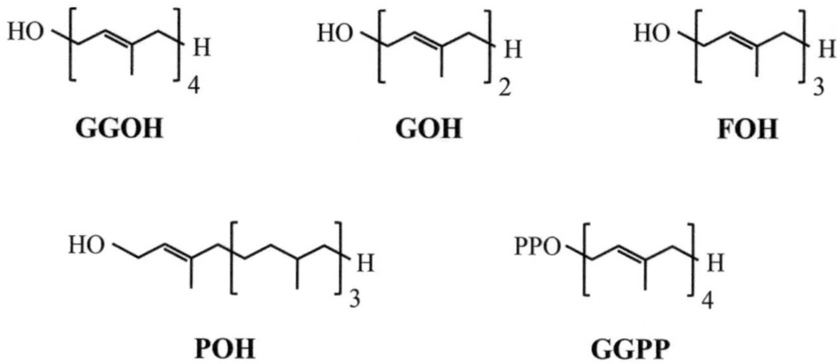

Figure 6. Chemical structures of isoprenoid groups; geranylgeraniol (GGOH), geraniol (GOH), farnesol (FOH), phytol (POH), and geranylgeranyl diphosphate (GGPP).

To further investigate the role of isoprenoids in steroidogenesis, other structurally related isoprenoids, such as geraniol (GOH) and farnesol (FOH) that have two and three isoprene units, respectively, as well as phytol (POH) and geranylgeranyl diphosphate (GGPP), were also examined (**Figure 6**). Accordingly, testosterone and progesterone levels were markedly increased upon treatment with POH and GGPP in I-10 cells. In contrast, FOH increased the levels of progesterone but not testosterone, whereas GOH did not affect steroidogenesis in I-10 cells [33]. These results indicate that most of the tested isoprenoids, and particularly POH, can stimulate the steroidogenic pathway in I-10 cells to the same extent as GGOH.

In summary, the novel role of MK-4 in stimulating steroidogenesis in I-10 cells through regulation of cAMP/PKA signaling may depend on GGOH and other structurally related isoprenoids. In addition, we found that MK-4, but not GGOH, enhanced glucose-stimulated insulin secretion (GSIS) by altering cAMP levels in INS-1 insulinoma cells (unpublished data). Some studies have reported that low testosterone levels could predict the development of type 2 diabetes and cardiovascular disease and have been linked to the increased risk of mortality in men [26–30]. It remains to be established if there is a direct connection between these diseases and vitamin K and what the different functions of MK-4 and GGOH may be. Taken together, these findings provide novel mechanistic insights in the process of steroidogenesis and GSIS and may be useful for the development of therapeutic strategies for men.

Author details

Hsin-Jung Ho, Hitoshi Shirakawa* and Michio Komai

*Address all correspondence to: shirakah@m.tohoku.ac.jp

Laboratory of Nutrition, Graduate School of Agricultural Science, Tohoku University, Sendai, Japan

References

[1] Davidson R.T., Foley A.L., Engelke J.A., Suttie J.W. Conversion of dietary phylloquinone to tissue menaquinone-4 in rats is not dependent on gut bacteria. J. Nutr. 1998;128(2):220–223.

[2] Thijssen H.H., Vervoort L.M., Schurgers L.J., Shearer M.J. Menadione is a metabolite of oral vitamin K. Br. J. Nutr. 2006;95(2):260–266.

[3] Nakagawa K., Hirota Y., Sawada N., Yuge N., Watanabe M., Uchino Y., et al. Identification of UBIAD1 as a novel human menaquinone-4 biosynthetic enzyme. Nature. 2010;468(7320):117–121. doi: 10.1038/nature09464.

[4] Shiraki M., Shiraki Y., Aoki C., Miura M. Vitamin K2 (menatetrenone) effectively prevents fractures and sustains lumbar bone mineral density in osteoporosis. Bone Miner. Res. 2000;15(3):515–521.

[5] Ohsaki Y., Shirakawa H., Hiwatashi K., Furukawa Y., Mizutani T., Komai M. Vitamin K suppresses lipopolysaccharide-induced inflammation in the rat. Biosci. Biotechnol. Biochem. 2006;70(4):926–932.

[6] Ohsaki Y., Shirakawa H., Miura A., Giriwono P.E., Sato S., Ohashi A., et al. Vitamin K suppresses the lipopolysaccharide-induced expression of inflammatory cytokines in cultured macrophage-like cells via the inhibition of the activation of nuclear factor kB through the repression of IKKα/β phosphorylation. J. Nutr. Biochem. 2010;21(11):1120–1126. doi: 10.1016/j.jnutbio.2009.09.011.

[7] Otsuka M., Kato N., Shao R.X., Hoshida Y., Ijichi H., Koike Y., et al. Vitamin K2 inhibits the growth and invasiveness of hepatocellular carcinoma cells via protein kinase A activation. Hepatology. 2004;40(1):243–251.

[8] Tokita H., Tsuchida A., Miyazawa K., Ohyashiki K., Katayanagi S., Sudo H., et al. Vitamin K2-induced antitumor effects via cell-cycle arrest and apoptosis in gastric cancer cell lines. Int. J. Mol. Med. 2006;17(2):235–243.

[9] Yokoyama T., Miyazawa K., Yoshida T., Ohyashiki K. Combination of vitamin K2 plus imatinib mesylate enhances induction of apoptosis in small cell lung cancer cell lines. Int. J. Oncol. 2005;1(26):33–40.

[10] Tabb M.M., Sun A., Zhou C., Grün F., Errandi J., Romero K., et al. Vitamin K2 regulation of bone homeostasis is mediated by the steroid and xenobiotic receptor SXR. J. Biol. Chem. 2003;278(4):43919–43927.

[11] Ichikawa T., Horie-Inoue K., Ikeda K., Blumberg B., Inoue S. Steroid and xenobiotic receptor SXR mediates vitamin K2-activated transcription of extracellular matrix-related genes and collagen accumulation in osteoblastic cells. J. Biol. Chem. 2006; 281(25):16927–16934.

[12] MacCorquodale D.W., Cheney L.C., Binkley S.B., Holcomb W.F., McKee R.W., Thayer Sidney A., et al. The constitution and synthesis of vitamin K1. J. Biol. Chem. 1939;131:357–370.

[13] Binkley S.B., McKee R.W., Thayer Sidney A., Doisy Edward A. The constitution of vitamin K2. J. Biol. Chem. 1940;133:721–729.

[14] Shirakawa H., Katsurai T., Komai M. Conversion of menaquinone-4 in animal organs and it functions. Jpn. Oil Chem. Soc. 2014;14(12):547–553.

[15] Huber A.M., Davidson K.W., O'Brien-Morse M.E., Sadowski J.A. Tissue phylloquinone and menaquinones in rats are affected by age and gender. J. Nutr. 1999;129(5):1039–1044.

[16] Payne A.H., Hardy M.P., editors. The Leydig Cell in Health and Disease. first ed. New Jersey: Human Press; 2007. 476 p. doi: 10.1007/978-1-59745–453-7.

[17] Shores M.M., Matsumoto A.M., Sloan K.L., Kivlahan D.R. Low serum testosterone and mortality in male veterans. Arch. Intern. Med. 2006;166(15):1660–1665.

[18] Huhtaniemi I. Late-onset hypogonadism: current concepts and controversies of pathogenesis, diagnosis and treatment. Asian J. Androl. 2014;16(2):192–202. doi: 10.4103/1008-682X.122336.

[19] Akishita M., Hashimoto M., Ohike Y., Ogawa S., Iijima K., Eto M., Ouchi Y. Low testosterone level as a predictor of cardiovascular events in Japanese men with coronary risk factors. Atherosclerosis. 2010;210(1):232–236. doi: 10.1016/j.atherosclerosis. 2009.10.037.

[20] Morris P.D., Channer K.S. Testosterone and cardiovascular disease in men. Asian J. Androl. 2012;14(3):428–435. doi: 10.1038/aja.2012.21.

[21] Beatrice A.M., Dutta D., Kumar M., Kumbenahalli Siddegowda S., Sinha A., Ray S., et al. Testosterone levels and type 2 diabetes in men: current knowledge and clinical implications. Diabetes Metab. Syndr. Obes. 2014;7:481–486. doi: 10.2147/DMSO.S50777.

[22] Tsang C.K., Kamei Y. Novel effect of vitamin K(1) (phylloquinone) and vitamin K(2) (menaquinone) on promoting nerve growth factor-mediated neurite outgrowth from PC12D cells. Neurosci. Lett. 2002;323(1):9–12.

[23] Otsuka M., Kato N., Shao R.X., Hoshida Y., Ijichi H., Koike Y., et al. Vitamin K2 inhibits the growth and invasiveness of hepatocellular carcinoma cells via protein kinase A activation. Hepatology. 2004;40(1):243–251.

[24] Ichikawa T., Horie-Inoue K., Ikeda K., Blumberg B., Inoue S. Vitamin K2 induces phosphorylation of protein kinase A and expression of novel target genes in osteoblastic cells. J. Mol. Endocrinol. 2007;39(4):239–247.

[25] Shirakawa H., Ohsaki Y., Minegishi Y., Takumi N., Ohinata K., Furukawa Y., et al. Vitamin K deficiency reduces testosterone production in the testis through down-

regulation of the Cyp11a a cholesterol side chain cleavage enzyme in rats. Biochim. Biophys. Acta 2006;1760(10):1482–1488.

[26] Ito A., Shirakawa H., Takumi N., Minegishi Y., Ohashi A., Howlader Z.H., et al. Menaquinone-4 enhances testosterone production in rats and testis-derived tumor cells. Lipids Health Dis. 2011;10:158.

[27] Takumi N., Shirakawa H., Ohsaki Y., Ito A., Watanabe T., Giriwono P.E., et al. Dietary vitamin K alleviates the reduction in testosterone production induced by lipopolysaccharide administration in rat testis. Food Funct. 2011;2(7):406–411. doi: 10.1039/c1fo10058k.

[28] Suhara Y., Hirota Y., Nakagawa K., Kamao M., Tsugawa N., Okano T. Design and synthesis of biologically active analogues of vitamin K2: evaluation of their biological activities with cultured human cell lines. Bioorg. Med. Chem. 2008;16(6):3108–3117.

[29] Fernandes N.V., Yeganehjoo H., Katuru R., DeBose-Boyd R.A., Morris L.L., Michon R. Geranylgeraniol suppresses the viability of human DU145 prostate carcinoma cells and the level of HMG CoA reductase. Exp. Biol. Med. (Maywood). 2013;238(11):1265–1274.

[30] Ohizumi H., Masuda Y., Nakajo S., Sakai I., Ohsawa S., Nakaya K. Geranylgeraniol is a potent inducer of apoptosis in tumor cells. J. Biochem. 1995;117(1):11–13.

[31] Masuda Y., Maeda S., Watanabe A., Sano Y., Aiuchi T., Nakajo S., et al. A novel 21-kDa cytochrome c-releasing factor is generated upon treatment of human leukemia U937 cells with geranylgeraniol. Biochem. Biophys. Res. Commun. 2006;346(2):454–460.

[32] Giriwono P.E., Shirakawa H., Ohsaki Y., Hata S., Kuriyama H., Sato S., et al. Dietary supplementation with geranylgeraniol suppresses lipopolysaccharide-induced inflammation via inhibition of nuclear factor-kB activation in rats. Eur. J. Nutr. 2013;52(3):1191–1199. doi: 10.1007/s00394-012-0429-y.

[33] Ho H.J., Shirakawa H., Yoshida R., Ito A., Maeda M., Goto T., et al. Geranylgeraniol enhances testosterone production via the cAMP/protein kinase A pathway in testis-derived I-10 tumor cells. Biosci. Biotechnol. Biochem. 2016;80(4):691-697. doi: 10.1080/09168451.2015.1123612.

Chapter 11

Vitamin K2 Facilitating Inter-Organ Cross-Talk

Jan O. Gordeladze, Håvard J. Haugen,
Gaute Floer Johnsen and Mona Møller

Additional information is available at the end of the chapter

http://dx.doi.org/10.5772/67153

Abstract

This chapter features how vitamin K2 is instrumental in bringing about inter-organ communication, thus facilitating (a) a synthesis/secretion of the endocrine, humoral factors from various organs and (b) physiological responses to the said factors by a multitude of organ systems of the body, thus creating a 'lattice' of reciprocal regulatory loops in order to ensure endocrine homeostasis.

Keywords: vitamin K2, MK-4, MK-7, PXR/SXR, FoxO and FoxA families of transcription factors, PI3K/Akt cascade, endocrine homeostasis, deiodinase (DIO_2), NF-κB, interleukins, IGFs, MMPs, FGFs, irisin, osteonectin, UCP1, VEGF, GIP/GLP

1. Introduction

Recently, an abstract entitled 'Epigenetic factors involved in musculo-skeletal interaction - How skeletal muscle cells and osteoblasts derived from stem cells communicate with special reference to histone deacetylases (HDACS), transcription factors (TFs), microRNAs, and vitamin K2' was presented at an OMICS conference in Chicago, USA [1].

This work shows how the interactive axis consisting of 'Epigenator-Initiator-Maintainer' components determines the ultimate phenotypes of the cellular functions or the chain of reactions within different body organs and tissues. The initiator signals (i.e. histone modifications, phenotype modifications through transcriptional control, via microRNA species) constitute forces, slowly tilting the cell phenotype towards a more or less stable profile.

Despite this tendency, phenotypic characteristics could be subjected to alterations, that is, either weakened or reinforced, or even altered, resulting from underlying or developing diseases and/or gene therapy. The present work encompasses some data showing manipulations

INTECH
open science | open minds

© 2017 The Author(s). Licensee InTech. This chapter is distributed under the terms of the Creative Commons Attribution License (http://creativecommons.org/licenses/by/3.0), which permits unrestricted use, distribution, and reproduction in any medium, provided the original work is properly cited. (cc) BY

of HDACs, transcription factors (TFs), as well as microRNAs and the impact of vitamin K2 (MK-7) on mineralizing cells (osteoblasts) and striated muscle cells exposed to either normal growth conditions or mediators of inflammation (i.e. Th-cells, macrophages, or interleukins), indicating that it is possible to engineer cells displaying an adapted phenotype where: (a) towards mineralization is reinforced, (b) untoward mineral deposition is halted and finally (c) mutual musculoskeletal interactions are 'reinforced'.

With the aid of various algorithms, one may reveal regulatory loops involving both TFs and microRNAs. The subjects TFs and microRNA species appeared to be part of an intricate hierarchical structure encompassing several classes of HDACs, including the Sirtuins, known to respond to cellular energy status (i.e. NADH/NAD^+ ratios). Finally, it was demonstrated that vitamin K2 (MK-7, via binding to the transcription factor SXR) interfered with a plethora of signalling pathways (such as the FoxA and FoxO families of transcription factors), the downstream of the signalling mechanisms represented by the PI3-kinase system (i.e. Akt/PKB and SGK, respectively), thus potentiating the cross-talk signals or suppressing the mineralizing character. It was concluded that vitamin K2 plays a pivotal role by optimizing the endocrine interaction between osteoblasts and striated muscle cells, facilitating a 'win-win' situation. Furthermore, we have shown that vitamin K2 may confer the ability for cross-talk between striated muscle cells and bones to include cells, such as insulin-producing β-cells, thyroid follicular cells, PTH-producing parathyroid cells and hepatocytes, in the absence or presence of inflammatory cells or their secreted cytokines/interleukins ± TNFα.

Sarcopenia (reduced muscle mass and/or function) and osteoporosis (bone brittleness) have generally been known for their relations to the locomotive syndrome and are linked to old age. Contrastingly, an increased muscle mass correlates with an enhanced bone mass and thus with a reduced fracture incidence. Genetic, as well as endocrine and mechanical factors, inflammation and nutritional states concurrently impinge on muscle tissue and bone metabolism.

Furthermore, a plethora of genes like myostatin and α-actinin-3 associate with both conditions. Factors such as vitamin D, growth hormones (like GH and IGF-1) and testosterone and pathological conditions with excess cortisol, as well as type I diabetes (T1DM), affect both muscle and bone tissues. It was shown that the genes Tmem119, osteoglycin and FAM5C may be critical for the commitment of myoprogenitor cells to the osteoblast lineage. Furthermore, osteoglycin and FAM5C might serve as muscle-derived humoral osteogenic factors. Others, encompassing myostatin, osteonectin, as well as IGF-1, irisin and osteocalcin, may also be associated with reciprocal metabolic interactions between muscle and bone [2].

2. Genetic factors and muscle/bone phenotypes

Genes such as myostatin, α-actinin-3, proliferator-activated receptor gamma coactivator 1-alpha (PGC-1α) and myocyte enhancer factor 2C (MEF-2C) are included in GWAS (genome-wide association study) as believed to be involved in a concurrent loss of muscle and bone tissue [3]. Myostatin, on the other hand, has been shown to be a negative regulator of muscle mass. Alpha-actinin-3 has been demonstrated to be abundantly expressed in fast-twitch skeletal muscle fibres and may also affect their differentiation towards fast-twitch fibres. Finally, it was shown [4] that a lack of α-actinin-3 may lead to a reduction in bone mineral density (BMD) in both humans and rodents.

PGC-1α seems to be instrumental in the modulation of mitochondrial biogenesis [5], and a further study established that PGC-1α elicited by physical activity seems to be crucial for oxidative metabolism in skeletal muscle fibres [6]. Furthermore, it was demonstrated that mitochondrial biogenesis induced by an enhancement in PGC-1α levels facilitates Wnt-mediated induction of osteoblastic differentiation of mesenchymal C3H10T1/2 cells [7]. These findings indicate that PGC-1α serves as a 'commitment' factor or inductor of stem cells to produce osteoblastic cells. Another essential factor is MEF-2C, which interacts with other myogenic regulatory factors, like Myf5 and MyoD, which in a synergistic fashion activates specific muscle-phenotypic genes. Animals devoid of MEF-2C in osteocytes make less sclerostin, a humoral factor acting as an inhibitor of the Wnt family of signalling molecules involved in osteoblast differentiation and bone formation. Thus, the MEF-2C-sclerostin signalling inhibits the formation of excessive bone mass and a 'healthy' turnover, which normally ensures minimal bone brittleness.

Qiu et al. [8] demonstrates that NF-κb-mediated signalling modulates myostatin transcription in myoblasts during cirrhosis-induced hyperammonaemia. This suggests that NF-κB antagonists are useful to reverse cirrhosis induced by sarcopenia. This observation also indicates that vitamin K2-induced modulation of NF-κB may determine the levels of humoral factors, which reciprocally regulate muscle and bone physiology. We have found [Gordeladze et al., 2015, unpublished] that preadipocytes, with mutated and superactive Gsα-induced adenylate cyclase activity, in the presence of vitamin K2 (MK-7), produce more beige-like adipocytes (see **Figure 1**) than large white adipocytes (ref), with an enhancement in PGC-1α levels (ref). Hence, it may be asserted that vitamin K2 facilitates Wnt-mediated induction of osteoblastic differentiation by

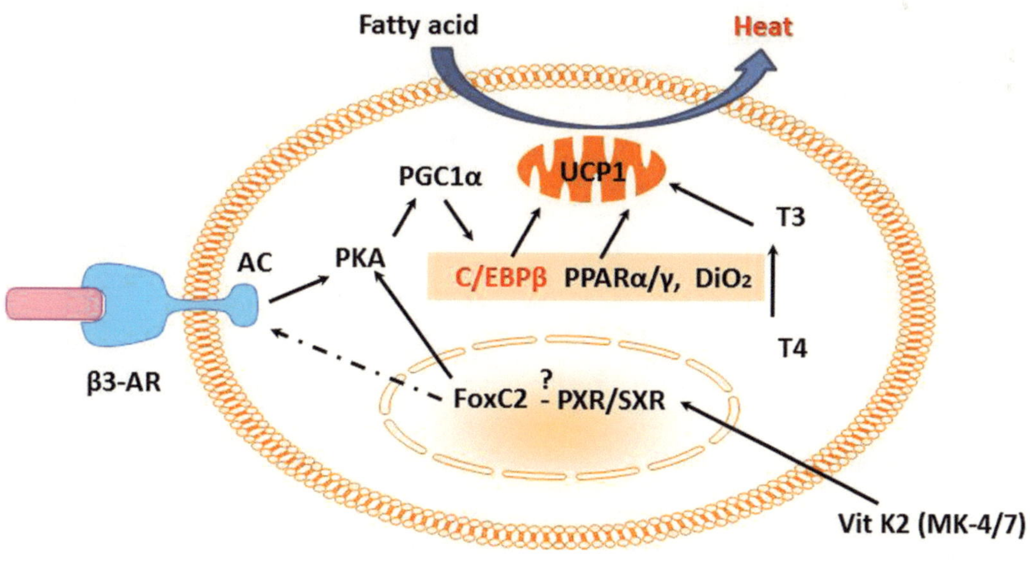

Figure 1. Putative working model showing how vitamin K2 may affect the hormonal signalling systems and transcription factors responsible for the transition of «white» adipocytes to «beige» adipocytes, thus blocking fat deposition and enhancing the production of heat from fatty acids.

enhancing the β-adrenoceptor and PKA-mediated signalling through PGC-1α of mesenchymal cells/stem cells, in order to fortify metabolic mechanisms 'ruled' by c/EBPβ, PPARα/γ and DiO_2.

However, the interaction between striated muscle cells and bones is illustrated in a better way in **Figure 2**. Here, it is shown that striated muscle cells communicate with 'the environment', consisting of other organs, such as white adipose tissue, liver, pancreas and bones through a multitude of endocrine/hormonal factors (see **Figure 2**(left)). However, if we just look closer at the reciprocal interactions between striated muscle cells and bone cells (osteoblasts, osteocytes and osteoclasts), there are still a large number of humoral factors, such as IGF-1, myostatin, osteoglycin, FAM5C, Irisin, Osteonectin, FGF2, IL-6, IL-7, IL-15, MMP-2, Sclerostin, Osteocalcin, MGP, VEGF and HGF (see **Figure 2**(right)), which 'capture' the two organs in a reciprocal regulatory 'looping system'. We have recently shown, for most part, the 'cross-talk' exchanged between these organs, particularly within the muscle-bone axis.

Vitamin K2 serves as a 'coupling agent', fine-tuning muscle-bone interactions, while concomitantly preserving its precision and strengthening it in the presence of inflammatory interleukins and INFα, and/or Th-1 and Th-17 cells.

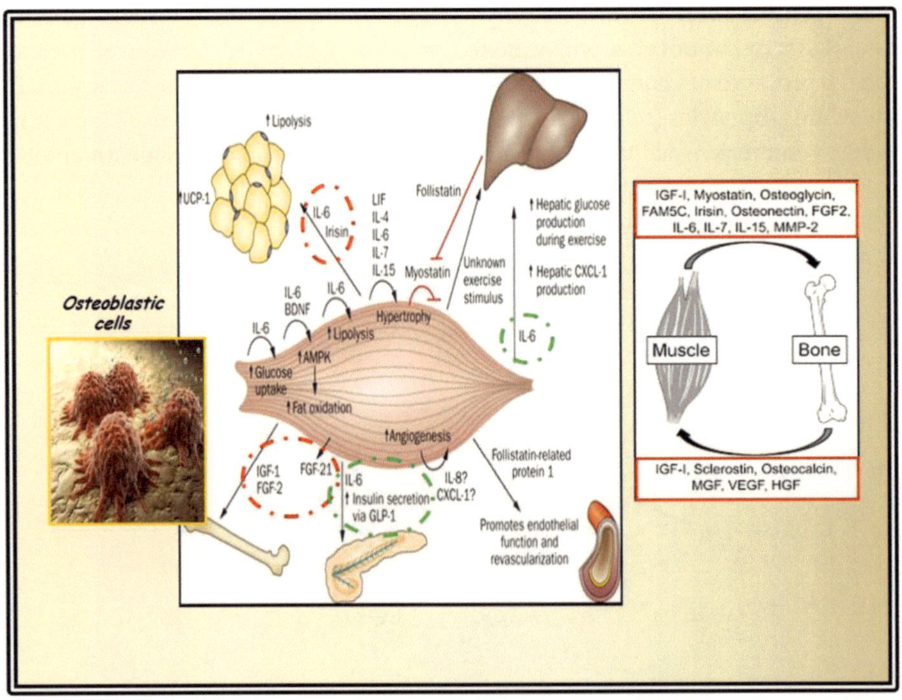

Figure 2. Endocrine communication between striated muscle cells and other organ systems, such as white adipose tissue, liver, bone and pancreas. Pay special attention to the plethora of hormones/cytokines being exchanged between the muscle and the bone. Ref.: Left: Benatti, F. B. & Pedersen, B. K. Nat. Rev. Rheumatol. 11, 86–97 (2015); published online 25 November 2014; doi:10.1038/nrrheum.2014.193. Right: Kawao N, Kaji H. J Cell Biochem. 2015 May;116(5):687–95. doi: 10.1002/jcb.25040. PMID: 25521430.

Since most of our cells throughout the body express the transcription factor PXR/SXR, binding vitamin K2, it may be asserted that vitamin K2 not only affects the phenotype of muscle and bone cells, shown here to interact in a reciprocal endocrine fashion but also (a) plays a major role in the determination and stabilization of the phenotype of a plethora of specialized cells in our body and (b) plays a pivotal role in the reciprocal interaction of various organ systems in our body to ensure optimal organ functions ('inter-organ cross-talk'). An interesting and elegant article written by Lara Pizzorno (see reference on last page) underscores the different effects of vitamin K2 in a comprehensive manner, supporting the notion that vitamin K2 is an essential biological factor supporting disease-free old age, which may be construed as if vitamin K2 is one important alimentary ingredient ensuring 'longevity'.

Longevity requires, of course, optimized and 'healthy' organ functioning throughout the body. Therefore, it is mandatory for the different organs of the body to communicate with each other and together form a 'cyn-organic' lattice where each organ communicates with the most part or all the others. Vitamin K2 may be one factor contributing to this inter-organ 'cross-talk', and there are several ways this little molecule exerts its integrative power. In this respect, the present book's chapter entitled 'Vitamin K2—small molecule with a large biological impact' featuring the molecular mechanisms, by which K2 exerts its actions, describes in detail how $DIO_{1,2,3}$ impacts the regulation of cell proliferation, lipogenesis, lipolysis, cholesterol metabolism, carbohydrate metabolism, muscle contraction, thermogenesis, cell communication, exocytosis, cell cycle regulation and growth regulation. Of particular interest is the type 2 deiodinase,(DIO_2), which, via higher brain centres, the pancreas, striated muscle cells, the liver as well as white and brown adipose tissues, converts T4 to its active form, T3, ensuring an integrated metabolic and hormonal homeostasis and 'steady state' or endocrine equilibrium between the different organ systems of the body. In this context, we have shown (ref) that vitamin K2 is able to sustain the cell phenotypes of different organ systems, such as bones, striated muscles and others in the presence of sub-chronical inflammation, as induced by the presence of either Th-1 cells, Th-17 cells, as well as a mixture of TNFα or inflammatory interleukins (e.g. IL-1 and IL-17).

Others have more directly proven that vitamin K2, via binding to PXR, affects both triglyceride turnover and gluconeogenesis in the liver (ref). The authors of this chapter describe how the MK-PXR complex via CD36, CPTA1 and SCD1 stimulates ketogenesis and hampers triglyceride production. Furthermore, they also show how the MK-PXR complex, via a cluster of transcription factors (FoxO1, CREB, PGC-1α and HNF4), stimulates the enzymes PEPCK1 and G6Pase in order to facilitate the metabolic conversion of lactate and amino acids to glucose (i.e. gluconeogenesis). This mechanism is remarkably like the one sketched in **Figure 1**, only that there are other members of the Fox family of transcription factors involved!

3. Novel findings related to the biology of vitamin K2

A summary of a literature survey, extracting articles from PubMed, featuring new research on the biological impact of vitamin K2 published in 2015 and 2016, gave the following results:

a. Shetty A et al. Urol Oncol. 2016 Sep 27. pii: S1078-1439(16)30108-9. doi: 10.1016/j.urolonc.2016.05.027. **Hepatoma-derived growth factor: A survival-related protein in prostate oncogenesis and a potential target for vitamin K2.**

Excerpt from the article: Hepatoma-derived growth factor (HDGF) is correlated with a poor prognosis in cancers. Forced overexpression of HDGF sustained viability of RWPE-1 human prostate cancer cells, while a knockdown of HDGF hampered their proliferation. Furthermore, using a HDGF antibody along with vitamin K2 reduced cell proliferation and inhibited NF-κB expression.

b. Sanguineti R et al. J Biol Regul Homeost Agents. 2016 Jul-Sep;30(3):713–726. **Vitamins D3 and K2 may partially counterbalance the detrimental effects of pentosidine in ex vivo human osteoblasts.**

Excerpt from the article: Osteoporosis is a multifaceted metabolic disorder, construed as inadequate bone strength. It was recently demonstrated that advanced glycation end products (AGEs) (e.g. pentosidine = PENT) may serve as indicators of senile bone brittleness. Optimal responses to hormones, like 1,25-dihydroxyvitamin D_3, serve as prerequisites for optimal, time-dependent osteoblast functioning (the oscillation between bone building and bone resorption phases). Vitamin K2 (MK-4/7) may enhance net vitamin D-induced bone formation.

Ex vivo human osteoblasts, incubated with vitamins D_3 and K2 and exposed to PENT for 72 hours, were assessed for gene expression, that is, ALP, COLIα1 and osteocalcin, as well as the RANKL/OPG ratio to assess net bone formation (as a result of the continuous remodelling process). The expression of RAGE, which is a well-characterized receptor of advanced glycation end products (AGEs), was also monitored. PENT + vitamins significantly inhibited ALP expression/secretion but did not show any impact on gene expression, which may be construed as hampered functional osteoblastic activity. Furthermore, PENT + vitamins enhanced gene expression of collagen, while protein secretion remained unchanged. Intracellular levels of collagen were partially lowered, while a fall in BGP gene expression, as well as intracellular protein levels, was seen with PENT exposure. The ratio RANKL/OPG was augmented, favouring bone resorption. Expression of the RAGE gene was, however, lowered, indicating that the detrimental AGE accumulation in the bone was attenuated and/or counteracted by the vitamins' D_3 and K2 exposure.

c. Ronn SH et al. Eur J Endocrinol. 2016 Sep 13. pii: EJE-16-0498. [Epub ahead of print] PLoS One. 2016 Aug 29;11(8):e0161886. doi: 10.1371/journal.pone.0161886. eCollection 2016. **Vitamin K2 (menaquinone-7) prevents age-related deterioration of trabecular bone microarchitecture at the tibia in postmenopausal women.**

Excerpt from the article: Vitamin K2 allegedly protects one against bone loss and fractures; however, its effect on bone quality has hitherto not been investigated. This article explores the effect of MK-7 on undercarboxylated osteocalcin (ucOC), bone mass, as well as bone quality. A randomized, 1-year, placebo-controlled, double-blind clinical trial was conducted, where the effect of some 400 μg of MK-7 (including supplementation with calcium and vitamin D) was assessed.

UcOC was diminished in the MK-7 group as compared with the placebo group subsequent to 3 months of treatment. Furthermore, the trabecular number in tibia was unchanged in the MK-7 group, while lowered in the placebo group. The trabecular spacing remained unchanged in the MK-7 group, while it increased in the placebo group. Finally, trabecular thickness stayed unchanged in the MK-7 group, while it increased in the placebo group. And one could not find significant differences between the groups at the radius or in BMD at any site.

The alterations in microarchitecture of bone tissue within the placebo group were in line with the age-related deteriorations of the trabecular bone structure, including trabecula loss, with a concomitant enhanced average thickness of the ones remaining. This may indicate that MK-7 sustains proper tibial trabecular bone microstructures.

d. Duan F et al. PLoS One. 2016 Aug 29;11(8):e0161886. doi: 10.1371/journal.pone. 0161886. eCollection 2016. **Vitamin K2 induces mitochondria-related apoptosis in human bladder cancer cells via ROS and JNK/p38 MAPK signal pathways**.

Excerpt from the article: Vitamin K2's impact on apoptosis in cancerous cells has been elucidated in many studies. In the present study, it was shown that vitamin K2 induced apoptosis in cancer cells of the bladder through mitochondrial pathways, that is (a) loss of mitochondrial membrane potential and (b) release of cytochrome C and the activation of the caspase-3 cascade. Additionally, c-Jun N-terminal kinase (JNK) and p38 MAPK phosphorylation were enhanced in vitamin K2-treated cells. The generation of reactive oxygen species (ROS) was detected in bladder cancer cells; however, the treatment with vitamin K2 and the antioxidant N-acetyl cysteine (NAC) blocked (a) the vitamin K2-triggered apoptosis, (b) the loss of mitochondria membrane potential and (c) the activation of JNK and p38 MAPK.

Hence, it seems that vitamin K2 brings about apoptosis in bladder cancer cells through: (a) ROS-mediated JNK/p38 MAPK and (b) mitochondrial pathways.

e. Yu YX et al. Acta Pharmacol Sin. 2016 Sep;37(9):1178–89. doi: 10.1038/aps. 2016.68. Epub 2016 Aug 8. **Vitamin K2 suppresses rotenone-induced microglial activation in vitro**.

Excerpt from the article: It has been shown that environmental factors (e.g. rotenone and others) cause neuroinflammation, contributing to the development of Parkinson's disease. Here, one explored molecular mechanism, pertaining to the repression by vitamin K2 (MK-4), on in vitro rotenone induced microglial activation.

The cell line (BV2) was exposed to rotenone in the presence or absence of MK-4. The levels of TNF-α or IL-1β were assessed, using ELISA technology. BV2 cells treated with rotenone with or without MK-4 were subjected to analyses of (a) the mitochondrial membrane potential, (b) ROS production, (c) immunofluorescence or (d) immunoblot assays. The neuroblastoma cells were exposed to conditioned media of BV2 cells, having been exposed to rotenone ± MK-4, and cell viability was assessed.

In rotenone-treated BV2 cells, MK-4 dose dependently counteracted upregulation of iNOS and COX-2 expression in the cells, while also the production of TNF-α and IL-1β. MK-4 significantly blocked rotenone-induced translocation to the nucleus of NF-κB. MK-4 significantly also attenuated rotenone-provoked p38 activation, ROS production and caspase-1 activation.

Vitamin K2 can directly suppress rotenone-induced activation of microglial BV2 cells in vitro by repressing ROS production and p38 activation.

f. Vissers LE et al. Atherosclerosis. 2016 Sep;252:15–20. doi: 10.1016/j.atherosclerosis. 2016.07.915. Epub 2016 Jul 25. **The relationship between vitamin K and peripheral arterial disease (PAD).**

Excerpt from the article: Vitamin K1 (phylloquinone) and vitamin K2 (menaquinones) are believed to diminish the risk of cardiovascular diseases by reducing or blocking calcification of the vascular bed.

The association between the intake of vitamins K1 and K2 with PAD was analysed in a prospective cohort of some 36,500 participants. The occurrence of PAD was obtained from national registries. Baseline intakes of K1 and K2s were calculated using standard food-frequency questionnaires.

During 12 years of follow-up, some 500 cases of PAD were documented. Menaquinone (K2) intake was associated with a reduced risk of PAD. A stronger association was observed in participants suffering from either hypertension or diabetes, respectively. Phylloquinone (K1) intake was not associated with PAD risk.

g. Villa, JK et al. Crit Rev Food Sci Nutr. 2016 Jul 20:0. [Epub ahead of print]. **Effect of vitamin K in bone metabolism and vascular calcification: a review of mechanisms of action and evidences.**

Excerpt from the article: Osteoporosis is associated with a public health concern, featuring the enhanced risk of incurring fractured bones, as well as vascular calcification. The family of vitamin K represents unique benefits on these issues, even though its aspects are far from exhaustively studied. The two important forms of vitamin K are phylloquinone = vitamin K1 and menaquinone = vitamin K2 (MK-4 and MK-7). In the present work, we investigated, in particular, the effect of vitamin K2 in bones and blood vessels. In addition to the known effects of vitamin K2, this particular form has been demonstrated to support bone formation via stimulating osteoblast differentiation, as well as osteocalcin carboxylation and enhanced ALP activity, IGF-1, as well as other growth-promoting protein (DF-15 and STCal-2) levels.

Furthermore, vitamin K2 lowers the osteoblast levels of proapoptotic proteins like Fas and Bax, while also hampering osteoclast differentiation via enhanced levels of osteoprotegerin (OPG) and reducing the amount of 'receptor activator of nuclear factor kappa-B ligand' NF-κB. In blood vessels, vitamin K2 diminishes the production of hydroxyapatite by carboxylating matrix-Gla protein, as well as the Gla-rich protein. Vitamin K2 hampers the death (apoptosis) of smooth vascular muscle cells, as well as reducing phenotype alteration of vascular smooth muscle cells to osteoblastic cells. The standard dosage of vitamin K2 studies of man amounts to 45 mg/day, which allows beneficial effects on bone and vasculature, especially in post-menopausal women with osteoporosis.

h. Ochsner et al., Endocrinol. 2016 Aug;30(8):937–48. doi: 10.1210/me.2016-1095. Epub 2016 Jul 13. **A reference transcriptome for constitutive androstane receptor and pregnane X receptor xenobiotic signaling.**

Excerpt from the article: PXR = PXR/NR1I3 and constitutive androstane receptor (CAR = CAR/NR1I2) both belong to the nuclear receptor (NR) superfamily of ligand-regulated TF2.

They are both mediators of endocrine-disrupting chemical signalling. In this work a "reference transcriptome" was generated, a 'reference transcriptome' encompassing members of frequently and differentially expressed of genes across some 160 experiments compiled from some 20 datasets, describing perturbations of both CAR- and PXR-based signalling pathways. Omitting the genes encoding members of the xenobiotic 'stress response', the ranking of genes especially involved in the metabolism of carbohydrate sheds light on the role of xenobiotics and thus vitamin K2, in the metabolic syndrome.

i. Zhang Y et al. Int J Biol Sci. 2016 Apr 28;12(7):776–85. doi: 10.7150/ijbs.15248. eCollection 2016. **Vitamin K2 ameliorates damage of blood vessels by glucocorticoid: a potential mechanism for its protective effects in glucocorticoid-induced osteonecrosis of the femoral head in a rat model.**

Excerpt from the article: Glucocorticoids have been reported to lower the blood vessels' number and also lower the blood supply in the femoral head. This is known to serve as an important mechanism, by which glucocorticoids induce osteonecrosis of the femoral head (ONFH). In order to prevent manifest drug-induced ONFH, bone formation with concomitant angiogenesis would be a beneficial treatment.

The present study investigates whether vitamin K2 could stimulate the formation of new blood vessels in the presence of glucocorticoids, both in vitro and in vivo. The effect of vitamin K2 on parameters, such as viability, migration, in vitro tube formation and key genes (like VEGF and PDGFB) incubated with or without dexamethasone were elucidated. VEGF, TGF-β and BMP-2, angiogenesis-related proteins secreted by osteoblastic cells (MG63 cell batches), were also detected.

Additionally, blood vessels in the femoral head of rats (given with or without a glucocorticoid and vitamin K2) were assessed. It showed that vitamin K2 fully protected the endothelial cells used from apoptosis, induced by dexamethasone. Furthermore, endothelial cell migration and in vitro tube formation were promoted. Furthermore, angiogenesis-related proteins were also stimulated by vitamin K2. In vivo studies revealed enhanced blood vessel volume with CD31-positive cells in rats, co-treated with vitamin K2.

In general, vitamin K2 demonstrated an ability to promote and sustain angiogenesis *in vitro* and to ameliorate/rescue vessels of the femoral head in glucocorticoid-treated rats *in vivo*. This indicates that vitamin K2 holds a role as a promising drug to be used in the prevention of steroid-induced ONFH.

j. Mi et al., Int J Exp Pathol. 2016 Apr;97(2):187–93. doi: 10.1111/iep.12178. Epub 2016 Jun 3. **Establishing a rat model for the study of vitamin K deficiency.**

Excerpt from the article: The main vitamin K-deficient model (i.e. the minidose warfarin exposure) differs from the pathological model of vitamin K deficiency, being a shortage of vitamin K. The present work aimed to establish a new method to provoke vitamin K deficiency in rats by combining vitamin K-deficient diets with intragastrical administration of gentamicin. In the diet- and gentamicin-provoked vitamin K-deficient animals, all rats suffered hepatic vitamin K1 and K2 loss and consequently an extended period of APTT. And, within the 21-day treatment group, one could also measure prolonged PT, as well as the decreasing FIX activities. Furthermore, in the 28-day treatment group, unmeasurable vitamin K1 and K2, enhanced PT and APTT values, as well as a decrease in FII, FVII, FIX and FX activities sustained the hypothesis that serious vitamin K deficiencies in the experimental animals had taken place. It was therefore hypothesized that the diet- and gentamicin-induced vitamin K-deficient model system could be used to study vitamin K's status. This vitamin K-deficient 28-day model could therefore be applied to research related to both the status of vitamin K and vitamin K-dependent coagulation per se. Hence, it was asserted that the combination of a vitamin K-deficient diet with the administration of gentamicin yields an ideal model to study vitamin K deficiency.

k. Okuyama et al. 2016;98(3–4):134–70. doi: 10.1159/000446704. Epub 2016 Jun 2. **Medicines and vegetable oils as hidden causes of cardiovascular disease and diabetes**.

Excerpt from the article: A coupling has been observed between cardiovascular diseases (CVD) and diabetes mellitus type 2 (T2DM); however, a causal relationship has hitherto not been made. Irrespective of aetiology, cholesterol-lowering medication (statins) has been recommended for both patient categories.

Statin-induced suppression of prenyl moieties in the pathway of cholesterol biosynthesis has been associated with both atherosclerosis and heart failure. However, both vegetable and hydrogenated oils appear to shorten the survival period incurred by stroke-prone spontaneously hypertensive animals through (a) downregulated platelet number, (b) enhanced haemorrhagic incidences and (c) aberrant kidney functioning. These phenomena are not related to their molecular constituents, and the present oils and drugs (i.e. statins and warfarin) share common mechanisms blocking vitamin K2-dependent biochemical processes.

l. Zhou C. Biochim Biophys Acta. 2016 Sep;1859(9):1112–20. doi: 10.1016/j.bbagrm.2016.02.015. Epub 2016 Feb 26. **Novel functions of PXR in cardiometabolic disease**.

Excerpt from article: Cardiometabolic disease is a worldwide epidemic, and our chemical environment is rapidly changing to the worse, resulting in an enhanced frequency and morbidity of chronic human diseases. But, mechanisms of how exposure to chemical compounds brings about cardiometabolic ailments are not well understood. A plethora of chemicals have, however, been shown to serve as activators of the pregnane X receptor (PXR), which, apart from serving as a nuclear receptor functioning as a xenobiotic sensor, also binds vitamin K2.

Recent investigations have unravelled new and intriguing roles of PXR in modulating potentially debilitating conditions such as obesity, insulin sensitivity, lipid homeostasis, atherogenesis and vascular functioning. The present reports indicate that PXR signalling significantly impacts pathophysiological conditions like cardiometabolic diseases in humans. This discovery of 'new' effects of PXR in the mentioned disease complexes not only heavily contributes to our comprehension of 'gene versus environment interactions', predisposing man to chronic disabilities, but also sheds light on future treatment modalities and the importance of optimal intake of vitamin K2, which binds to PXR, enabling a plethora of tissues in the body to function properly under strain from extrinsic and intrinsic challenges. The present article is part of a special issue called 'Xenobiotic nuclear receptors: New Tricks for An Old Dog'. Editor: Dr. Wen Xie.

m. Liu et al. PLoS One. 2016 Feb 18;11(2):e0149639.doi: 10.1371/journal.pone. 0149639. eCollection 2016. **Role of UBIAD1 in intracellular cholesterol metabolism and vascular cell calcification.**

Excerpt from an article: Vascular calcification is a risk factor coupled to mortality suffering from chronic kidney disease. Cholesterol turnover is linked to this vascular calcification process. In the present work, the role of the UbiA prenyltransferase domain containing 1 (UBIAD1) in cholesterol turnover and vascular cell sclerosis (i.e. calcification) was scrutinized.

Human umbilical vein smooth muscle cells (HUVSMCs) were exposed to (a) a traditional growth medium (with 1.4 mmol/L Pi) or (b) calcification medium (3.0 mmol/L Pi). When treated with medium (b), HUVSMCs were incubated with cholesterol or menaquinone-4 (a product synthesized by UBIAD1). Matrix calcium quantitation, alkaline phosphatase (ALP) levels and cellular cholesterol menaquinone-4 (vitamin K2) levels were analysed.

To make the story short, it was concluded that (a) high intracellular cholesterol content contributes to phosphate-induced vascular cell differentiation and calcification, (b) UBIAD1 or menaquinone-4 (vitamin K2) decreased vascular cell differentiation and calcification through its potent role of reciprocally modulating cellular cholesterol.

n. Litwa E. et al. J Steroid Biochem Mol Biol. 2016 Feb;156:43–52. doi: 10.1016/j.jsbmb. 2015.11.018. Epub 2015 Nov 28. **RXRα, PXR and CAR xenobiotic receptors mediate the apoptotic and neurotoxic actions of nonylphenol in mouse hippocampal cells.**

Excerpt from the article: Here, the authors investigated the role of RXR (the retinoid receptor), PXR and CAR (the constitutive androstane receptor) as to the apoptotic and toxic impacts of nonylphenol (NP) in neuronal mouse cell cultures. The present work showed that NP stimulated caspase-3, while inducing lactate dehydrogenase (LDH) activity in hippocampal cells, reactions which were paralleled by an enhancement of protein levels of RXRα, PXR as well as CAR. NP enhanced RXR, PXR and CAR mRNA synthesis, and the present effects preceded an enhancement in corresponding protein levels. The staining techniques applied here also indicated an NP-induced translocation of receptor-specific immunofluorescence from cytoplasm to the nucleus.

The use of specific siRNAs further indicated that RXR-, PXR- and CAR-siRNA-transfected cells had turned less vulnerable to NP-induced stimulation of caspase-3 and LDH, hence underscoring their involvement in RXRα/PXR/CAR, signalling pathways in the apoptotic-neurotoxic reactions induced by NP.

o. Sy C. et al. Mar Drugs. 2015 Nov 19;13(11):7020–39. doi: 10.3390/md13117020. **Interactions between carotenoids from marine bacteria and other micronutrients: impact on stability and antioxidant activity.**

Excerpt from the article: Spore forming and pigmented marine bacteria, like the Bacillus indicus HU36, serve as natural sources of oxygenated carotenoids. In the present study, the stability, as well as the antioxidant activity (i.e. resistance to lipid peroxidation), of HU36 carotenoids with bacterial MK-7 was investigated.

Unexpectedly, MK-7 substantially improved the ability of HU36 carotenoids to block Fe(II)-induced lipid peroxidation, even though MK-7 remained unconsumed in the incubation medium. Hence, it was asserted that their presence modifies the antioxidant properties exerted by the carotenoids, probably by allowing them to scavenge radicals. The HU36 carotenoids and phenol-derived antioxidants showed synergism in the inhibition of linoleic acid peroxidation. This reaction might have risen from antioxidants, which interacted via heme-iron-based electron transfer.

p. Puri A. et al. J Food Sci Technol. 2015 Dec;52(12):8228–35. doi: 10.1007/s13197-015-1903-3. Epub 2015 Jun 23. **Effect of sequential bio-processing conditions on the content and composition of vitamin K2 and isoflavones in fermented soy food.**

Excerpt from the article: The present paper features the effect of sequential addition of *Bifidobacterium bifidum, Bacillus subtilis,* as well as *Rhizopus oligosporus* on the contents and composition of vitamin K2 contained within fermented soya foods. Soya beans, fermented with *B. bifidum,* were treated as described: The fermented bacterial bulk mass was re-fermented with a co-culture of *B. subtilis* and *R. oligosporus.* This study indicated that the co-fermentation of soya beans with different microbe combinations, however, in a given, predefined sequence, may enhance nutritional value better than mono-culture fermentations, which is a result of the positive correlation between applied enzymes (lipase, phytase, β-glucosidase), menaquinone-7 (vitamin K2) and soya isoflavone contents.

q. Poon CC. et al. Eur J Pharmacol. 2015 Nov 15;767:30–40. doi: 10.1016/j.ejphar. 2015.09.048. Epub 2015 Oct 8. **In vitro vitamin K(2) and 1α,25-dihydroxyvitamin D(3) combination enhances osteoblasts anabolism of diabetic mice.**

Excerpt from the article: The present study features the anabolic effect and cellular action of vitamin K2 or/and 1α,25-dihydroxyvitamin D_3 ($1,25(OH)_2D_3$) on iliac crest osteoblasts from C57BL/KsJ lean (+/+), as well as obese and diabetic (db/db) mice. A lower ALP (alkaline phosphatase) activity and a reduced expression of anabolic bone markers and osseous formation-related transcription factors (i.e. OC = osteocalcin, Dlx5, Runx2, ATF4 and OSX) were consistently observed in the osteoblastic cells of db/db mice, as compared with lean mice. Significantly higher deposits of Ca^{2+} by osteoblastic cells were seen in lean mice, as compared with db/db mice. Concomitant administration of vitamin K2 (MK-7) and active vitamin D ($1,25(OH)_2D_3$) brought about an increase in calcium deposits by osteoblasts in all animals. Furthermore, vitamins K2 and D successively (for 3 weeks) augmented levels of the anabolic bone turnover markers, including bone formation-related transcription factors. The combined treatment with vitamin K2 plus vitamin $1,25(OH)_2D_3$ substantially stimulated migration, as well as the reappearance of both surface microvilli and ruffles expressed by osteoblasts of the db/db mice. Hence, the present data indicate that the vitamin K2 plus vitamin D_3 combination might serve as a novel therapeutic strategy for the treatment of diabetes-associated osteoporosis.

r. Qin XY. J Nutr Sci Vitaminol (Tokyo). 2015;61(4):285–90. doi: 10.3177/jnsv.61.285. **Carboxylic derivatives of vitamin K2 inhibit hepatocellular carcinoma cell growth through caspase/transglutaminase-related signaling pathways.**

Excerpt from the article: The chemoprevention of liver cancer (e.g. Hepatocellular carcinoma = HCC) has turned out to be one of the most challenging aspects of medical research. Vitamin K2 (MK-4, MK-7) has especially been launched for its established chemopreventive effect in the treatment of HCC, while inconsistent or contradicting results reported in clinical trials have emerged. The study described in the present paper was undertaken to add to our understanding of the anti-HCC cell proliferative effect of vitamin K2 and its derivatives, taking its chemical structure into account. However, no marked effects were observed with the original vitamin K2, while vitamin K2 derivatives, bearing isoprene units as well as a carboxyl-terminated side chains, appeared to inhibit the growth of HCC cells in a dose-dependent manner without damaging normal hepatocytes. Traditional loss-of-function analyses concluded beyond doubt that the anti-HCC cell effect witnessed by the vitamin K2 derivatives was not conferred by the vitamin K2 binding protein, 'Bcl-2 homologous antagonist/killer', known as 'Bak', but rather associated with the caspase-transglutaminase related pathway of signalling.

s. Hey H. Ugeskr Laeger. 2015 Aug 3;177(32):V12140700. **[Vitamin K2 influences several diseases]**. [Article in Danish]

Excerpt from the article: In the present paper, the evidence of the biological effects of vitamin K2 is discussed. Deficiency of vitamin K2 is indeed a factor in a plethora of chronic diseases, such as diabetes, osteoporosis, cancer, inflammations and cardiovascular ailments. Vitamin K2 deficiency is very commonly associated with the diseases mentioned above, even though vitamin K2 is rarely included in the treatment regimens used by clinicians. A multitude of randomized clinical investigations have demonstrated that patients with ailments, such as osteoporosis, cancer, cardiovascular diseases, Alzheimer's disease, adiposity and many others, can benefit from vitamin K2 supplementation. However, further studies are needed in order to ascertain the effect of vitamin K2 supplements in patients with diabetes and inflammatory bowel diseases.

As a last comment, the readers are highly recommended to read the article by Lara Pizzorno (http://weblisting.freetemplatespot.com/lmreview.com/), a compilation of beneficial features of vitamin K2 as: 'Essential for the Prevention of Age-Associated Chronic Diseases' amongst other articles on the subject of 'Longevity'.

Author details

Jan O. Gordeladze[1,2]*, Håvard J. Haugen[2], Gaute Floer Johnsen[2] and Mona Møller[3]

*Address all correspondence to: j.o.gordeladze@medisin.uio.no

1 Institute of Basic Medical Science, University of Oslo, Norway

2 Department of Odontology, University of Oslo, Norway

3 Kappa Bioscience, Oslo, Norway

References

[1] Jan O Gordeladze et al. Epigenetic factors involved in musculo-skeletal interaction: how skeletal muscle cells and osteoblasts derived from stem cells communicate with special reference to histone deacetylases (HDACS), transcription factors (TFs), microRNAs and vitamin K2. Orthop Muscular Syst 2016;5:2(Suppl). doi:10.4172/2161-0533.C1.027.

[2] Kawao N, Kaji H. Interactions between muscle tissues and bone metabolism. J Cell Biochem. 2015;116(5):687–95. doi:10.1002/jcb.25040. PMID: 25521430.

[3] Karasik D, Cohen-Zinder M. The genetic pleiotropy of musculoskeletal aging. Front Physiol. 2012;3:303. doi:10.3389/fphys.2012.00303.

[4] Yang N, Schindeler A, McDonald MM, Seto JT, Houweling PJ, Lek M, Hogarth M, Morse AR, Raftery JM, Balasuriya D, MacArthur DG, Berman Y, Quinlan KG, Eisman JA, Nguyen TV, Center JR, Prince RL, Wilson SG, Zhu K, Little DG, North KN. α-Actinin-3 deficiency is associated with reduced bone mass in human and mouse. Bone. 2011;49(4):790–8. doi:10.1016/j.bone.2011.07.009. Epub 19 July 2011.

[5] Baldelli S, Lettieri Barbato D, Tatulli G, Aquilano K, Ciriolo MR. The role of nNOS and PGC-1α in skeletal muscle cells. J Cell Sci. 2014;127(Pt 22):4813–20. doi:10.1242/jcs.154229. Review.

[6] Handschin C, Spiegelman BM. PGC-1 coactivators and the regulation of skeletal muscle fiber-type determination. Cell Metab. 2011;13(4):351; author reply 352. doi:10.1016/j.cmet.2011.03.008. No abstract available.

[7] An JH, Yang JY, Ahn BY, Cho SW, Jung JY, Cho HY, Cho YM, Kim SW, Park KS, Kim SY, Lee HK, Shin CS. Enhanced mitochondrial biogenesis contributes to Wnt induced osteoblastic differentiation of C3H10T1/2 cells. Epub 2010 Apr 14. Bone. 2010 Jul;47(1):140-50. doi: 10.1016/j.bone.2010.04.593

[8] Qiu J, Thapaliya S, Runkana A, Yang Y, Tsien C, Mohan ML, Narayanan A, Eghtesad B, Mozdziak PE, McDonald C, Stark GR, Welle S, Naga Prasad SV, Dasarathy S. Hyperammonemia in cirrhosis induces transcriptional regulation of myostatin by an NF-κB-mediated mechanism. Proc Natl Acad Sci U S A. 2013;110(45):18162–7. doi:10.1073/pnas.1317049110. Epub 21 Oct 2013.

Section 5

Implications of Vitamin K2 in Non-Humans

Chapter 12

Vitamin K2 in Animal Health: An Overview

Jayde O'Neil, Bethany Scarrott,

Ragnhild Aven Svalheim, Jonathan Elliott and

Stephen J. Hodges

Additional information is available at the end of the chapter

http://dx.doi.org/10.5772/63901

Abstract

The role of vitamin K in animal health has not received much attention. Vitamin K studies have, for the most part, addressed the use of animals in the investigation of vitamin K physiology and pathophysiology, often using the rodent as a model system. However, vitamin K performs the same role in animals as it does in man and there are areas, such as animal nutrition, where a better understanding of animal requirements in general, and with ageing, could benefit animal health and continued well-being.

Keywords: vitamin K2, vitamin K3, menadione, coprophagy, UBIAD1

1. Introduction

The post-translational gamma-carboxylation of proteins by vitamin K is common across the animal kingdom, yet the organisms most studied for the relationship between this vitamin in health and disease are predominantly man and the rat.

There are several forms of vitamin K: a single vitamer from plants, vitamin K1, or phylloquinone, and a family of K2 vitamins, the menaquinones, which are distinguished by the number of isoprene units in their side chain at position 3 of the 2-methyl-1,4-naphthoquinone moiety. All the long side-chain menaquinones are derived from bacterial sources. Menaquinone-4 is found less frequently in bacteria, those species where it has been reported as a major menaquinone are usually extremophiles and microaerophiles [1–3], and it may be the evolutionary hinge between vitamin K1 using phytoplankton and photosynthetic cyanobacteria [4]. In contrast to bacteria-mediated generation of biologically functional vitamin K2, it is also

INTECH
open science | open minds

© 2017 The Author(s). Licensee InTech. This chapter is distributed under the terms of the Creative Commons Attribution License (http://creativecommons.org/licenses/by/3.0), which permits unrestricted use, distribution, and reproduction in any medium, provided the original work is properly cited. (cc) BY

possible for the specific form of vitamer, menaquinone-4, to be synthesized in animal tissues from vitamin K3 [5–7]. It is also worth noting that, in the rat, vitamin K1 can be converted to menaquinone-4 without the need for bacterial mediators [8–10].

The general understanding of vitamin K in the physiology and pathophysiology of animal health and well-being is less well developed than it is for man. As with man, overt vitamin K deficiency is rare in animals and several examples will be identified. It is clearly not possible to cover all animal species in this overview, and we have selected a reasonably broad range of important agricultural and companion animals.

With respect to captive collection species, although not discussed here as there is very limited information available, the general concepts of vitamin K in the health and well-being of these animals are open to review. This is particularly pertinent to the conservation programmes to support endangered animal breeding projects, as vitamin K2 is beginning to be highlighted as an essential factor in embryogenesis.

For the purposes of this overview, animals can be broadly divided into herbivores and omnivore/carnivores. Common sense would suggest that the former group would be considered to derive most of their vitamin K from the plant vitamer, vitamin K1. We will demonstrate that this is not a simple extrapolation to food source and that vitamin K2 congeners can, in some cases, be at least as dominant a source of vitamin K. The latter grouping of animals is diverse, and we will develop the hypothesis that the vitamin K2 congeners are a dominant form of vitamin K that makes a significant contribution to maintaining the vitamin K status of these animals.

Investigation into the feeding practices for domesticated animals, both companion and agricultural animals, demonstrates that there are several examples where nutritional developments of animal feeds have defaulted into the use of vitamin K3, even though this naphthoquinone compound is unable to participate directly in the gamma-carboxylation of vitamin K-dependent proteins [11]. The value of vitamin K3 in animal nutrition may be derived, in part, from its role as a bacterial growth factor [12–14], where it is converted into vitamin K2 [15]. Many bacteria, particularly the gram-positive bacteria, contain vitamin K2 as their major quinone [16–18]. There are indications that vitamin K2 derived from intestinal bacteria contribute to the vitamin K status of some animals, such as ruminants, more than other animals, and this will be discussed in more detail in the relevant sections.

The following sections will discuss where animals derive their source, or sources, of vitamin K and, where known, what vitamin K-related health pathologies have been described in animals.

2. Avian vitamin K requirements

Given that the original identification of vitamin K was made in the chicken by Henrik Dam [19, 20], it is appropriate to begin this overview by considering avian species. The early work of Dam demonstrated that chicks fed a diet designed to be deficient in sterols, following the nonpolar solvent extraction of their feed, developed subdural or muscular hemorrhage and presented with a retardation in their blood clotting time. Subsequent experiments demonstrated that Inclusion of other essential nutrients, which were known at that time, did not prevent hemorrhage, suggesting the existence of a new vitamin.

Extending his investigations, Dam determined that geese and ducks were also susceptible to the development of clotting problems, but pigeons and canaries appeared to be much less prone to develop hemorrhagic problems [21].

Due to the enormous value of avian species in human nutrition, vitamin K sufficiency is an important subject. Furthermore, the more recent suggestion of a role for vitamin K in skeletal biology also promotes the understanding of adequate vitamin K nutrition in avian species, such as the turkey, which is prone to suffer from skeletal health problems [22] and which has considerable economic consequences [23]. Similarly, broiler chickens, with their rapid growth requirements, also have skeletal health problems [24]. The vitamin K in the feed for hens to offset skeletal health problems has been investigated, and there is some benefit from 8 to 10 mg/Kg feed of vitamin K3 [25, 26]. A more recent study [27] has suggested that in the chicken, the liver and the pancreas process vitamin K, and in particular menaquinone-4, in different ways. The observation of a 10-fold greater proportion of menaquinone-4 epoxide in the chicken pancreas as compared to the liver suggested that the vitamin K 2,3-epoxide reductase complex (VKOR) may not be as efficient in the avian pancreas as it is in the liver.

The production demands on laying hens also puts considerable strains on the bird's skeleton, which has also been shown to benefit from additional vitamin K3 in the diet [28]. From the discussion on the need to modify vitamin K3 into an active form of vitamin K, there is a strong likelihood of commercial avian health needs being met by the conversion of vitamin K3 into vitamin K2, as either menaquinone-4 or, with the potential for coprophagy, longer chain menaquinones from fecal bacteria.

Non-commercial avian species have received little attention for their vitamin K requirements. However, it should be noted that wild birds can often be the casualties of deliberate or accidental poisoning from anticoagulant agents that inhibit the vitamin K cycle [29, 30].

2.1. Rodentia and Lagomorpha

2.1.1. The rat

The wild the rat, as an omnivore, will derive their vitamin K from a variety of sources such as plant material and dead animal tissues and also from bacteria in decaying plant and animal materials. In the laboratory, the rat has been the most extensively investigated animal for vitamin K physiological and pathophysiological requirements. It has provided the standard, from which much human investigation has been initiated.

One of the fundamental elements of undertaking a rodent experiment on vitamin K has been the need to derive an animal with a vitamin K deficiency that survives long enough for investigation. Outside the use of a vitamin K-specific anticoagulants, such as warfarin and brodifacoum, this has been difficult to achieve nutritionally derived vitamin K deficiency [31,32]; Dr

Cees Vermeer, personal communication. With fastidious experimentation, it is possible to reduce vitamin K tissue levels in rats, and with certain diets, this can affect the intestinal vitamin K2-producing flora [33].

The reported high fecal coprophagy in the rat demonstrates that inducing vitamin K deficiency is difficult, even when animals are contained in elevated, wire-bottom metabolic cages with additional tail-cups on the rats to collect feces [34].

There is also tissue variation in the activity of VKOR in the rat [35], which can have implications, in some circumstances, on the functional efficacy of carboxylation of vitamin K-dependent proteins in extrahepatic tissues.

Polymorphism within the gene for VKOR has been associated with increased rodent resistance to anticoagulant rodenticides. This has profound potential commercial implications as the initial shift to more potent agents from the original warfarin used to control rodents has shown potential problems, due to environmental persistence, while the need to control rodents remains necessary [36].

2.1.2. The mouse

In the wild, the mouse has the same exposure to sources of vitamin K as described for the rat, and, like the rat, mice in the wild are also showing increasing vitamin K anticoagulant resistance.

The mouse has been widely used as a laboratory model for many diseases due to a fast breeding rate and the potential for genetic homogeneity. The additional development of creating transgenic mice that over- or under-express a specific gene (or genes) together with site-directed, or conditional, gene manipulation has increased their importance in medical and scientific research enormously. These technologies have been exploited to demonstrate the essential nature of the VKOR enzyme [37].

The enzyme UbiA prenyltransferase domain-containing protein 1 (UBIAD1) is expressed across vertebrates and has been shown to synthesize the vitamin K2 congener menaquinone-4 [38]. Deletion of this gene in mice has been found to be lethal, preventing development beyond embryonic day 7.5 [39], suggesting a fundamentally important role for vitamin K2 in embryonic development.

Mice have the same limitations for vitamin K research as described for the rat. Inducing nutritional deficiency and preventing coprophagy is difficult, but can be achieved with some success using strict practices.

2.1.3. The rabbit

Rabbits and hares are prolific herbivores, and therefore, a large amount of their vitamin K requirement is derived from vitamin K1. Although, as with rodents, rabbits consume substantial fecal matter; as long ago as 1882 it was reported that rabbits produce two types of fecal matter, one more liquid than the other [40]. Subsequently, the 'soft' feces were found to have a similar protein, fiber and nutrient composition to the caecal contents [41] and that coprophagy is also normal for the wild rabbit [42]. The caecum of the rabbit is a perfect environment for the incubation of numerous species of bacteria [43] and that the nursing doe probably carries the responsibility for ensuring her kittens have the appropriate intestinal/caecal microbial flora established early in life [44]. Indeed, an incorrect microbiota platform may lead to pathological conditions in the rabbit [45].

The sequelae from the discussion on coprophagy in the rabbit are that the recycling of bacterial contents from feces and maybe, importantly, the caecal content microbiota could contribute significantly to the animal's vitamin K status, in addition to their nutritional vitamin K1 intake. This has not been directly measured in the rabbit, but while there is ongoing debate about the vitamin K2 nutritional benefits from large bowel absorption of these fat-soluble vitamins, the small intestine absorption of vitamin is known [46].

There is no information on the hare that describes how it meets its full requirements for vitamin K, and presumably, the rabbit serves as an appropriate example to draw similar conclusions.

2.2. Agricultural animals

Agricultural species are important for essential food and other product production, and the health and well-being of livestock is important for their growth and reproduction. We will consider several major animal species, but with centuries of breeding and a wide diversity in breeds, we take a generic position in descriptions, except where specific health problems have been identified. The horse is included in this discussion as, while there is a substantial element of meat production from the horse and related species, it also has value in equine sport activities and a rapid growth rate is a selected feature for many of these activities.

2.2.1. Ovine vitamin K requirements

With over 200 sheep breeds that have been reared over thousands of years for different features, such as wool, milk, skins, meat and even the ability to clear vegetation, a great deal of genetic diversity has been engineered into the different flocks. As committed grazing herbivores, most of their vitamin K needs will be met by vitamin K1. However as a ruminant, there will be an added contribution through vitamin K2 from the microflora that undertake additional fermentation in their rumen. There is more direct published literature on vitamin K in the cow and the relationship to rumination, this is considered in Section 2.3.4.

Grazing can bring sheep into contact with toxic plants, such as giant fennel (*Ferula communis*) which contains a 4-hydroxycoumarin compound, a relative of the widely used human anticoagulant pharmaceutical and rodenticide warfarin. The giant fennel is widely distributed around the Mediterranean basin and has been associated with a hemorrhagic syndrome in several species of livestock, including sheep and goats [47]. In vivo experiments in sheep show that extracts from the plant can cause a range of symptoms, including hemorrhage [48, 49]. The active coumarol agents inhibit the VKOR enzyme in much the same way as warfarin, and there is a report of species susceptibility differences [50]. This syndrome can be successfully treated with vitamin K [51].

The new-born lamb, such as human neonates, can to some extent experience postpartum hemorrhage. In the case of the Rambouillet breed of sheep, there is a potentiation of this problem [52]. This problem has been causally linked to a genetic defect in the vitamin K-dependent carboxylase enzyme [53, 54].

2.2.2. *Caprine vitamin K requirements*

As with sheep, the goat has a long history of domestication by man and there are now many breeds of goat that have been generated, as with sheep, for a diverse range of reasons. In contrast to sheep, goats are browsers and will also crop tree branches and shrubs. While the goat is credited with eating anything, this is not the case, being inquisitive animals they test many things with their lips and mouth, and may then ingest inedible objects. The goat has the same chance of eating toxic plants as sheep, such as giant fennel.

The absence of the literature on hemorrhagic disease suggests that millennia of breeding programmes have not drawn in a vitamin K-related genetic hemorrhagic disease or, if it did, the mutation was lethal before the animal reached reproductive maturity.

2.2.3. *Porcine vitamin K requirements*

The history of the pig is even older than that of sheep and goats with indications that the pig was first domesticated 9–10,000 years ago in the Middle East [55]. Despite this heritage, there are many fewer breeds of pig than there are for sheep and goats. The pig, wild boar and hogs are natural scavengers and will consume anything edible, including decaying matter and fecal waste; the latter ability has been utilized by humans as part of local sewage management in the form of the 'pig-privy' [56]. This suggests that in the wild and in free-living domestic pig populations, long-chain vitamin K2 will be at least as important, if not more important, as vitamin K1 in maintaining the vitamin K status of the pig.

The vitamin K status of boar and hogs has not been the subject of investigation. However, husbandry of the domestic pig, due to its economic importance, has been considered extensively.

A comparative investigation into the coagulation status of several animal species found that the pig most closely mirrored the human neonate [57]. Therefore, the description of postpartum hemorrhage in the piglet due to vitamin K deficiency [58] is not too surprising as it is seen in the new-born human infant, where the recommendation is for vitamin K prophylaxis at birth. Increased hygienic conditions for farrowing sows housed in elevated sties preventing coprophagy and antibiotic usage were identified as potential factors in vitamin K deficiency in the pig. Around this time, supplementation with 2.2 mg/Kg feed with vitamin K3 was suggested for pigs if a vitamin K deficiency was suspected [59].

2.2.4. *Bovine vitamin K requirements*

Bovine species have as long an association with man as other domesticated animals. Cave paintings demonstrate that ancient man was aware of the auroch in his environment and was probably a major predator of these animals. Domestication from the auroch has been suggested to date back to around 10,000 years in Asia and in the Near East. The modern-day descendents are the Zebu cattle in Asia and the common taurine cattle breeds in Europe. Genetic diversity from the DNA of skeletal remains of the European ancestors suggests that their diversity is so limited as to indicate that the modern taurine bovid has been derived from as few as 80 original cows [60].

The bovids, as ruminants, are as likely to derive their vitamin K requirements from rumen microflora as they are from plant sources. This has support when bovine liver menaquinone stores are considered against equine liver stores. The long-chain vitamin K2 congeners in bovine liver contain measurable menaquinones, up to menaquinone-13, while equine liver does not and vitamin K1 from plants is the principle form of vitamin K found in the liver of the horse [61–63].

A recent study looking at muscle levels of vitamin K in two bovine breeds, the Norwegian Red and the Jersey, raised under identical conditions in one location [64], found that the tissue distribution in muscle is distinctly different to that reported for the liver and different between the two breeds. This study found that there were two dominant forms of vitamin K, vitamin K1 and menaquinone-4. The former correlating with the muscle fat content, there being no equivalent relationship for menaquinone-4. These findings mirror a previous food screen report [65]. Also, different muscles in the two breeds found varying levels of vitamins K. The possibility of a relationship between the enzyme UBIAD1 and menaquinone-4 levels was not considered in the study, but this could, at least in part, explain some of the vitamin K2 findings.

Possibly, the single most relevant historical relationship between bovine species and vitamin K is the discovery of a vitamin K anticoagulant from the ingestion of spoilt sweet clover feed. This eventually led to the discovery and development of one of the world's most prescribed pharmaceutical drugs, warfarin [66–71].

2.2.5. Equine vitamin K requirements

The finding that horse liver contains predominantly vitamin K1 relates to the dietary source for this animal, but which also supports the contention that ruminants gain broader vitamin K support from bacterial vitamin K2 sources. In a large, rapidly growing animal like the horse, an adequate vitamin K supply may need to meet more than coagulation requirements. An indication of this may be suggested from a 4-week-old Standardbred colt initially presenting with vitamin K deficiency bleeding that continued to fail to thrive after normalization of the coagulation defect by the administration of vitamin K [72]. While there may have been other underlying pathologies, this observation of a vitamin K deficiency bleeding may have been the result of a genetic mutation in the breeding line as the stallion that sired the described foal had also previously sired a colt with coagulopathy health problems [72]. An inability to follow up the research prevented a definitive description of a vitamin K-dependent deficiency disease in these horses.

A study, with an interest in race horse growth, and in particular bone health [73], investigated the best form of vitamin K to administer orally to horses. Circulating vitamin K1 increased in relation to the administered dose; however, menaquinone-4 administration did not show corresponding plasma level increases, the suggestion being that the horses in this study did not absorb this vitamer. Interestingly, administration of vitamin K3 did cause menaquinone-4 plasma levels to rise, which may relate to UBIAD1 synthesis of menaquinone-4 or intestinal flora.

Another consideration in equine health is for vitamin K3 administered parenterally; this form of intervention was been found to cause pronounced renal toxic effects [74, 75]. Reports of parenteral administration of vitamin K1 or K2 prophylaxis, or remedial coagulopathy, interventions were not obvious from screening the literature.

3. Piscine vitamin K requirements

The world growth in aquaculture has increased enormously over the last decade. From 2005 to 2013, the global worth of this industry has more than doubled to over 150 billion USD [76]. Any captive species will be dependent on being fed an appropriate diet for health in order to produce good quality product for an ever-demanding consumer population.

Fish, like other animals, are not able to *de novo* synthesize vitamin K and have to obtain it from their diet. Vitamin K deficiency in fish results in several familiar health problems that are found in terrestrial animals, including increased blood coagulation time, reduced growth, anemia, hemorrhage, weak bones, and occurrence of spinal curvature, short tails and increased mortality, together with problems specifically related to fish, such as loss of fin tissue [77–79].

The role of vitamin K in salmonids began to be investigated in the 1960s [80, 81], and today, minimum requirements for vitamin K supplement in fish feed are largely based on the effect of vitamin K on blood coagulation. Although estimates of dietary vitamin K requirement differ a great deal among fish species, and the quantitative requirement of vitamin K for most fish is still largely unknown. In addition, dietary studies on fish entail problems that are not encountered in terrestrial animals, such as vitamin leaching into their environment from supplemented feed [82].

Studies on zebrafish and Japanese puffer fish have found genes for vitamin K-dependent factors (VII, IX, X and prothrombin). Also, continuous exposure to warfarin causes spontaneous bleeding in zebrafish [83–85]. In common carp, warfarin prolonged the prothrombin time and activated partial thromboplastin time, whereas supplementation of menadione prevented increase in prothrombin time. For the large yellow croaker, the blood coagulation time generally decreased with increasing dietary menadione levels [86].

Menadione can be alkylated enzymatically to menaquinone-4 in tissues [87]. This conversion has been recorded in several fish species such as Atlantic salmon [88–90], Atlantic cod [91], cultured sardines [92], mummichog [93], ayu [94] and large yellow croaker [86].

The enzyme UBIAD1 that converts vitamin K1 into menaquinone-4 has been the subject of an investigation in a zebrafish mutant (*reddish*^{s587}*:reh*). The fish develop normally for 24–36 h, but by 48 h they present with cranial hemorrhage [95]. The UBIAD1 gene was found to be expressed as early as the single-cell stage. With continued development, the *reh* zebrafish mutant presents with higher expression of UBIAD1 in the vasculature than cardiac tissues and gene expression decreases with time. Introduction of an antisense morpholino oligonucleotide targeted splice to knockdown wild-type UBIAD1, or the administration of warfarin, produced similar vascular models to the reh mutant, with the warfarin challenge having notably less impact on cardiac tissue compared to the vascular effects. The defect in these fish could be salvage by the introduction of wild-type zebrafish or human UBIAD1 mRNA, but not *reh* UBIAD1 mRNA. Similarly, the knockdown or warfarin-treated larvae could be rescued by the administration of vitamin K1 or menaquinone-4, vitamin K1 being used as the source of naphthoquinone for the UBIAD1 conversion to menaquinone-4. The overall finding that the UBIAD1 gene and enzyme expression has an important role in vascular endothelial cell survival has implications across all tissues and in cancer.

Bone and spinal deformities are a major problem in commercial fish farming. Deformities are not only an economic problem for fish farms, but also raise ethical and welfare issues for the aquaculture industry. The importance of vitamin K in fish skeletal health has increased interest in vitamin K requirement for normal bone development in fish. There are a few studies that have dealt specifically with the effect of vitamin K deficiency on fish bone health [78, 88, 90, 96, 97]. Tissue-specific gene expression of the vitamin K-dependent proteins, such as osteocalcin and matrix Gla protein (MGP), has been shown in the vertebrae of Atlantic salmon; however, dietary vitamin K was not found to regulate the expression of MGP [90, 98]. Interestingly, neither juvenile Atlantic salmon [90] or Atlantic salmon smolts [88] showed any sign of deformities on a diet lacking vitamin K supplementation.

Mummichog given feed that was not enriched with vitamin K grew thin, weak bones. Vitamin K deficiency induced bone structure abnormalities such as vertebral fusion and row irregularity, both in early development and during later growth [78, 96]. Furthermore, the offspring of vitamin K-deprived fish had higher incidences of abnormal vertebral formation 5 days after hatching when compared to larvae from fish fed a vitamin K-enriched diet [78]. In haddock, vitamin K appears to be necessary for bone mineralization [99]. However, vitamin K does not affect the number of osteoblast in haddock, while bone deformities coincided with an increased amount of osteoid and a decrease in bone mineral content. In the Senegalese sole given feed enriched with vitamin K, there was a notable improved larval growth performance and post-larval skeletal quality. Also, vitamin K modulated expression of protein involved in several biological processes including muscle contraction and development, cytoskeletal network, skin development, energy metabolism, protein chaperoning and folding, and bone development [97].

It now seems that vitamin K supply may be less than optimal for bone development, but sufficient to maintain normal growth and hemostasis [82, 87].

Fish feed is commonly enriched with vitamin K3 (menadione) in the form of water soluble salts, normally menadione sodium bisulphite (MSB) and menadione nicotinamide bisulphite (MNB) [82].

Using menadione in fish feed is, however, not without problems; too high a dosage, in particular MSB, has proven to cause reduced growth [91, 100]. Nevertheless, it remains one of the most common vitamin K supplements in fish feed.

4. Companion animals

The species and breeds within species that have become companion animals have expanded considerably. Rabbits, rats and various exotic animals are increasingly being kept as domestic pets. In this section, we focus on cats and dogs as they dominate the sector.

4.1. Canine vitamin K requirements

The dog has been domesticated for possibly even longer than agricultural animals [101–103]. The possible breeding matrix that has led to the huge array of current domestic dogs is known to carry several genetic defects [104], and while there are several well-described coagulopathies in the dog, vitamin K-specific deficiency is not a widely reported genetic mutation. One case study [105], considered that a black Labrador retriever admitted for a ovariohysterectomy later presented clinically with a vitamin K deficiency coagulopathy. After ruling out several other options, including fat malabsorption problems, ingested coumarin-based rodenticide, other xenobiotics, liver disease, and noting that the problem was resolved and managed by vitamin K administration, the authors suggested this was a possible case of vitamin K deficiency. This type of disease, therefore, remains a rare disease in the dog.

Conversely, scavenging for food has caused numerous cases of accidental coumarin-based anticoagulation poisonings in the dog that carry through all geographical regions and is repeatedly reported over time [106–108]. The problem has probably become exacerbated with the introduction of more persistent anticoagulant rodenticides.

Some early studies [109, 110] found that the vitamin K1 supplemented dog stored most of the administered tritiated vitamer in the liver. It was also noted that a proportion was converted into menaquinones, particularly menaquinone-4. In the light of the recent studies on UBIAD1 [38], this may have an explainable origin.

Incomplete gamma-carboxylation of vitamin K-dependent proteins has been implicated in human joint diseases [111, 112]. Some dog breeds have a predisposition to clinically significant arthritic diseases; however, studies on the potential to alter the course of arthritic disease in dogs with vitamin K have not been undertaken.

The commercial dog food manufacturers have an open policy on nutritional information in their products. Many do not explicitly refer to vitamin K content of their products. One major supplier notes that in their dry food product, vitamin K 'activity' is supplied as vitamin K3.

4.2. Feline vitamin K requirements

Cats are scavengers, they hunt birds and small animals including rodents, these activities bring them into contact with rodenticides [113], and therefore, it is not surprising that the principle reason for veterinary use of vitamin K clinically is as a rescue medication due to accidental rodenticide intoxication. As with the dog, the problem has probably been exacerbated by the increasing use of the more persistent anticoagulants used as rodenticides that have replaced warfarin in order to overcome rodent warfarin resistance.

Starting in the 1950s in South West UK, a breed of Rex cat was developed out of some accidental breeding with feral tom cats, which led to some reverse mating into their own genetic line with the intent to maintain the Rex breed. One of these lines of development led to the Devon Rex cat. In 1990, three Devon Rex cats were described with a vitamin K deficiency character [114], after exclusion of other factors such as accidental anticoagulant ingestion, liver disease, intestinal malabsorption problems and treatment with vitamin K to correct their deficiency. The nature of the defect in the Devon Rex was investigated in the Netherlands, and this cat was found to have a decreased ability to gamma-carboxylate vitamin K-dependent clotting factors due to a decrease binding of reduced vitamin K and the clotting factors to the carboxylase enzyme [115].

With increasing age, cats also develop diseases that cause vitamin K deficiency coagulopathies, such as liver disease, inflammatory bowel disease and secondary malabsorption syndrome [116–118].

It is possible to induce a vitamin K deficiency through diet, presumably as the cat is not particularly associated with coprophagic behavior. In an early study of queens and their kittens fed either a commercial tuna- or a salmon-based fish diet, there was a notable increase in blood clotting times [119]. Where the information is available, current commercial cat diets provide vitamin K 'activity' in the form of vitamin K3, principally in the dry food products.

The cat is also prone to present clinically with chronic kidney disease (CKD) [120], and the reported prevalence is high, particularly in aged cats. It is interesting that the pathophysiology of feline CKD has been proposed to be sufficiently similar to human disease that the cat could provide a natural model to investigate human CKD [121]. In human CKD, there are several reports of an association with low vitamin K status [122] and a recent multi-ethnic study demonstrated an inverse association between estimated glomerular filtration rate and a functional marker of vitamin K deficiency, namely de-phospho-undercarboxylated matrix Gla protein [123].

Our pilot studies looking at circulating vitamin K in the healthy aged cats found that menaquinone-4 was the dominating form of vitamin K. This observation would suggest that the vitamin K3 in cat diet is converted through UBIAD1 to menaquinone-4, although this has not been specifically demonstrated. There is a line of thought that vitamin K may also be provided to the cat through colonic bacterial supply. The absence of long-chain menaquinones in our study suggest that this is unlikely and evidence supporting colonic absorption of fat-soluble vitamins K in general is limited [124].

Systemic inflammation is widely accepted as a dominant driver in the aetiology of CKD, and this is an active area of therapeutic interest [125]. The re-emerging observation of direct anti-inflammatory activity for vitamins K1 and K2 and in particular their common 7-carbon carboxylic acid catabolite [126–128] suggests that a low vitamin K status in CKD may also translate to a weakened anti-inflammatory potential in the CKD patient and the CKD cat. The role of vitamin K in the homeostatic physiology of the kidney and the pathophysiology of feline CKD has not been the subject of focused study.

As with the ageing dog, cats are also likely to present with degenerative joint diseases [129] and the high co-morbidity relationship with CKD suggests that vitamin K deficiency in the ageing cat is possible [130]. The function of vitamin K in the aetiology of feline degenerative joint diseases remains to be investigated.

5. Concluding remarks

Vitamin K in animals as a general subject has not been systematically investigated. There are, however, common vitamin K-related health issues that animals share with man. These are predominantly associated with coagulopathies, but diseases of the musculoskeletal system and kidney injury may have considerable overlaps to the mutual benefit of man and animals.

The requirements for animal health and well-being are poorly defined for vitamin K, with, in some cases, misconceptions about the contributions to vitamin K status from colonic bacterial sources. Animals may have a distinctly greater reliance on vitamin K2, in large part due to their diet, which is regulated by the feed that is given to them. Furthermore, the upper alimentary canal supply of menaquinones may be of central benefit to support the vitamin K status of several species such as ruminants and possibly in particular rabbit kittens.

An exciting emerging area is the molecular regulation embryological development and growth by the *in situ* generation of the vitamin K2 congener menaquinone-4, through the prenyltransferase UBIAD1. The work on the zebrafish has the potential to radiate across all species, and this may be of particular importance in conservation breeding programmes.

There is reason to suspect that in some metabolic and inflammatory diseases, there is a pronounced vitamin K deficiency. The prospect of intervening in some of these pathologies with vitamin K, such as kidney disease, is already being proposed for human patients. Defining the vitamin K requirements in different animals, beyond simple hepatic coagulation factor needs, may suggest newer approaches to veterinary medicine that could be investigated.

Acknowledgements

This work was facilitated, in part, by the award of a senior fellowship to SJH by The Northern Norway Regional Health Authority.

http://dx.doi.org/10.5772/63901

Author details

Jayde O'Neil[1], Bethany Scarrott[1], Ragnhild Aven Svalheim[2], Jonathan Elliott[1] and Stephen J. Hodges[1,3*]

*Address all correspondence to: shodges@rvc.ac.uk

1 Department of Comparative Biomedical Sciences, The Royal Veterinary College, London, UK

2 Nofima AS, Muninbakken, Breivika, Tromsø, Norway

3 Anesthesia and Critical Care Research Group, The University Hospital of North Norway and UIT-The Arctic University of Norway, Tromsø, Norway

References

[1] Marcelli SW, Chang HT, Chapman T, Chalk PA, Miles RJ, Poole RK. The respiratory chain of Helicobacter pylori: identification of cytochromes and the effects of oxygen on cytochrome and menaquinone levels. FEMS Microbiol Lett. 1996;138(1):59–64.

[2] Antunes A, Rainey FA, Wanner G, Taborda M, Pätzold J, Nobre MF, da Costa MS, Huber R. A new lineage of halophilic, wall-less, contractile bacteria from a brine-filled deep of the Red Sea. J Bacteriol. 2008;190(10):3580–7.

[3] Srinivas TN, Anil Kumar P, Madhu S, Sunil B, Sharma TV, Shivaji S. Cesiribacter andamanensis gen. nov., sp. nov., isolated from a soil sample from a mud volcano. Int J Syst Evol Microbiol. 2011;61(7):1521–7. [Erratum: Int J Syst Evol Microbiol. 2011;61(10): 2561]

[4] Mimuro M, Tsuchiya T, Inoue H, Sakuragi Y, Itoh Y, Gotoh T, Miyashita H, Bryant DA, Kobayashi M. The secondary electron acceptor of photosystem I in Gloeobacter violaceus PCC 7421 is menaquinone-4 that is synthesized by a unique but unknown pathway. FEBS Lett. 2005;579(17):3493–6.

[5] Taggart WV, Matschiner JT. Metabolism of menadione-6,7-3H in the rat. Biochemistry. 1969;8(3):1141–6.

[6] Dialameh GH, Taggart WV, Matschiner JT, Olson RE. Isolation and characterization of menaquinone-4 as a product of menadione metabolism in chicks and rats. Int J Vitam Nutr Res. 1971;41(3):391–400.

[7] Komai M, Shirakawa H, Kimura S. Newly developed model for vitamin K deficiency in germfree mice. Int J Vitam Nutr Res. 1988;58(1):55–9.

[8] Thijssen HH, Drittij-Reijnders MJ. Vitamin K distribution in rat tissues: dietary phylloquinone is a source of tissue menaquinone-4. Br J Nutr. 1994;72(3):415–25.

[9] Davidson RT, Foley AL, Engelke JA, Suttie JW. Conversion of dietary phylloquinone to tissue menaquinone-4 in rats is not dependent on gut bacteria. J Nutr. 1998;128(2):220–3.

[10] Ronden JE, Drittij-Reijnders MJ, Vermeer C, Thijssen HH. Intestinal flora is not an intermediate in the phylloquinone-menaquinone-4 conversion in the rat. Biochim Biophys Acta. 1998;1379(1):69–75.

[11] Buitenhuis HC, Soute BA, Vermeer C. Comparison of the vitamins K1, K2 and K3 as cofactors for the hepatic vitamin K-dependent carboxylase. Biochim Biophys Acta. 1990;1034(2):170–5.

[12] Peterson WH, Peterson MS. Relation of bacteria to vitamins and other growth factors. Bacteriol Rev. 1945;9(2):49–109.

[13] Glick MC, Zilliken F, György P. Supplementary growth promoting effect of 2-methyl-1,4-naphthoquinone of Lactobacillus bifidus var. pennsylvanicus. J Bacteriol. 1959;77(2):230–6.

[14] Starr SE, Killgore GE, Dowell VR Jr. Comparison of Schaedler agar and trypticase soy-yeast extract agar for the cultivation of anaerobic bacteria. Appl Microbiol. 1971;22(4): 655–8.

[15] Bentley R, Meganathan R. Biosynthesis of vitamin K (menaquinone) in bacteria. Microbiol Rev. 1982;46(3):241–80.

[16] Collins MD, Pirouz T, Goodfellow M, Minnikin DE. Distribution of menaquinones in actinomycetes and corynebacteria. J Gen Microbiol. 1977;100(2):221–30.

[17] Collins MD, Jones D. Distribution of isoprenoid quinone structural types in bacteria and their taxonomic implication. Microbiol Rev. 1981;45(2):316–54.

[18] Shah HN, Collins MD. Genus Bacteroides. A chemotaxonomical perspective. J Appl Bacteriol. 1983;55(3):403–16.

[19] Dam H. Cholesterinsoffwechsel in Hühnereirn Hühnchen. Biochem. Z 1929:215;475–492.

[20] Dam H. The antihaemorrhagic vitamin of the chick. Biochem J. 1935;29(6):1273–85.

[21] Dam H, Schønheyder F, Lewis L. The requirement for vitamin K of some different species of animals. Biochem J. 1937;31(1):22–7.

[22] Sullivan, T. and Al-Ubaidi YY. Linear growth and mineralization of bones in broad breasted bronze turkeys. Poultry Sci. 1963;42:46–49.

[23] Sullivan, T. Skeletal problems in poultry: estimated annual cost and descriptions. Poultry Sci. 1994;73:879–882.

[24] Fleming RH. Nutritional factors affecting poultry bone health. Proc Nutr Soc. 2008;67(2):177–83.

[25] Fleming RH, McCormack HA, Mcteir L & Whitehead CC. Bone structure and strength at different ages in laying hens and effects of dietary particulate limestone, vitamin K and ascorbic acid. Br Poult Sci 1998;39, 434–440.

[26] Zhang C, Li D, Wang F, Dong T. Effects of dietary vitamin K levels on bone quality in broilers. Arch Tierernahr. 2003;57(3):197–206.

[27] Askim M, Haug A, Gadeholt G. Dietary intake of menaquinone-4 may determine hepatic and pancreatic menaquinone-4 in chickens. Food Nutr Res. 2012;56.

[28] Fleming RH, McCormack HA, Mcteir L, Whitehead CC. Effects of dietary particulate limestone, vitamin K3 and fluoride and photostimulation on skeletal morphology and osteoporosis in laying hens. Br Poult Sci 2003;44:683–689.

[29] Stone WB, Okoniewski JC, Stedelin JR. Poisoning of wildlife with anticoagulant rodenticides in New York. J Wildl Dis. 1999;35(2):187–93.

[30] Guitart R, Sachana M, Caloni F, Croubels S, Vandenbroucke V, Berny P. Animal poisoning in Europe. Part 3: wildlife. Vet J. 2010;183(3):260–5.

[31] Matschiner JT, Taggart WV. Bioassay of vitamin K by intracardial injection in deficient adult male rats. J Nutr. 1968;94(1):57–9.

[32] Craciun AM, Groenen-van Dooren MM, Thijssen HH, Vermeer C. Induction of prothrombin synthesis by K-vitamins compared in vitamin K-deficient and in brodifacoum-treated rats. Biochim Biophys Acta. 1998;1380(1):75–81.

[33] Mathers JC, Fernandez F, Hill MJ, McCarthy PT, Shearer MJ, Oxley A. Dietary modification of potential vitamin K supply from enteric bacterial menaquinones in rats. Br J Nutr. 1990;63(3):639–52.

[34] Barnes RH and Fiala G. Effects of the prevention of coprophagy in the rat. I. Growth studies. J Nutr. 1958;64(4):533–40.

[35] Hammed A, Matagrin B, Spohn G, Prouillac C, Benoit E, Lattard V. VKORC1L1, an enzyme rescuing the vitamin K 2,3-epoxide reductase activity in some extrahepatic tissues during anticoagulation therapy. J Biol Chem. 2013;288(40):28733–42.

[36] Esther A, Endepols S, Freise J, Klemann N, Runge M, Pelz HJ. Rodenticide resistance and consequences. Bundesgesundheitsblatt Gesundheitsforschung Gesundheitsschutz. 2014;57(5):519–23.

[37] Spohn G, Kleinridders A, Wunderlich FT, Watzka M, Zaucke F, Blumbach K, Geisen C, Seifried E, Müller C, Paulsson M, Brüning JC, Oldenburg J. VKORC1 deficiency in mice causes early postnatal lethality due to severe bleeding. Thromb Haemost. 2009;101(6):1044–50.

[38] Nakagawa K, Hirota Y, Sawada N, Yuge N, Watanabe M, Uchino Y, Okuda N, Shimomura Y, Suhara Y, Okano T. Identification of UBIAD1 as a novel human menaquinone-4 biosynthetic enzyme. Nature. 2010;468:117–121.

[39] Nakagawa K, Sawada N, Hirota Y, Uchino Y, Suhara Y, Hasegawa T, Amizuka N, Okamoto T, Tsugawa N, Kamao M, Funahashi N, Okano T. Vitamin K2 biosynthetic enzyme, UBIAD1 is essential for embryonic development of mice. PLoS One. 2014;9(8):e104078.

[40] Morot C. Des pelotes stomacal des leporides. Mem. Soc. Centr. Med.Vet. 1882;12, Ser. 1.

[41] Eden A. Coprophagy in the rabbit: origin of the night feces. Nature 1940;145,628–629.

[42] Southern HN. Coprophagy in the wild rabbit. Nature 1940;145,262.

[43] Monteils V, Cauquil L, Combes S, Godon JJ, Gidenne T. Potential core species and satellite species in the bacterial community within the rabbit caecum. FEMS Microbiol Ecol. 2008;66(3):620–9.

[44] Kovács M, Szendro Z, Milisits G, Bóta B, Bíró-Németh E, Radnai I, Pósa R, Bónai A, Kovács F, Horn P. Effect of nursing methods and faeces consumption on the development of the bacteroides, lactobacillus and coliform flora in the caecum of the newborn rabbits. Reprod Nutr Dev. 2006;46(2):205–10.

[45] Bäuerl C, Collado MC, Zúñiga M, Blas E, Pérez Martínez G. Changes in cecal microbiota and mucosal gene expression revealed new aspects of epizootic rabbit enteropathy. PLoS One. 2014;9(8):e105707.

[46] Hollander D, Rim E. Vitamin K2 absorption by rat everted small intestinal sacs. Am J Physiol. 1976;231(2):415–9.

[47] Appendino G, Tagliapietra S, Gariboldi P, Nano GM, Picci V. Oxygenated prenylated coumarins from F. communis. Phytochemistry 1988:27;3619–3624.

[48] Tligui N, Ruth GR. Ferula communis variety brevifolia intoxication of sheep. Am J Vet Res. 1994;55(11):1558–63.

[49] Tligui N, Ruth GR, Felice LJ. Plasma ferulenol concentration and activity of clotting factors in sheep with Ferula communis variety brevifolia intoxication. Am J Vet Res. 1994;55(11):1564–1569.

[50] Louvet MS, Gault G, Lefebvre S, Popowycz F, Boulven M, Besse S, Benoit E, Lattard V, Grancher D. Comparative inhibitory effect of prenylated coumarins, ferulenol and ferprenin, contained in the 'poisonous chemotype' of Ferula communis on mammal liver microsomal VKORC1 activity. Phytochemistry. 2015;118:124–30.

[51] Lamnaouer D. Anticoagulant activity of coumarins from Ferula communis L. Therapie. 1999;54(6):747–51.

[52] Baker DC, Robbe SL, Jacobson L, Manco-Johnson MJ, Holler L, Lefkowitz J. Hereditary deficiency of vitamin-K-dependent coagulation factors in Rambouillet sheep. Blood Coagul Fibrinolysis. 1999;10(2):75–80.

[53] Johnson JS, Laegreid WS, Basaraba RJ, Baker DC. Truncated gamma-glut amyl carboxylase in rambouillet sheep. Vet Pathol. 2006;43(4):430–7.

[54] Johnson JS, Soute BA, Olver CS, Baker DC. Defective gamma-glutamyl carboxylase activity and bleeding in Rambouillet sheep. Vet Pathol. 2006;43(5):726–32.

[55] Rosenberg M and Redding RW. "Ancestral Pigs: A New (Guinea) Model for Pig Domestication in the Near East". In: Ancestors for the Pigs: Pigs in Prehistory. S.M. Nelson ed., pp. 65–76. Masca Research Papers in Science and Archaeology Vol. 15. University of Pennsylvania Museum.

[56] Nemeth D. "Privy-Pigs Prehistory? A Korean Analogy for Neolithic Chinese Substance Practices" In: Ancestors for the Pigs: Pigs in Prehistory. S.M. Nelson ed., pp. 11–25.

[57] Massicotte P, Mitchell L, Andrew M. A comparative study of coagulation systems in newborn animals. Pediatr Res. 1986;20(10):961–5.

[58] Schendel HE, Johnson BC. Vitamin K deficiency in the baby pig. J. Nutr. 1962 76:124–130.

[59] Bowland JP Progress in swine nutrition for the veterinarian. Can Vet J. 1973;14(1):3–11.

[60] Bollongino R, Burger J, Powell A, Mashkour M, Vigne JD, Thomas MG. Modern taurine cattle descended from small number of near-eastern founders. Mol Biol Evol. 2012;29(9): 2101–4.

[61] Matschiner JT, Taggart WV, Amelotti JM. The vitamin K content of beef liver. Detection of a new form of vitamin K. Biochemistry. 1967;6(5):1243–8.

[62] Matschiner JT, Amelotti JM. Characterization of vitamin K from bovine liver. J Lipid Res. 1968;9(2):176–9.

[63] Duello TJ, Matschiner JT. Identification of phylloquinone in horse liver. Arch Biochem Biophys. 1970;138(2):640–5.

[64] Rødbotten R, Gundersen T, Vermeer C, Kirkhus B. Vitamin K2 in different bovine muscles and breeds. Meat Sci. 2014;97(1):49–53.

[65] Elder SJ, Haytowitz DB, Howe J, Peterson JW, Booth SL. Vitamin K contents of meat, dairy, and fast food in the U.S. diet. J Agric Food Chem. 2006;54(2):463–7.

[66] Campbell HA, and Link KP. Studies on the Hemorrhagic Sweet Clover Disease. IV. The isolation and crystallization of the hemorrhagic agent. J. Biol. Chem. 1941:138, 21–33.

[67] Overman RS, Stahmann MA, Sullivan WR, Huebner CF, Campbell HA, Link P. Studies on the hemorrhagic sweet clover disease. VII. The effect of 3,3′-methylenebis (4-hydroxycoumarin) on the prothrombin time of the plasma of various animals. J. Biol. Chem. 1942;142, p. 941.

[68] Overman RS, Stahmann MA, Huebner CF, Sullivan WR, Spero L, Doherty DG, Ikawa M, Graf L, Roseman S, Link KP. Anticoagulant activity and structure in the 4-hydroxycoumarin group. J. Biol. Chem. 1944;153:5–24.

[69] Schofield FW. Damaged sweet clover: the cause of a new disease in cattle simulating hemnorrhagic septicemia and blackleg. J. Am. Vet. Med. Assoc. 1924;64:553.

[70] Roderick LMA. The pathology of sweet clover disease in cattle. J. Am. Vet. Med. Assoc. 1929;74:314.

[71] Stahmann, MA, Huebner, CF, Link KP. Studies on the hemorrhagic sweet clover disease. XIII.J. Biol. Chem. 1941;138:513–527.

[72] McGorum BC, Henderson IS, Stirling D, Wallace R, Haggart C, Thomas AE. Vitamin K deficiency bleeding in a Standardbred colt. J Vet Intern Med. 2009;23(6):1307–10.

[73] Terachi T, Inoue Y, Ashihara N, Kobayashi M, Ando K, Matsui T. Plasma vitamin K concentration in horses supplemented with several vitamin K homologs. J Anim Sci. 2011;89(4):1056–61.

[74] Rebhun WC, Tennant BC, Dill SG, King JM. Vitamin K3-induced renal toxicosis in the horse. J Am Vet Med Assoc. 1984;184(10):1237–9.

[75] Maxie G, van Dreumel T, McMaster D, Baird J. Ontario. Menadione (vitamin K(3)) toxicity in six horses. Can Vet J. 1992;33(11):756–7.

[76] ftp://ftp.fao.org/FI/STAT/summary/a-6.pdf

[77] Taveekijakarn P, Miyazaki T, Matsumoto M, Arai S. Studies on vitamin K deficiency in amago salmon, *Oncorhynchus rhodurus* (Jordan & McGregor). J Fish Dis 1996;19:209–214.

[78] Udagawa M. The Effect of parental vitamin K deficiency on bone structure in mummichog Fundulus heteroclitus. J World Aquacult Soc 2004;35:366–371.

[79] Lall SP, Lewis-McCrea, LM. Role of nutrients in skeletal metabolism and pathology in fish—an overview. Aquaculture 2007;267:3–19.

[80] Doolittle RF, Surgenor DM. Blood coagulation in fish. Am J Physiol 1962;203:964–970.

[81] Poston HA. Effect of dietary vitamin K and sulfaguanidine on blood coagulation time, microhematocrit, and growth of immature brook trout. Progr Fish-Cult 1964;26:59–64.

[82] Krossøy C, Waagbø R, Ørnsrud R. Vitamin K in fish nutrition. Aquaculture Nutrition 2011;17:585–594.

[83] Jagadeeswaran P, Sheehan JP. Analysis of blood coagulation in the zebrafish. Blood Cells, Molecules, and Diseases 1999;25:239–249.

[84] Jagadeeswaran P, Gregory M, Johnson S, Thankavel B. Haemostatic screening and identification of zebrafish mutants with coagulation pathway defects: an approach to identifying novel haemostatic genes in man. Br J Haematol 2000;110:946–956.

[85] Hanumanthaiah R, Thankavel B, Day K, Gregory M, Jagadeeswaran P. Developmental expression of vitamin K-dependent gamma-carboxylase activity in zebrafish embryos: effect of warfarin. Blood Cell Mol Dis 2001;27:992–999.

[86] Cheng L, Zhang W, Lin S, Xu W, Mai K. Effects of dietary vitamin K on growth performances, blood coagulation time and menaquinone-4 (MK-4) concentration in tissues of juvenile large yellow croaker Pseudosciaena crocea. Aquacult Res. 2015;46:1269–1275.

[87] Udagawa M. Physiological role of vitamin K in fish. Japan Agric Res Quart 2000;34:279–284.

[88] Graff IE, Waagbø R, Fivelstad S, Vermeer C, Lie Ø, Lundebye AK. A multivariate study on the effects of dietary vitamin K, vitamin D3 and calcium, and dissolved carbon dioxide on growth, bone minerals, vitamin status and health performance in smolting Atlantic salmon *Salmo salar* L. J Fish Dis 2002;25:599–614.

[89] Graff IE, Krossøy C, Gjerdevik K, Julshamn K. Influence of dietary menadione nicotinamide bisulphite (vitamin K3) and phylloquinone (vitamin K1) on Atlantic salmon (*Salmo salar* L.) tissue levels, determined by high-performance liquid chromatography with fluorescence detection. Aquacult Nutr 2010;16:637–647.

[90] Krossøy C, Waagbø R, Fjelldal PG, Wargelius A, Lock E-J, Graff IE, Ørnsrud R. Dietary menadione nicotinamide bisulphite (vitamin K3) does not affect growth or bone health in first-feeding fry of Atlantic salmon (*Salmo salar* L.). Aquacult Nutr 2009;15:638–649.

[91] Grahl-Madsen E, Lie Ø. Effects of different levels of vitamin K in diets for cod (*Gadus morhua*). Aquaculture 1997;151:269–274.

[92] Udagawa M, Nakazoe J-i, Murai T. Tissue distribution of phylloquinone and menaquinone-4 in sardine, Sardinops Melanostictus. Comp Biochem Physiol Part B: Comp Biochem. 1993;106:297–301.

[93] Udagawa M, Hirose K. Physiological abnormality and the tissue vitamin K level of mummichog Fundulus heteroclitus fed with a vitamin K free diet. Fisheries Sci. 1998;64:612–616.

[94] Udagawa M, Nakazoe J-I, Murai T. Comparison of the composition of vitamin K in different forms between cultured and wild ayu (*Plecoglossus altivelis*). Fisheries Sci. 1993;65:331–332.

[95] Hegarty JM, Yang H, Chi NC. UBIAD1-mediated vitamin K2 synthesis is required for vascular endothelial cell survival and development. Development. 2013;140(8):1713–9.

[96] Udagawa M. The effect of dietary vitamin K (phylloquinone and menadione) levels on the vertebral formation in mummichog Fundulus heteroclitus. Fisheries Sci. 2001;67:104–109.

[97] Richard N, Fernández I, Wulff T, Hamre K, Cancela L, Conceição LE, Gavaia PJ. Dietary supplementation with vitamin K affects transcriptome and proteome of Senegalese

sole, improving larval performance and quality. Marine biotechnology (New York, N.Y.) 2014:16;522–537.

[98] Krossøy C, Ørnsrud R, Wargelius A. Differential gene expression of bgp and mgp in trabecular and compact bone of Atlantic salmon (*Salmo salar* L.) vertebrae. J Anatomy 2009;215:663–672.

[99] Roy PK, Lall, SP. Vitamin K deficiency inhibits mineralization and enhances deformity in vertebrae of haddock (*Melanogrammus aeglefinus* L.). Comp Biochem Physiol Part B, Biochem Mol Biol 2007;148:174–183.

[100] Grisdale-Helland B, Helland SJ, Åsgård T. Problems associated with the present use of menadione sodium bisulfite and vitamin A in diets for Atlantic salmon. Aquaculture 1991;92:351–358.

[101] Freedman AH, Gronau I, Schweizer RM, Ortega-Del Vecchyo D, Han E, Silva PM, Galaverni M, Fan Z, Marx P, Lorente-Galdos B, Beale H, Ramirez O, Hormozdiari F, Alkan C, Vilà C, Squire K, Geffen E, Kusak J, Boyko AR, Parker HG, Lee C, Tadigotla V, Wilton A, Siepel A, Bustamante CD, Harkins TT, Nelson SF, Ostrander EA, Marques-Bonet T, Wayne RK, Novembre J. Genome sequencing highlights the dynamic early history of dogs. PLoS Genet. 2014(1):e1004016.

[102] Duleba A, Skonieczna K, Bogdanowicz W, Malyarchuk B, Grzybowski T. Complete mitochondrial genome database and standardized classification system for Canis lupus familiaris. Forensic Sci Intnl: Genetics 2015;19:123–129.

[103] Skoglund P, Ersmark E, Palkopoulou E, Dalén L. Ancient wolf genome reveals an early divergence of domestic dog ancestors and admixture into high-latitude breeds. Curr Biol. 2015;25(11):1515–9.

[104] Mellersh C. DNA testing and domestic dogs. Mamm Genome. 2012;23(1–2):109–23.

[105] Mason DJ, Abrams-Ogg A, Allen D, Gentry PA, Gadd KR. Vitamin K-dependent coagulopathy in a black Labrador Retriever. J Vet Intern Med. 2002;16(4):485–8.

[106] Mount ME, Kass PH. Diagnostic importance of vitamin K1 and its epoxide measured in serum of dogs exposed to an anticoagulant rodenticide. Am J Vet Res. 1989;50(10): 1704–9.

[107] Robben JH, Mout HC, Kuijpers EA. Anticoagulant rodenticide poisoning in dogs in The Netherlands. Tijdschr Diergeneeskd. 1997;122(17):466–71.

[108] Waddell LS, Poppenga RH, Drobatz KJ. Anticoagulant rodenticide screening in dogs: 123 cases (1996–2003). J Am Vet Med Assoc. 2013;242(4):516–21.

[109] Duello TJ, Matschiner JT. Characterization of vitamin K from pig liver and dog liver. Arch Biochem Biophys. 1971 May;144(1):330–8.

[110] Duello TJ, Matschiner JT. The relationship between the storage forms of vitamin K and dietary phylloquinone in the dog. Int J Vitam Nutr Res. 1971;41(2):180–8.

[111] Hamidi MS, Cheung AM. Vitamin K and musculoskeletal health in postmenopausal women. Mol Nutr Food Res. 2014;58(8):1647–57.

[112] Shea MK, Kritchevsky SB, Hsu FC, Nevitt M, Booth SL, Kwoh CK, McAlindon TE, Vermeer C, Drummen N, Harris TB, Womack C, Loeser RF. Health ABC Study. The association between vitamin K status and knee osteoarthritis features in older adults: the Health, Aging and Body Composition Study. Osteoarthritis Cartilage. 2015;23(3): 370–8.

[113] Kohn B, Weingart C, Giger U. Haemorrhage in seven cats with suspected anticoagulant rodenticide intoxication. J Feline Med Surg. 2003;5(5):295–304.

[114] Maddison JE, Watson AD, Eade IG, Exner T. Vitamin K-dependent multifactor coagulopathy in Devon Rex cats. J Am Vet Med Assoc. 1990;197(11):1495–7.

[115] Soute BA, Ulrich MM, Watson AD, Maddison JE, Ebberink RH, Vermeer C. Congenital deficiency of all vitamin K-dependent blood coagulation factors due to a defective vitamin K-dependent carboxylase in Devon Rex cats. Thromb Haemost. 1992;68(5):521–5.

[116] Edwards DF, Russell RG. Probable vitamin K-deficient bleeding in two cats with malabsorption syndrome secondary to lymphocytic-plasmacytic enteritis. J Vet Intern Med. 1987;1(3):97–101.

[117] Lisciandro SC, Hohenhaus A, Brooks M. Coagulation abnormalities in 22 cats with naturally occurring liver disease. J Vet Intern Med. 1998;12(2):71–5.

[118] Center SA, Warner K, Corbett J, Randolph JF, Erb HN. Proteins invoked by vitamin K absence and clotting times in clinically ill cats. J Vet Intern Med. 2000;14(3):292–7.

[119] Strieker MJ, Morris JG, Feldman BF, Rogers QR. Vitamin K deficiency in cats fed commercial fish-based diets. J Small Anim Pract. 1996;37(7):322–6.

[120] Brown CA, Elliott J, Schmiedt CW, Brown SA. Chronic kidney disease in aged cats: clinical features, morphology, and proposed pathogeneses. Vet Pathol. 2016;53(2):309–26.

[121] Shipov A, Segev G, Meltzer H, Milrad M, Brenner O, Atkins A, Shahar R. The effect of naturally occurring chronic kidney disease on the micro-structural and mechanical properties of bone. PLoS One. 2014;9(10):e110057.

[122] McCabe KM, Adams MA, Holden RM. Vitamin K status in chronic kidney disease. Nutrients. 2013;5(11):4390–8.

[123] Wei FF, Drummen NE, Schutte AE, Thijs L, Jacobs L, Petit T, Yang WY, Smith W, Zhang ZY, Gu YM, Kuznetsova T, Verhamme P, Allegaert K, Schutte R, Lerut E, Evenepoel P, Vermeer C, Staessen JA. Vitamin K dependent protection of renal function in multi-ethnic population studies. EBioMedicine. 2016;4:162–9.

[124] Shearer MJ. Vitamin K metabolism and nutriture. Blood Rev. 1992;6(2):92–104.

[125] Machowska A, Carrero JJ, Lindholm B, Stenvinkel P. Therapeutics targeting persistent inflammation in chronic kidney disease. Transl Res. 2016;167(1):204–13.

[126] Weiser H, Kormann A. Pharmacological properties of vitamin K1 and analogues: Antidote functions, analgesic and antiphlogistic properties. Unpublished results presented as a poster at the 17th Steenbock Symposium June 1987, Current Advances in Vitamin K Research, University of Wisconsin, Madison, USA. 1988.

[127] Reddi K, Henderson B, Meghji S, Wilson M, Poole S, Hopper C, Harris M, Hodges SJ. Interleukin 6 production by lipopolysaccharide-stimulated human fibroblasts is potently inhibited by naphthoquinone (vitamin K) compounds. Cytokine. 1995;7(3): 287–90.

[128] Soper RJ, Oguz C, Emery R, Pitsillides AA, Hodges SJ. Vitamin K catabolite inhibition of ovariectomy-induced bone loss: structure-activity relationship considerations. Mol Nutr Food Res. 2014;58(8):1658–66.

[129] Lascelles BD, Henry JB 3rd, Brown J, Robertson I, Sumrell AT, Simpson W, Wheeler S, Hansen BD, Zamprogno H, Freire M, Pease A. Cross-sectional study of the prevalence of radiographic degenerative joint disease in domesticated cats. Vet Surg. 2010;39(5): 535–44.

[130] Marino CL, Lascelles BD, Vaden SL, Gruen ME, Marks SL. Prevalence and classification of chronic kidney disease in cats randomly selected from four age groups and in cats recruited for degenerative joint disease studies. J Feline Med Surg. 2014;16(6):465–72.

Section 6

Biosynthesis of Vitamin K2 and Various Metabolic Effects

Chapter 13

Medicinal Chemistry of Vitamin K Derivatives and Metabolites

Shinya Fujii and Hiroyuki Kagechika

Additional information is available at the end of the chapter

http://dx.doi.org/10.5772/63511

Abstract

Vitamin K acts as a cofactor for γ-glutamyl carboxylase. Recently, various biological activities of vitamin K have been reported. Anti-proliferative activities of vitamin K, especially in vitamin K_3, are well known. In addition, various physiological and pharmacological functions of vitamin K_2, such as transcription modulators as nuclear steroid and xenobiotic receptor (SXR) ligands and anti-inflammatory effects, have been revealed in the past decade. Characterization of vitamin K metabolites is also important for clinical application of vitamin K and its derivatives. In this chapter, recent progress on the medicinal chemistry of vitamin K derivatives and metabolites is discussed.

Keywords: vitamin K derivative, metabolite, antitumor activity, anti-inflammatory activity, steroid and xenobiotic receptor/pregnane X receptor

1. Introduction

Vitamin K is a specific cofactor for γ-glutamyl carboxylase (GGCX), which catalyzes formation of γ-carboxyglutamyl (Gla) residues in vitamin K–dependent proteins (**Figure 1**) [1]. Various other biological activities of vitamin K and its derivatives have also been reported. For example, vitamin K_3 (menadione), a vitamin K homologue that was considered as a synthetic vitamin K, has antitumor activity [2–5], as does vitamin K_2 (menaquinone) [6, 7]. Among the homologues of vitamin K_2, menaquinone-4 (MK-4), which contains four isoprene units, has been intensively investigated. It binds to nuclear receptor human pregnane X receptor (PXR), which is also called steroid and xenobiotic receptor (SXR), and regulates transcription of osteoblastic genes [8, 9]. It also exhibits anti-inflammatory activity by suppressing the NF-kB pathway [10], and has an inhibitory effect on arteriosclerosis [11]. It binds 17β-hydroxysteroid dehydrogenase 4 and

INTECH
open science | open minds

© 2017 The Author(s). Licensee InTech. This chapter is distributed under the terms of the Creative Commons Attribution License (http://creativecommons.org/licenses/by/3.0), which permits unrestricted use, distribution, and reproduction in any medium, provided the original work is properly cited.

modulates estrogen metabolism [12]. Further, it enhances testosterone production [13, 14], and shows growth-inhibitory activity toward hepatocellular carcinoma (HCC) cells [6, 7]. These biological activities of vitamin K and its analogues are attractive targets of drug discovery, and the activities of vitamin K metabolites have also attracted much interest. A great many natural and synthetic biologically active 1,4-naphthoquinone derivatives (i.e., vitamin K derivatives) have been reported. In this chapter, we will focus on three medicinal-chemistry studies of vitamin K activities.

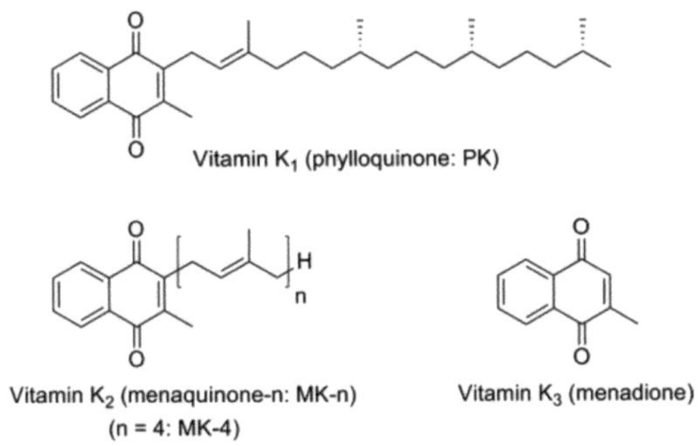

Figure 1. Structures of vitamin K homologues.

2. Menadione derivatives as antitumor agents

The antitumor activity of thioether derivatives is one of the most intensively investigated fields in the medicinal chemistry of menadione derivatives. Several series of naphthoquinone derivatives and benzoquinone derivatives bearing an alkyl, alkoxy, or alkylthio group as a side chain have been synthesized and biologically evaluated by assay of growth-inhibitory activity toward human hepatoma cell line HepB3. Almost all of the tested compounds, as well as the parent menadione, exhibited significant inhibitory activity, and the alkylthio derivatives were more potent than the corresponding alkyl and alkoxy derivatives. Among these compounds, a 2-hydroxyethylthio derivative Cpd 5 (compound 5; NSC 672121) exhibited the most potent activity (**Figure 2**) [15]. Subsequent studies revealed that Cpd 5 irreversibly inhibits growth-regulatory phosphatase Cdc25 by arylating a cysteine residue in the catalytic site, causing cell-cycle arrest [16–19].

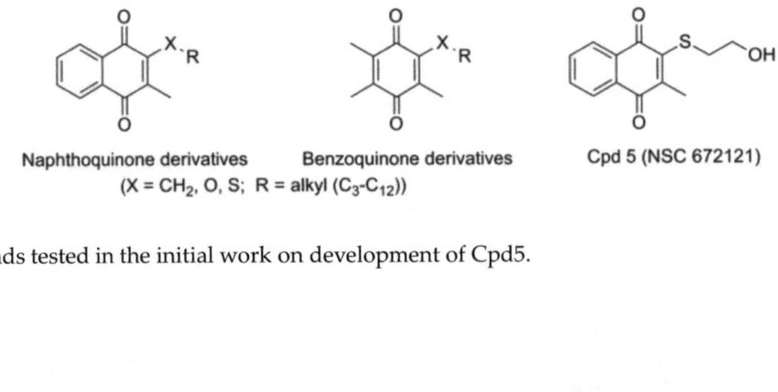

Figure 2. Compounds tested in the initial work on development of Cpd5.

Based on the finding that Cpd5 inhibits Cdc25 and exerts antitumor activities, various menadione derivatives have been developed as candidate antitumor compounds. Bis(2-hydroxyethylthio)naphthoquinone derivative NSC 95397 (**Figure 3**) showed potent Cdc25-inhibitory activity and inhibited proliferation of several cancer cell lines with greater potency than that of Cpd 5 [20]. Hydroxylated NSC 95397 derivatives exhibited enhanced Cdc25-inhibitory activity and inhibited growth of several cancer cell lines [21]. Fluorinated Cpd 5 was three times more potent than Cpd-5 itself in Hep3B growth inhibition and induced phosphorylation of ERK1/2, JNK1/2 and p38 in HepB3 cells [22]. Calculations suggested that fluorinated Cpd 5 cannot generate reactive oxygen species because of its modified redox profile, and therefore, the compound appears to function as a pure arylating agent [23]. Modification of the core structure afforded a maleimide derivative PM-20 with a submicromolar IC_{50} value for HepB3 growth inhibition. Structure-activity relationship study indicated that the biphenyl structure of PM-20 is essential for activity (**Figure 3**) [24].

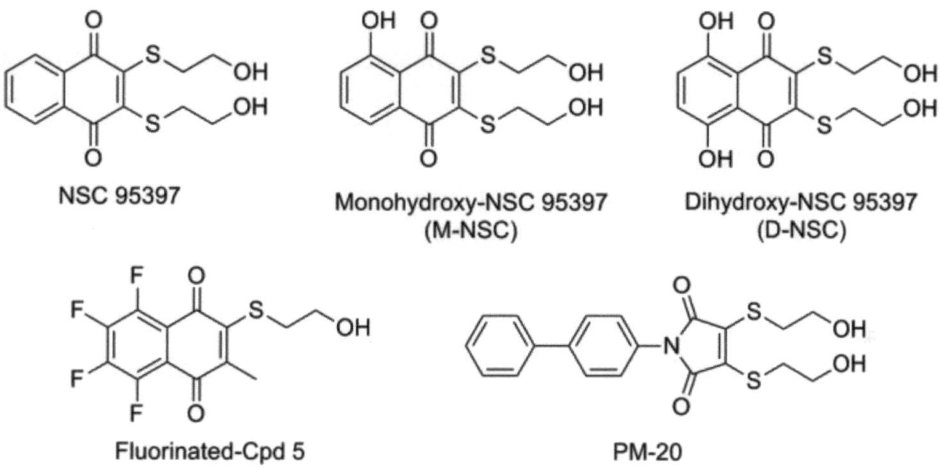

Figure 3. Structures of Cpd 5 derivatives bearing a 2-hydroxyethylthio moiety.

Modification of the hydroxyethyl side chain of Cpd-5 and NSC 95397 was also investigated. Carboxylic acid derivatives such as compounds **1, 3**, and **4** (**Figure 4**) were designed to interact with arginine residues in the catalytic site of Cdc25B, and indeed, they exhibited potent Cdc25B3-inhibitory activity [25, 26]. Though the cytotoxic activities of these carboxylic acid derivatives, especially dicarboxylic acid **4**, were low, prodrug-type benzyl ester derivatives exhibited enhanced growth-inhibitory activity toward HeLa cells. It was also found that Cpd 5 derivatives bearing a modified terminal, such as **6**, showed selective cytotoxicity toward neuroblastoma cell lines, whereas the parent menadione and Cpd 5 exhibited cytotoxicity toward both neuroblastoma cells and normal cell lines [27]. Aminoalkylmenadione derivatives such as **7** showed angiogenesis-inhibitory activity (**Figure 4**) [28].

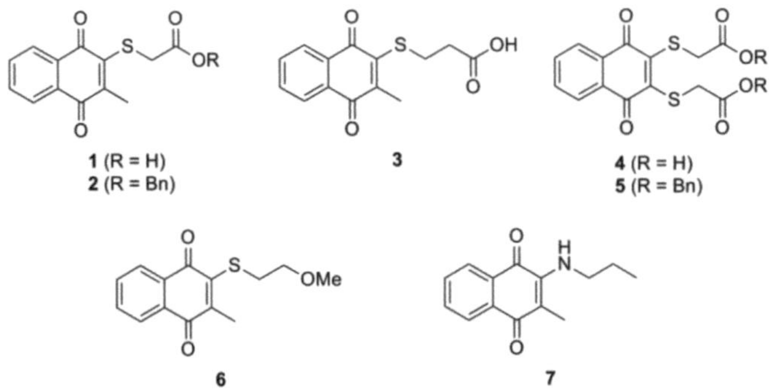

Figure 4. Examples of side chain-modified Cpd 5 derivatives.

A natural product, plumbagin (5-hydroxymenadione, **Figure 5**), shows anticancer and antiproliferative activities [29]. It suppresses the NF-kB activation pathway by modulating p65 and IkBα kinase activation to potentiate cytokine- and drug-induced apoptosis [30]. Structurally related naphthoquinone derivatives juglone and 1,4-naphthoquinone exerted similar TNFα-induced NF-kB inhibitory activities, whereas menadione did not [30]. Another natural product, lapachol, which has a hydroxyl group instead of the methyl group of MK-1, has anticancer activity [31]. A synthetic analogue **8** bearing two isoprene units also exerted antitumor activity (**Figure 5**) [32], and various biologically active lapachol derivatives have been developed [33]. The 2-hydroxy-1,4-naphthoquinone structure has distinct chemistry; for example, it has the characteristics of 1,2-naphthoquinone (e.g., lapachol can cyclize to form α-lapachone or β-lapachone), in contrast to 2-methyl-1,4-naphthoquinone.

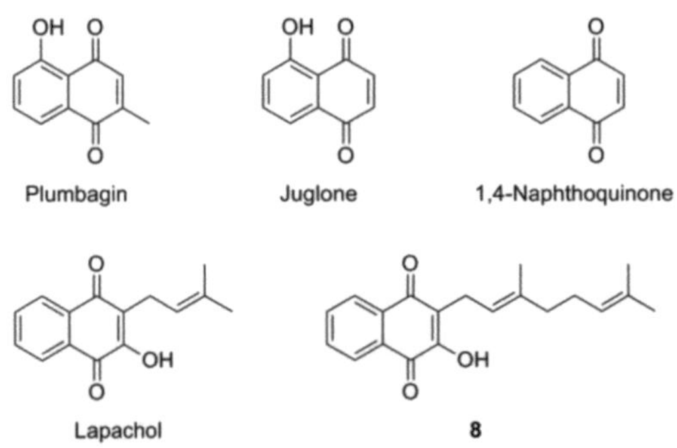

Figure 5. Some vitamin K–related naphthoquinone derivatives with antitumor activity.

3. Structure-activity relationship of MK-4 derivatives as nuclear SXR ligands

In the early twenty-first century, it was found that MK-4 binds a nuclear receptor, steroid, and xenobiotic receptor (SXR), which is a human homologue of pregnane X receptor (PXR), and regulates transcription of osteoblastic genes [8, 9]. Structure-activity relationships of MK-4 as an SXR ligand were intensively investigated by Suhara et al., using deuterated derivatives (**Figure 6**). Saturation of double bond(s) in the side chain significantly reduced the SXR agonistic activity. Triene derivative **9** bearing a 6,7-saturated side chain exerted only moderate activity, and diene **10**, monoene **11** (phylloquinone-d_7), and alkyl derivative **12** were inactive. Removal of methyl groups also reduced the activity, but demethylated compounds **13–16** still retained significant activity [34].

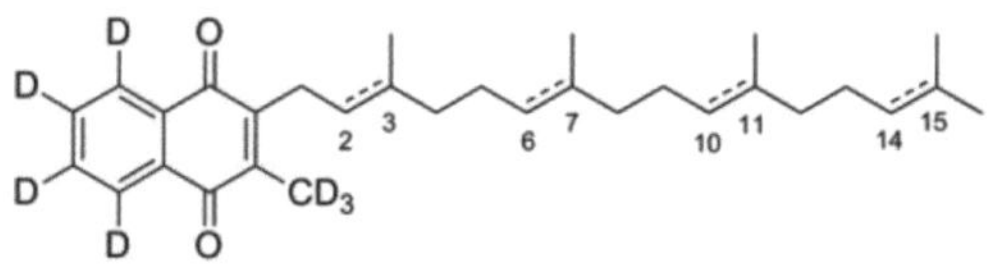

6,7-dihydro : **9** (triene)
6,7,10,11-tetrahydro : **10** (diene)
6,7,10,11,14,15-hexahydro : **11** (phylloquinone-d_7)
2,3,6,7,10,11,14,15-octahydro : **12** (alkyl)

des 3-Me : **13**
des 7-Me : **14**
des 11-Me : **15**
des 15-Me : **16**

Figure 6. Compounds used in SAR study of SXR.

The length of the side chain is important for the SXR activity of menaquinones. MK-1 bearing one prenyl group showed little ligand potency, while MK-2, MK-3, and MK-4 were more active. In the SXR-GAL4 one hybrid assay system, MK-3 was the most potent compound, and MK-2 and MK-4 showed somewhat lower activity. In the assay system using SXRE, MK-2, and MK-3 were the most potent compounds [35]. "Double side chain" vitamin K analogues bearing the same side chains at the 2-position and 3-position of the naphthoquinone ring were also designed and synthesized. MK-1-W and MK-2-W were as potent as MK-3 and MK-4, whereas MK-3-W, MK-4-W, and PK-W showed little activity (**Figure 7**) [35].

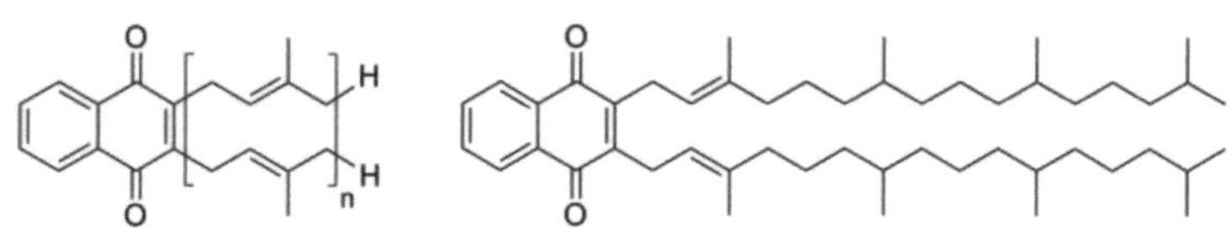

n = 1 : MK-1-W
n = 2 : MK-2-W
n = 3 : MK-3-W
n = 4 : MK-4-W

PK-W

Figure 7. Structures of double side chain vitamin K analogs.

Substitution at the terminal of the side chain of menaquinones significantly affects SXR ligand potency. Hydroxylated derivatives MK-2-ω-OH, MK-3-ω-OH, and MK-4-ω-OH showed little activity in the SXR-GAL4 one hybrid assay system, whereas compounds **17** and **18** bearing a terminal phenyl group exhibited more potent activity than the parent menaquinones (**Figure 8**). Compounds **17** and **18** also exhibited potent activity in the SXRE assay system [36]. Thus, a suitable hydrophobic side chain is essential for SXR activity of menaquinones.

Interestingly, Suhara et al. also found that menaquinone derivatives bearing a terminal hydrophobic substituent have the ability to induce selective neuronal differentiation of neuronal progenitor cells. The most potent compound **19** was twice as effective as the EtOH control, based on quantitation of Map2 mRNA (**Figure 8**) [37].

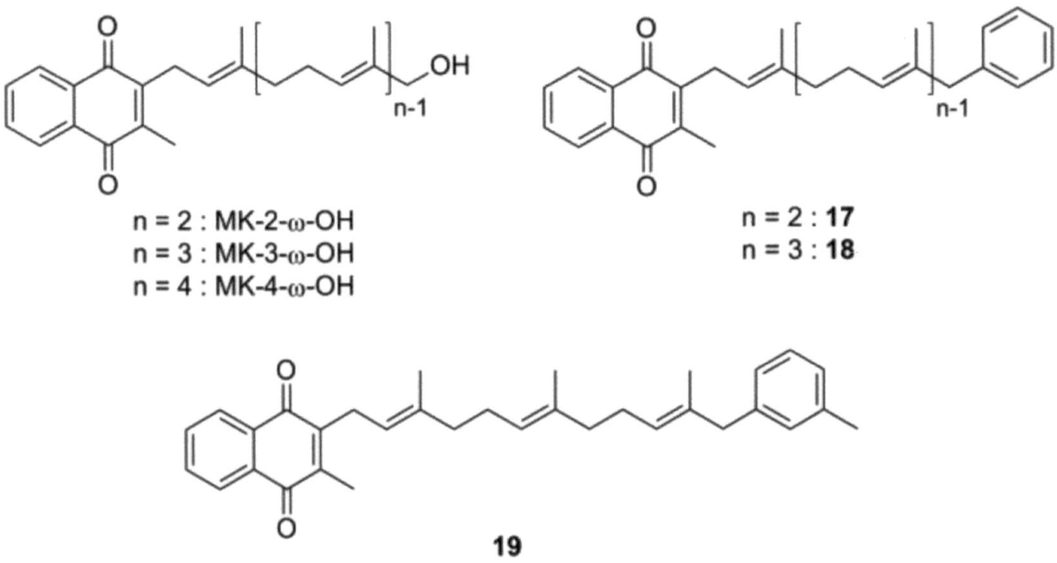

Figure 8. Structures of menaquinone derivatives with modified terminal.

4. Synthesis and biological activity of menaquinone metabolites

The biological activities of metabolites of vitamin K are also important. MK-4 is one of the most interesting vitamin K homologues because of its multifunctional properties, and ω-carboxyl homologues of MK-4 (MK-4-ω-COOH), K acid I, K acid II and their glucuronides have been identified as metabolites [38–42]. It is considered that MK-4 is initially metabolized to MK-4-ω-COOH by ω-oxidation, followed by β-oxidation to afford intermediary carboxylic acids (**Figure 9**) [43]. These carboxylic acids can be categorized into two groups; MK-n-ω-COOH derivatives bearing a α,β-unsaturated carboxy group and MK-n-(ω-2)-COOH derivatives bearing a γ,δ-unsaturated carboxy group. Chemical synthesis of these metabolites is essential for evaluation of their properties, and several synthetic routes have been reported.

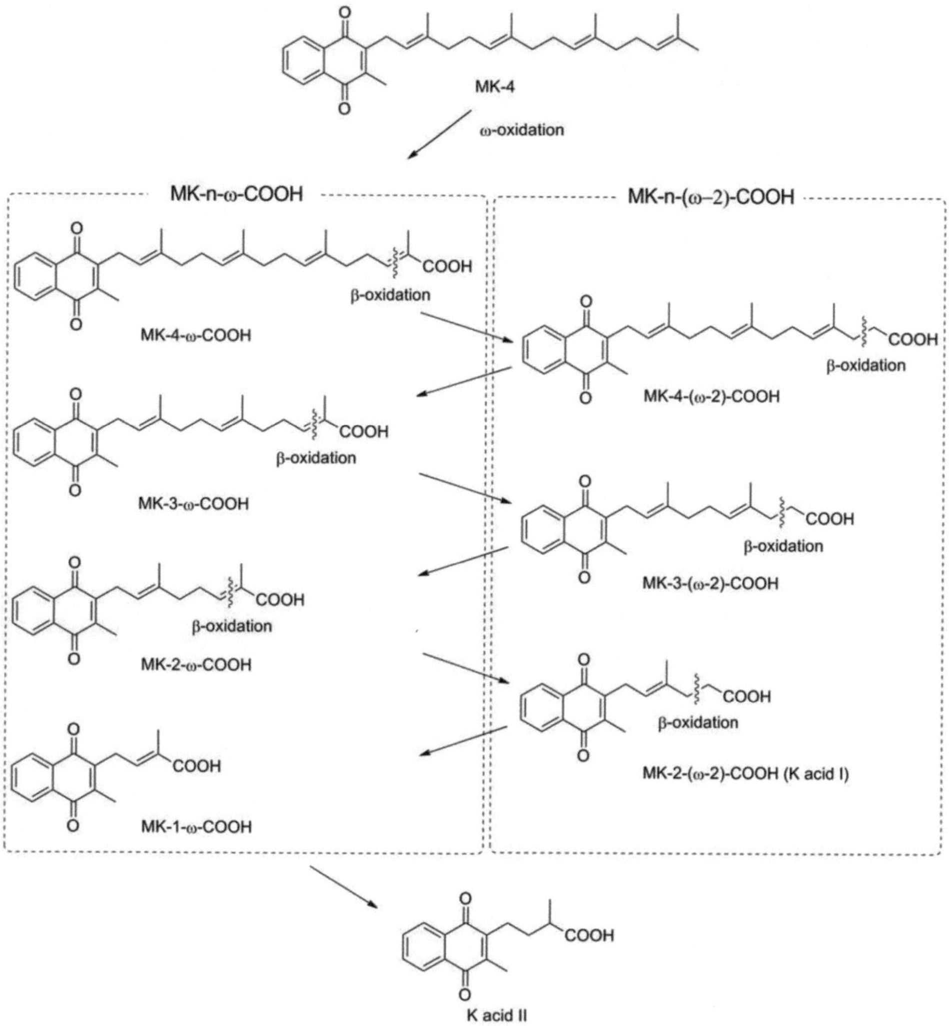

Figure 9. Putative catabolic pathways of MK-4.

4.1. Synthesis of menaquinone metabolites

The MK-4 metabolites K acid I and K acid II are also metabolites of phylloquinone (vitamin K_1). Several chemical syntheses of K acid I and K acid II have been reported. Watanabe et al. synthesized K acid I by direct addition of a carboxy side chain to the naphthoquinone framework using BF_3 etherate [44]. A route involving a malonyl derivative and decarboxylation was also investigated (**Figure 10**) [45]. They also synthesized K acid II. Addition of a side chain moiety by Friedel-Crafts acylation, followed by Clemmensen reduction, afforded naphthylcarboxylic acid, and oxidation of the naphthol moiety using Fremy's salt gave K acid II. Direct alkylation of naphthoquinone using peroxide also afforded K acid II (**Figure 11**) [44].

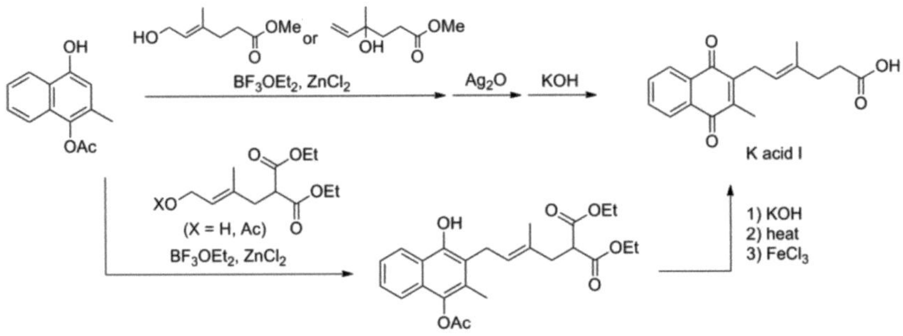

Figure 10. Synthetic route to K acid I (Watanabe et al.).

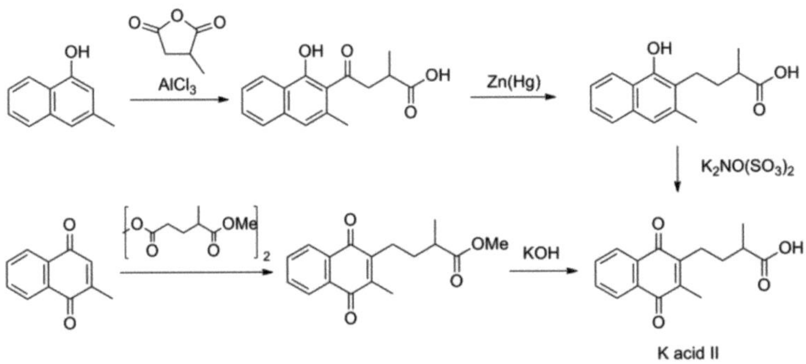

Figure 11. Synthetic route of K acid II (Watanabe et al.).

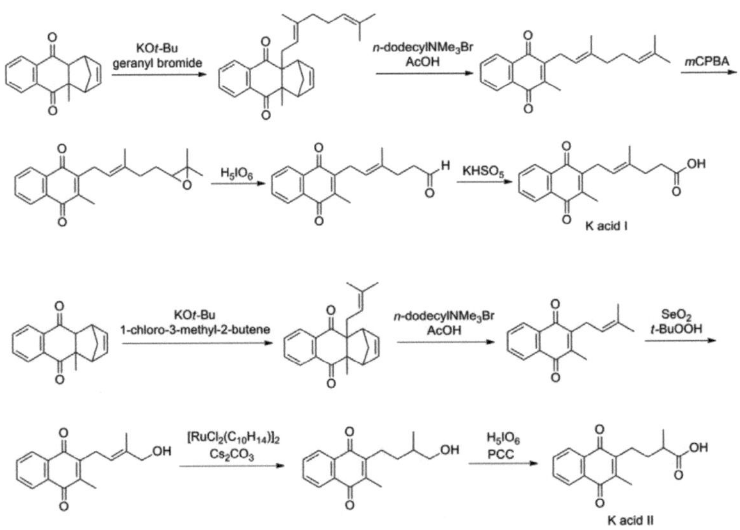

Figure 12. Synthetic routes of K acid I and K acid II (Teitelbaum et al.).

Teitelbaum et al. synthesized K acid I and K acid II by oxidation of MK-2 and MK-1, respectively. They prepared intermediary MK-n using a menadione-cyclopentadiene adduct as the same starting material (**Figure 12**) [46].

Okamoto et al. synthesized MK-1-ω-COOH by using Wittig reaction as a key step. To prepare the intermediary aldehyde, they employed alkylation and oxidative cleavage (**Figure 13**) [47].

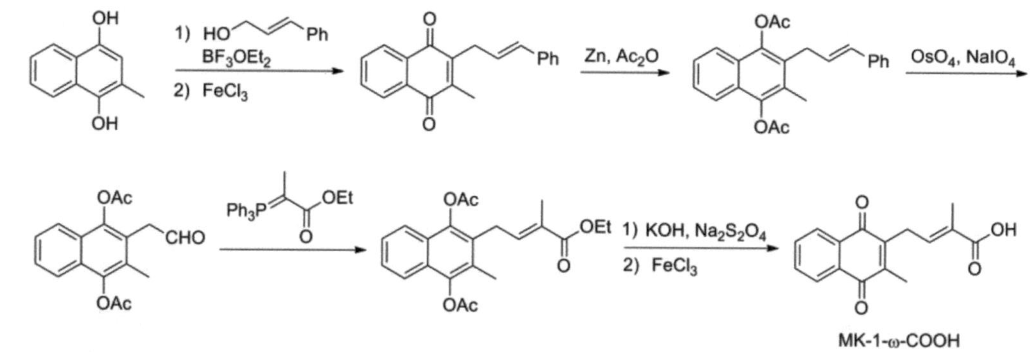

Figure 13. Synthesis of MK-1-ω-COOH (Okamoto et al.).

Terao et al. synthesized MK-3-(ω-2)-COOH and MK-4-(ω-2)-COOH using Claisen rearrangement as a key reaction. Claisen reaction of triethyl orthoacetate and MK-n derivative gave two-carbon-atom-extended carboxylic acid esters, and then hydration afforded MK-n-(ω-2)-COOH derivatives (**Figure 14**) [48].

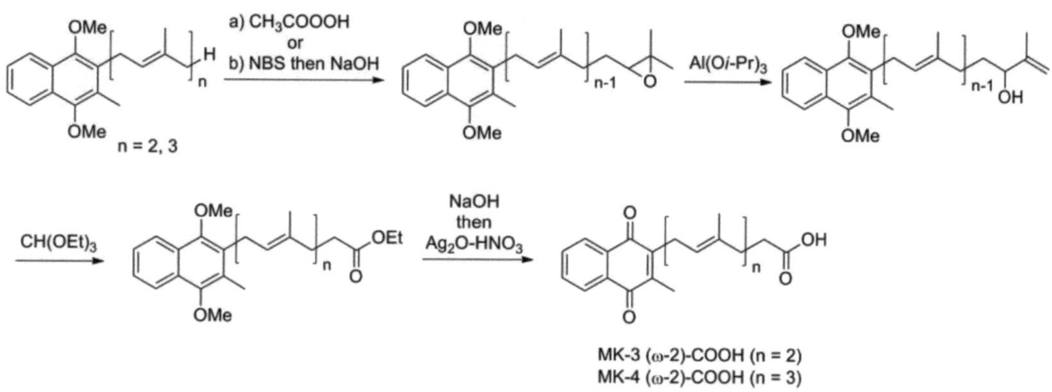

Figure 14. Synthesis of MK-n-(ω-2)-COOH derivatives (Terao et al.).

Masaki et al. employed sulfur-contractive anionic [2,3]-sigmatropic rearrangement for side chain elongation. Treatment of allyl sulfide with base afforded two-carbon-atom-extended carboxylic acid esters in one pot (**Figure 15**). MK-2-(ω-2)-COOH and MK-3-(ω-2)-COOH were obtained in this way [49].

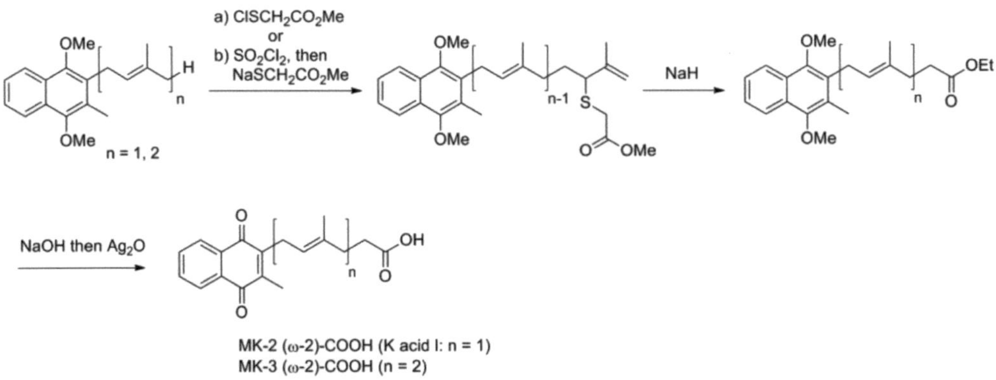

Figure 15. Synthesis of MK-n-(ω-2)-COOH derivatives (Masaki et al.).

Fujii et al. reported systematic synthesis of menaquinone metabolites. MK-n-ω-COOH derivatives were synthesized by oxidation of the terminal carbon of MK-n derivatives. Stereoselective oxidation with selenium oxide, followed by stepwise oxidation, gave MK-n-ω-COOH derivatives. K acid II was synthesized by hydrogenation of MK-1-ω-COOH (**Figure 16**) [50]. MK-n-(ω-2)-COOH derivatives were synthesized by oxidative cleavage of MK-n derivatives. Epoxidation of terminal olefin followed by perchloric acid treatment afforded 1,2-diols. Oxidative cleavage of the diol moiety followed by oxidative reactions gave MK-n-(ω-2)-COOH derivatives (**Figure 17**) [50]. These synthetic schemes correspond to the putative catabolic pathways of menaquinones, that is, ω-oxidation and β-oxidation.

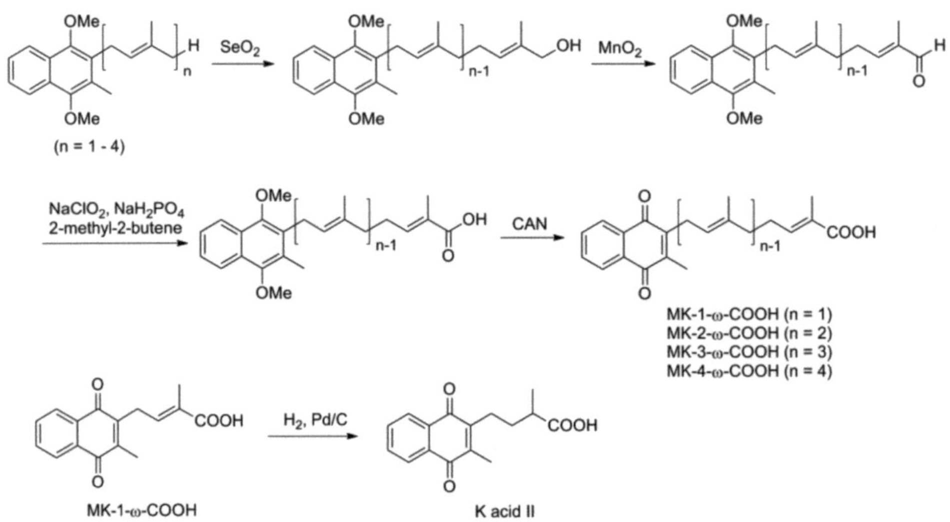

Figure 16. Synthesis of MK-n-ω-COOH derivatives (Fujii et al.).

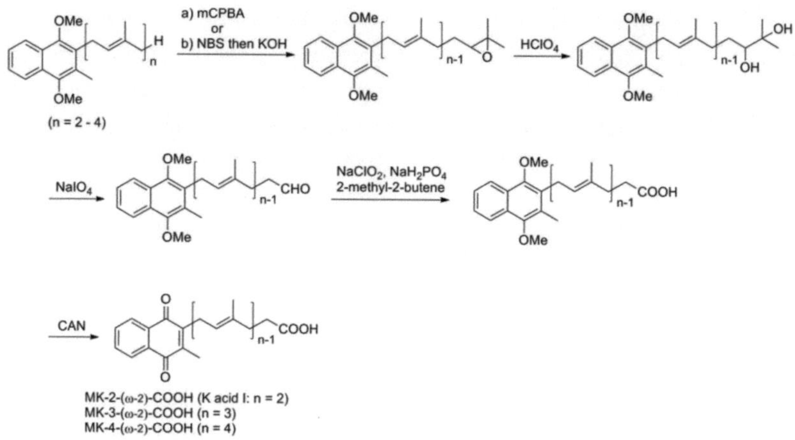

Figure 17. Synthesis of MK-n-(ω-2)-COOH derivatives (Fujii et al.).

Suhara et al. designed and synthesized ω-hydroxy derivatives (ω-alcohols) and ω-formyl derivatives (ω-aldehydes) as menaquinone metabolite analogs. ω-Oxidized side chain moieties were prepared from corresponding isoprene derivatives, and the side chain parts were introduced into the naphthalene core. Oxidation to quinone form afforded ω-alcohols, and then PDC oxidation afforded ω-aldehydes (**Figure 18**) [51, 52].

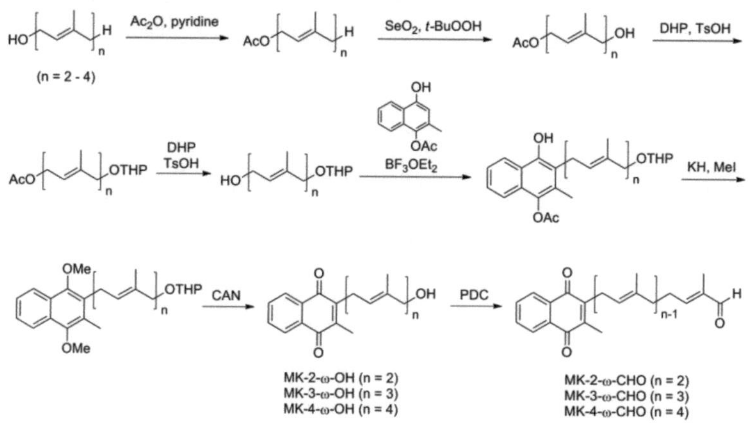

Figure 18. Synthesis of MK-n-ω-alcohols and MK-n-ω-aldehydes (Suhara et al.).

4.2. Biological activities of menaquinone metabolites

These menaquinone carboxylic acid derivatives and related quinone carboxylic acids, including ubiquinone derivatives and tocopheryl derivatives, show lysosomal membrane-stabilizing activity [45, 47]. Appropriate hydrophobicity of the side chain appears to be essential for this activity. Some of these compounds also exert inhibitory effects on the generation of the slow-reacting substance of anaphylaxis [48].

MK-4 has various biological activities, such as anti-inflammatory activity and antitumor activity, and these activities of the menaquinone metabolites were also investigated. All tested menaquinone metabolites inhibited LPS-induced production of proinflammatory cytokines in RAW264.7 cells [50]. It is suggested that naphthoquinone structure is essential for the anti-inflammatory activity of menaquinone derivatives. Regarding antitumor activity, several carboxylic acids, such as MK-2-ω-COOH, significantly inhibited proliferation of JHH7 and HepG2 hepatocellular carcinoma cell lines. On the other hand, MK-2-ω-COOH did not inhibit proliferation of normal hepatic cells. Anti-proliferative activity may be associated with caspase/transglutaminase-related pathways [53].

The ω-alcohols and ω-aldehydes showed apoptosis-inducing activity toward human leukemia cell line HL-60 and human osteosarcoma cell line MG-63. The ω-aldehydes were more potent than the corresponding ω-alcohols [51, 52]. The vitamin K potency of MK-4-ω-OH, that is, its coenzyme activity for GGCX, was also evaluated. MK-4-ω-OH showed a larger V_{max}/K_m value than that of intact MK-4, indicating that MK-4-ω-OH has greater coenzyme activity than MK-4 [52].

5. Future perspective

Vitamin Ks are attractive lead compounds for drug discovery. One of the most promising applications is as candidate antitumor agents, though the mechanism of action of Cpd 5 could be different from that of intact vitamin Ks. In addition, bone homeostasis and neural effects are also possible targets of vitamin K derivatives. Vitamin K may also be used as a food supplement, and therefore, characterization of its metabolites is important. It is noteworthy that some menaquinone metabolites have characteristic activities distinct from those of intact vitamin K_2. Though a clinical study of MK-4 as an agent to prevent recurrence of hepatocellular carcinoma was terminated [54], the metabolites and their analogs still represent potential drug candidates.

Author details

Shinya Fujii[1,2] and Hiroyuki Kagechika[1*]

*Address all correspondence to: kage.chem@tmd.ac.jp

1 Institute of Biomaterials and Bioengineering, Tokyo Medical and Dental University, Kanda-Surugadai, Chiyoda-ku, Tokyo, Japan

2 Institute of Molecular and Cellular Biosciences, the University of Tokyo, Yayoi, Bunkyo-ku, Tokyo, Japan

http://dx.doi.org/10.5772/63511

References

[1] Litwack G. ed. Vitamin K, vitamins and hormones, vol. 78, Elsevier Academic Press, New York, 2008.

[2] Prasad K. N., Edwards-Prasad J., Sakamoto A. Vitamin K_3 (menadione) inhibits the growth of mammalian tumor cells in culture. *Life Sci.*, 29, 1387–1392 (1981).

[3] Ngo E. O., Sun T.-P., Chang J.-Y., Wang C.-C., Chi K.-H., Cheng A.-L., Nutter L. M. Menadione-induced DNA damage in a human tumor cell line. *Biochem. Pharmacol.*, 42, 1961–1968 (1991).

[4] Wu F. Y.-H., Chang N.-T., Chen W.-J., Juan C.-C. Vitamin K_3-induced cell cycle arrest and apoptotic cell death are accompanied by altered expression of c-fos and c-myc in nasopharyngeal carcinoma cells. *Oncogene*, 8, 2237–2244 (1993).

[5] Juan C.-C., Wu F. Y.-H. Vitamin K_3 inhibits growth of human hepatoma HepG2 cells by decreasing activities of both p34-Cdc2 kinase and phosphatase. *Biochem. Biophys. Res. Commun.*, 190, 907–913 (1993).

[6] Otsuka M., Kato N., Shao R. X., Hoshida Y., Ijichi H., Koike Y., Taniguchi H., Moriyama M., Shiratori Y., Kawabe T., Omata M. Vitamin K_2 inhibits the growth and invasiveness of hepatocellular carcinoma cells via protein kinase A activation. *Hepatology*, 40, 243–251 (2004).

[7] Habu D., Shiomi S., Tamori A., Takeda T., Tanaka T., Kubo S., Nishiguchi S. Role of vitamin K_2 in the development of hepatocellular carcinoma in women with viral cirrhosis of the liver. *JAMA*, 292, 358–361 (2004).

[8] Tabb M. M., Sun A., Zhou C., Grün F., Errandi J., Romero K., Pham H., Inoue S., Mallick S., Lin M., Forman B. M., Blumberg B. Vitamin K_2 regulation of bone homeostasis is mediated by the steroid and xenobiotic receptor SXR. *J. Biol. Chem.*, 278, 43919–43927 (2003).

[9] Ichikawa T., Horie-Inoue K., Ikeda K., Blumberg B., Inoue S. Steroid and Xenobiotic receptor SXR mediates vitamin K_2-activated transcription of extracellular matrix-related genes and collagen accumulation in osteoblastic cells. *J. Biol. Chem.*, 281, 16927–16934 (2006).

[10] Ohsaki Y., Shirakawa H., Miura A., Giriwono P. E., Sato S., Ohashi A., Iribe M., Goto T., Komai M. Vitamin K suppresses the lipopolysaccharide-induced expression of inflammatory cytokines in cultured macrophage-like cells via the inhibition of the activation of nuclear factor kB through the repression of IKKα/β phosphorylation. *J. Nutr. Biochem.*, 21, 1120–1126 (2010).

[11] Spronk H. M. H., Soute B. A. M., Schurgers L. J., Thijssen H. H. W., De Mey J. G. R., Vermeer C. Tissue-specific utilization of menaquinone-4 results in the prevention of arterial calcification in warfarin-treated rats. *J. Vasc. Res.*, 40, 531–537 (2003).

[12] Otsuka M., Kato N., Ichimura T., Abe S., Tanaka Y., Taniguchi H., Hoshida Y., Moriyama M., Wang Y., Shao R.-X., Narayan D., Muroyama R., Kanai F., Kawabe T., Isobe T., Omata M. Vitamin K_2 binds 17β-hydroxysteroid dehydrogenase 4 and modulates estrogen metabolism. *Life Sci.* 76, 2473–2482 (2005).

[13] Shirakawa H, Ohsaki Y, Minegishi Y, Takumi N, Ohinata K, Furukawa Y, Mizutani T., Komai M. Vitamin K deficiency reduces testosterone production in the testis through down-regulation of the Cyp11a a cholesterol side chain cleavage enzyme in rats. *Biochim Biophys Acta*, 1760, 1482–1488 (2006).

[14] Ito A., Shirakawa H., Takumi N., Minegishi Y., Ohashi A., Howlader Z. H., Ohsaki Y., Sato T., Goto T., Komai M. Menaquinone-4 enhances testosterone production in rats and testis-derived tumor cells. *Lipids Health Dis.*, 10, 158 (2011).

[15] Nishikawa Y., Carr B. I., Wang M., Kar S., Finn F., Dowd P., Zheng Z. B., Kerns J., Naganathan S. Growth inhibition of hepatoma cells induced by vitamin K and its analogs. *J. Biol. Chem.*, 270, 28304–28310 (1995).

[16] Ni R., Nishikawa Y., Carr B. I. Cell growth inhibition by a novel vitamin K is associated with induction of protein tyrosine phosphorylation. *J. Biol. Chem.*, 273, 9906–9911 (1998).

[17] Nishikawa Y., Wang Z., Kerns J., Wilcox C. S., Carr B. I. Inhibition of hepatoma cell growth in vitro by arylating and non-arylating K vitamin analogs. Significance of protein tyrosine phosphatase inhibition. *J. Biol. Chem.*, 274, 34803–34810 (1999).

[18] Tamura K., Southwick E. C., Kerns J., Rosi K., Carr B. I., Wilcox C., Lazo J. S. Cdc25 inhibition and cell cycle arrest by a synthetic thioalkyl vitamin K analogue. *Cancer Res.*, 60, 1317–1325 (2000).

[19] Kar S. Lefterov I. M., Wang M., Lazo J. S., Scott C. N., Wilcox C. S., Carr B. I. Binding and inhibition of Cdc25 phosphatase by vitamin K analogues. *Biochemistry*, 42, 10490–10497 (2003).

[20] Lazo J. S., Nemoto K, Pestell K. E., Cooley K., Southwick E. C., Mitchell D. A., Furey W., Gussio R., Zaharevitz D. W., Joo B., Wipf P. Identification of a potent and selective pharmacophore for Cdc25 dual specificity phosphatase inhibitors. *Mol Pharmacol.*, 61, 720–728 (2002).

[21] Peyregne V. P., Kar S., Ham S. W., Wang M., Wang Z., Carr B. I. Novel hydroxyl naphthoquinones with potent Cdc25 antagonizing and growth inhibitory properties. *Mol Cancer Ther.*, 4, 595–602 (2005).

[22] Kar S., Wang M., Ham S. W., Carr B. I. Fluorinated Cpd 5, a pure arylating K-vitamin derivative, inhibits human hepatoma cell growth by inhibiting Cdc25 and activating MAPK. *Biochem. Pharmacol.*, 72, 1217–1227 (2006).

[23] Ham S. W., Choe J.-I., Wang M.-F., Peyregne V., Carr B. I. Fluorinated quinoid inhibitor: possible "pure" arylator predicted by the simple theoretical calculation. *Bioorg. Med. Chem. Lett.*, 14, 4103–4105 (2004).

[24] Kar S., Wang M., Yao W., Michejda C. J., Carr B. I. PM-20, a novel inhibitor of Cdc25A, induces extracellular signal-regulated kinase 1/2 phosphorylation and inhibits hepatocellular carcinoma growth in vitro and in vivo. *Mol. Cancer Ther.*, 5, 1511–1519 (2006).

[25] Brun M.-P., Braud E., Angotti D., Mondésert O., Quaranta M., Montes M., Miteva M., Gresh N., Ducommun B., Garbay C. Design, synthesis, and biological evaluation of novel naphthoquinone derivatives with CDC25 phosphatase inhibitory activity. *Bioorg. Med. Chem.*, 13, 4871–4879 (2005).

[26] Braud E., Goddard M.-L., Kolb S., Brun M.-P., Mondésert O., Quaranta M., Gresh N., Ducommun B., Garbay C. Novel naphthoquinone and quinolinedione inhibitors of CDC25 phosphatase activity with antiproliferative properties. *Bioorg. Med. Chem.*, 16, 9040–9049 (2008).

[27] Kitano T., Yoda H., Tabata K., Miura M., Toriyama M., Motohashi S., Suzuki T. Vitamin K_3 analogs induce selective tumor cytotoxicity in neuroblastoma. *Biol. Pharm. Bull.*, 35, 617–623 (2012).

[28] Kayashima T., Mori M., Mizutani R., Nishio K., Kuramochi K., Tsubaki K., Yoshida H., Mizushina Y., Matsubara K. Synthesis and biological evaluation of vitamin K derivatives as angiogenesis inhibitor. *Bioorg. Med. Chem.*, 18, 6305–6309 (2010).

[29] Krishnaswamy M., Purushothaman K. K. Plumbagin: a study of its anticancer, antibacterial and antifungal properties. *Indian J. Exp. Biol.*, 18, 876–877 (1980).

[30] Sandur S. K., Ichikawa H., Sethi G., Ahn K. S., Aggarwal B. B. Plumbagin (5-hydroxy-2-methyl-1,4-naphthoquinone) suppresses NF-kB activation and NF-kB-regulated gene products through modulation of p65 and IkBα kinase activation, leading to potentiation of apoptosis induced by cytokine and chemotherapeutic agents. *J. Biol. Chem.*, 281, 17023–17033 (2006).

[31] Rao K. V., McBride T. J., Oleson J. J. Recognition and evaluation of lapachol as an antitumor agent. *Cancer Res.*, 28, 1952–1954 (1968).

[32] Hartwell J. L., Abbott B. J. Antineoplastic principles in plants: recent developments in the field. *Adv. Pharmacol. Chemother.*, 7, 117–209 (1969).

[33] For recent examples: (i) Fiorito S., Epifano F., Bruyere C., Mathieu V., Kiss R. Genovese S. Growth inhibitory activity for cancer cell lines of lapachol and its natural and semi-synthetic derivatives. *Bioorg. Med. Chem. Lett.*, 24, 454–457 (2014). (ii) Wang S.-H., Lo C.-Y., Gwo Z.-H., Lin H.-J., Chen L.-G., Kuo C.-D., Wu J.-Y. Synthesis and biological evaluation of lipophilic 1, 4-naphthoquinone derivatives against human cancer cell lines. *Molecules*, 20, 11994–12015 (2015).

[34] Suhara Y., Hanada N., Okitsu T., Sakai M., Watanabe M., Nakagawa K., Wada A., Takeda K., Takahashi K., Tokiwa H., Okano T. Structure-activity relationship of novel menaquinone-4 analogues: modification of the side chain affects their biological activities. *J. Med. Chem.* 55, 1553–1558 (2012).

[35] Suhara Y., Watanabe M., Motoyoshi S., Nakagawa K., Wada A., Takeda K., Takahashi K., Tokiwa H., Okano T. Synthesis of new vitamin K analogues as steroid and xenobiotic receptor (SXR) agonists: insights into the biological role of the side chain part of vitamin K. *J. Med. Chem.* 54, 4918–4922 (2011).

[36] Suhara Y., Watanabe M., Nakagawa K., Wada A., Ito Y., Takeda K., Takahashi K., Okano T. Synthesis of novel vitamin K_2 analogues with modification at the ω-terminal position and their biological evaluation as potent steroid and xenobiotic receptor (SXR) agonists. *J. Med. Chem.* 54, 4269–4273 (2011).

[37] Suhara Y., Hirota Y., Hanada N., Nishina S., Eguchi S., Sakane R., Nakagawa K., Wada A., Takahashi K., Tokiwa H., Okano T. Synthetic small molecules derived from natural vitamin K homologues that induce selective neuronal differentiation of neuronal progenitor cells. *J. Med. Chem.* 58, 7088–7092 (2015).

[38] Wiss O., Gloor U. Absorption, distribution, storage, and metabolites of vitamins K and related quinones. *Vitam. Horm.*, 24, 575–586 (1966).

[39] Watanabe M., Toyoda M., Imada I., Morimoto H. Ubiquinone and related compounds. XXVI. The urinary metabolites of phylloquinone and α-tocopherol. *Chem. Pharm. Bull.*, 22, 176–182 (1974).

[40] McBurney A., Shearer M. J., Barkhan P. Preparative isolation and characterization of the urinary aglycones of vitamin K_1 (phylloquinone) in man. *Biochem. Med.*, 24, 250–267 (1980).

[41] Tadano K., Yuzuriha T., Sato T., Fujita T., Shimada K., Hashimoto K., Satoh T. Identification of menaquinone-4 metabolites in the rat. *J. Pharmacobiodyn.*, 12, 640–645 (1989).

[42] Harrington D. J., Soper R., Edwards C., Savidge G. F., Hodges S. J., Shearer M. J. Determination of the urinary aglycone metabolites of vitamin K by HPLC with redox-mode electrochemical detection. *J. Lipid. Res.*, 46, 1053–1060 (2005).

[43] Edson K. Z., Prasad B., Unadkat J. D., Suhara Y., Okano T., Guengerich F. P., Rettie A. E. Cytochrome P450-dependent catabolism of vitamin K: ω-hydroxylation catalyzed by human CYP4F2 and CYP4F11. *Biochemistry*, 52, 8276–8285 (2013).

[44] Watanabe M., Kawada M., Nishikawa M., Imada I., Morimoto H. Ubiquinone and related compounds. XXVII. Synthesis of urinary metabolites of phylloquinone and α-tocopherol. *Chem. Pharm. Bull.*, 22, 566–575 (1974).

[45] Watanabe M., Okamoto K., Imada I., Morimoto H. Ubiquinone and related compounds. XXXI. Synthesis of urinary metabolites of ubiquinone, phylloquinone, α-tocopherol and related compounds. *Chem. Pharm. Bull.*, 26, 774–783 (1978).

[46] Teitelbaum A. M., Scian M., Nelson W. L., Rettie A. E. Efficient syntheses of vitamin K chain-shortened acid metabolites. *Synthesis*, 47, 944–948 (2015).

[47] Okamoto K., Watanabe M., Kawada M., Goto G., Ashida Y., Oda K., Yajima A., Imada I., Morimoto H. Synthesis of quinones having carboxy- and hydroxy-alkyl side chains,

http://dx.doi.org/10.5772/63511

and their effects on rat-liver lysosomal membrane. *Chem. Pharm. Bull.*, 30, 2797–2819 (1982).

[48] Terao S., Shiraishi M., Kato K., Ohkawa S., Ashida Y., Maki Y. Quinones. Part 2. General synthetic routes to quinone derivatives with modified polyprenyl side chains and the inhibitory effects of these quinones on the generation of the slow reacting substance of anaphylaxis (SRS-A). *J. Chem. Soc. Perkin Trans 1*, 2909–2920. (1982).

[49] Masaki Y., Sakuma K., Kaji K. Regio- and stereoselective terminal allylic carboxymethylation of *gem*-dimethyl olefins. Synthesis of biologically important linear degraded terpenoids. *Chem. Pharm. Bull.*, 33, 1930–1940 (1985).

[50] Fujii S., Shimizu A., Takeda N., Oguchi K., Katsurai T., Shirakawa H., Komai M., Kagechika H. Systematic synthesis and anti-inflammatory activity of ω-carboxylated menaquinone derivatives – Investigation on identified and putative vitamin K metabolites. *Bioorg. Med. Chem.*, 23, 2344–2352 (2015).

[51] Suhara Y., Murakami A., Kamao M., Mimatsu S., Nakagawa K., Tsugawa N., Okano T. Efficient synthesis and biological evaluation of ω-oxygenated analogues of vitamin K_2: study of modification and structure-activity relationship of vitamin K_2 metabolites. *Bioorg. Med. Chem. Lett.*, 17, 1622–1625 (2007).

[52] Suhara Y., Hirota Y., Nakagawa K., Kamao M., Tsugawa N., Okano T. Design and synthesis of biologically active analogues of vitamin K_2: evaluation of their biological activities with cultured human cell lines. *Bioorg. Med. Chem.*, 16, 3108–3117 (2008).

[53] Qin X.-Y., Fujii S., Shimizu A., Kagechika H., Kojima S. Carboxylic derivatives of vitamin K_2 inhibit hepatocellular carcinoma cell growth through caspase/transglutaminase-related signaling pathways. *J. Nutrit. Sci. Vitaminol.*, 61, 285–290 (2015).

[54] Yoshida H., Shiratori Y., Kudo M., Shiina S., Mizuta T., Kojiro M., Yamamoto K., Koike Y., Saito K., Koyanagi N., Kawabe T., Kawazoe S., Kobashi H., Kasugai H., Osaki Y., Araki Y., Izumi N., Oka H., Tsuji K., Toyota J., Seki T., Osawa T., Masaki N., Ichinose M., Seike M., Ishikawa A., Ueno Y., Tagawa K., Kuromatsu R., Sakisaka S., Ikeda H., Kuroda H., Kokuryu H., Yamashita T., Sakaida I., Katamoto T., Kikuchi K., Nomoto M., Omata M. Effect of vitamin K_2 on the recurrence of hepatocellular carcinoma. *Hepatology*, 54, 532–540 (2011).

Chapter 14

From Protein Folding to Blood Coagulation: Menaquinone as a Metabolic Link between Bacteria and Mammals

Brian M. Meehan and Jonathan Beckwith

Additional information is available at the end of the chapter

http://dx.doi.org/10.5772/63342

Abstract

Menaquinones have long played a central role in bacterial metabolism due to their solubility in membranes and their ability to mediate electron transfer reactions between a large variety of enzymes. In addition to acting as important nodes in fermentation and respiration, menaquinones are critical to the formation of disulphide bonds in the periplasm. Their utility as molecular wires has also led to their incorporation into redox reactions in higher-order organisms, where they participate in numerous physiological processes, including blood coagulation. Through studying the menaquinone-dependent pathways in organisms across the phylogenetic spectrum, researchers have begun to uncover intriguing metabolic links and have identified novel compounds for modulating these vital pathways.

Keywords: menaquinone, vitamin K, vitamin K epoxide reductase, DsbB, disulphide bond formation

1. Rise of the quinones

As life began to emerge in the seas of primordial Earth, one of the first orders of business was the construction of a plasma membrane to protect and concentrate biomolecules in the cytoplasm of what would become the first cell. This served to increase the efficiency of the biochemical reactions necessary for growth and propagation. The constant influx of salt water across the plasma membrane, however, would have led to extremely high internal osmotic pressure in what were essentially bags of chemicals, necessitating the invention of efflux pumps to drive

INTECH
open science | open minds

© 2017 The Author(s). Licensee InTech. This chapter is distributed under the terms of the Creative Commons Attribution License (http://creativecommons.org/licenses/by/3.0), which permits unrestricted use, distribution, and reproduction in any medium, provided the original work is properly cited. (cc) BY

ions back out into the surrounding milieu. These early efflux pumps most likely exported protons at the expense of ATP, which the cells were forced to make through substrate-level phosphorylation, an inefficient process. However, leakage of protons back into the cell could drive the ATPase in the reverse direction, thus linking the extrusion of protons to the creation of cellular energy.

Aside from the issue of osmotic pressure, the plasma membrane also created a conundrum in that the organic molecules necessary to drive metabolism were prevented entry. Thus, cells developed membrane-associated transporters capable of importing such nutrients. Catabolism of these organic molecules provided the cells with necessary building blocks and resulted in the liberation of electrons, which could then be collected on redox-active molecules like NAD+ or FAD+, reducing them to NADH and $FADH_2$.

To regenerate the pools of electron carriers, these ancient microbes resorted to fermentation, the process by which electrons are dropped onto self-derived organic molecules. This freed up NAD+ and FAD+ to participate in more rounds of catabolism of carbon sources, thus driving metabolism. However, fermentation is an inefficient process and thus limited the growth rate and abundance of these microbes. Only when the enzymes involved in fermentation (i.e., nitrate reductase, fumarate reductase) evolved to associate with the plasma membrane did these ancient microbes begin to tap into the power they needed to flourish. Now, the process of passing electrons onto terminal acceptors could be coupled to the extrusion of protons into the extracellular space. With greater numbers of protons pumped out of the cell, their leakage back across the membrane could greatly increase the amount of ATP generated [1]. In essence, the cells could target these molecular machines to the membrane to produce the chemical energy necessary to fuel metabolism. The effect could be further amplified by linking these redox reactions together, but that required cofactors capable of accepting and donating electrons to act as molecular wires. The earliest such cofactors were likely iron-sulfur clusters and flavins, but these were not readily inserted into the highly lipophilic environment of the plasma membrane. Thus arose the quinones, which are fat-soluble redox molecules capable of associating with membrane-embedded enzymes. By linking together several modular redox complexes into an electron transport chain capable of extruding protons, quinones potentiated a huge leap forward in bioenergetics and greatly increased the capacity for complexity in biological systems.

To maximize the utility of having quinones available in their membranes, these ancient microbes needed to find a way to tap into more plentiful electron donors and acceptors. One potential source of electrons present in abundance on primordial Earth was water. However, the vast amounts of energy necessary to pull electrons from water presented a formidable obstacle to its utilization. Only when high-energy solar radiation came to be employed in the process known as photosynthesis were "cyanobacteria" successful in linking the fixation of CO_2 to hydrolysis [2]. Quinones were key to the evolution of photosystem I, yet another example of their power and adaptability in biological systems.

While the increased access to electrons represented a potential windfall to energy-starved cells, the development of photosynthesis was nonetheless catastrophic to life on Earth. Concomitant with the liberation of electrons from water by photosynthesis was the production of a new and toxic gas–oxygen. Among the many harmful effects of an oxygenated atmosphere were the generation of reactive oxygen species, which poisoned the metabolic pathways of early microbial cells by destroying important cofactors and enzymes [3]. In this noxious environment lay opportunity, however, as oxygen could serve as an extremely effective electron acceptor if cells could evolve mechanisms for reducing it as part of their electron transport chains. Once again quinones mediated a metabolic breakthrough, linking the reduction of oxygen to membrane components via cytochrome oxidases. This process of aerobic respiration led to an estimated 16-fold increase in the capacity to generate ATP [4] and may have opened the door to the development of complex eukaryotic life [5].

The metabolic flexibility of quinones means that their use is not limited simply to the respiratory chains of microbes or the photosynthetic centers of plants. Many higher-order organisms not only incorporated quinones into their respiratory chains, but have utilized these highly effective molecular wires in many different redox-dependent reactions. While the quinones are ancient, they remain very important to life on this planet.

2. Biosynthesis of menaquinone and phylloquinone

The major quinones found in nature are ubiquinone (UQ), menaquinone (MK), and phylloquinone (K1) (**Figure 1**). They differ not only in structure but also in their redox potentials, so the incorporation of one or the other as a cofactor allows for fine-tuning of electron transfer reactions. The distribution in nature of genes involved in menaquinone biosynthesis suggests that it was most likely the original quinone; this is supported by the observation that menaquinone is readily oxidized in aerobic environments, suggesting that it existed long before the appearance of oxygen [6]. There are many different species of menaquinone, though they differ only in the length of their isoprenyl side chains. These differences are reflected in the nomenclature of menaquinones, wherein the number of isoprene units is indicated (i.e., MK-4). Phylloquinone, usually considered distinct from the menaquinones, is merely MK-4 with a more heavily saturated lipophilic tail.

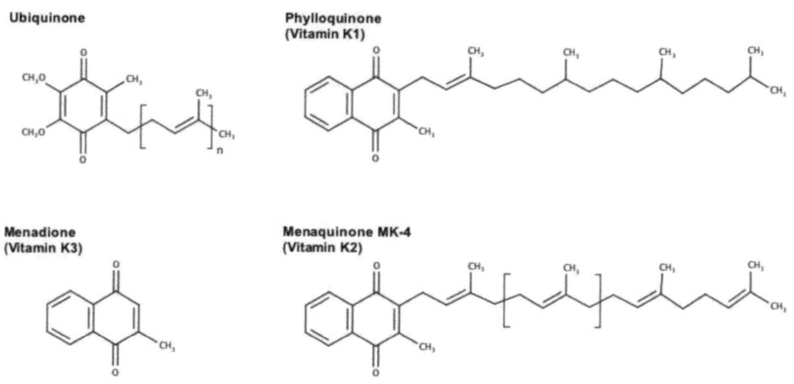

Figure 1. Structures of common quinones.

The canonical pathway for menaquinone biosynthesis in bacteria is well-established and has been reviewed in great detail elsewhere [7]. Of particular interest to this review, however, is the prenylation reaction mediated by MenA in which the lipophilic tail is attached to 1,4-dihydroxy-2-naphthoic acid (**Figure 2**). In essence, this is the critical step that links the redox-active quinone to the membrane. The lipophilic substrate of MenA is made up of repeating isoprenal subunits, the exact number of which is determined by the octaprenyl pyrophosphate (OPP) synthase encoded by the particular microbe. The ultimate chain length of these products is determined by a molecular ruler mechanism wherein bulky amino acid residues at the bottom of each of OPP's active sites block chain elongation [8], and it is this step that controls the identity of the primary MK produced by an organism. There is some evidence to suggest, however, that growth temperature also plays a role in the length and degree of saturation of the aliphatic side chain [9]. Phylloquinone biosynthesis in cyanobacteria is predicted to proceed via a pathway very similar to that of MK biosynthesis. However, the cyanobacterial MenA incorporates a mostly saturated phytyl tail at position C-3 rather than the partially unsaturated isoprenyl side chain associated with MK [10]. Recently, an alternative pathway

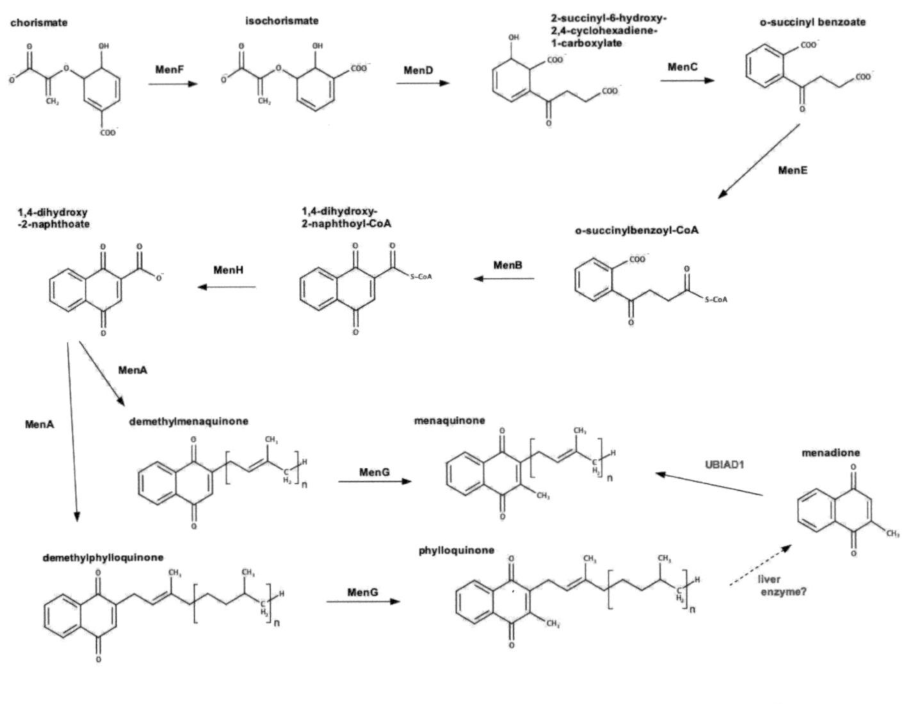

Figure 2. Biosynthesis of menaquinone and phylloquinone. The highly conserved sequence of reactions required for conversion of chorismate to menaquinone is shown. The pathway for phylloquinone biosynthesis is thought to proceed via the same steps, with the exception of the prenylation step mediated by MenA. The production of demethylphylloquinone versus demethylmenaquinone is determined by which substrate is provided to MenA by octaprenyl pyrophosphate synthase (OPP). A phytyl chain results in the production of demethylphylloquinone, whereas a more highly unsaturated isoprenyl chain results in demethylmenaquinone. In mice, metabolism of phylloquinone in the liver has been shown to release menadione, which in turn can be prenylated by the MenA homolog UBIAD1 to produce menaquinone.

http://dx.doi.org/10.5772/63342

for menaquinone biosynthesis has been described in several Achaea and Gram-negative bacteria, including *Helicobacter pylori,* Chlamydia species, and spirochetes [11]. While both pathways start with chorismate, the formation of the quinone proceeds via completely different reactions. The diversity of pathways for biosynthesis of MKs serves to underscore the importance of its role in metabolism.

Phylloquinone biosynthesis in plants is not as well understood as in cyanobacteria, though the pathway likely mirrors that of menaquinone. One striking difference is that the first four reactions proceeding from chorismate, mediated by the products of the *menF, D, C,* and *H* genes in bacteria, are accomplished in plants by one fusion protein known as PHYLLO [12]. Later steps are compartmentalized between the chloroplast and peroxisome, adding an additional level of complexity to production in plants [13].

Unlike the organisms mentioned above, higher-order organisms are incapable of *de novo* synthesis of MKs. It is therefore imperative that MKs be supplied through diet, thus classifying MKs as a "vitamin" for mammals. The identification of MKs as essential vitamins arose from the Nobel Prize-winning work of Edward Adelbert Doisy and Henrik Dam in 1943. Their recognition that a fat-soluble compound played a key role in blood coagulation led them to name it "vitamin K" ("K" for the German word "Koagulationsvitamin"). Purification and characterization revealed two forms of vitamin K—phylloquinone thus became known as vitamin K1 and menaquinone as vitamin K2 with regards to their requirement in mammals.

Though humans cannot synthesize vitamin K *de novo,* several homologs of menaquinone biosynthetic enzymes can be found in the human genome, including one for the prenylating enzyme MenA (UBIAD1). This enzyme was expressed in an insect cell line and was shown to be capable of converting menadione (K3) and K1 into MK-4 [14]. Such a result suggested that while humans cannot make their own MKs, they may be capable of converting biosynthetic intermediates into the final product. Indeed, recent results show that the phytyl tail of K1 is cleaved in the intestine of mice to make K3 and then prenylated by UBIAD1 in the cerebrum [15, 16]. Furthermore, human subjects fed K1 exhibited increased levels of K3, indicating that dietary K1 may play a major role in overall MK levels [16], and gnotobiotic rats fed K1 saw increased levels of MK4 in tissues, indicating that the conversion is not dependent on gut bacteria [17]. UBIAD1 has also been demonstrated to be essential for the embryonic development of mice [18], so vitamin K's route through the body is not completely understood and the complexities begin with the vitamin's source.

3. Sources of vitamin K

While the reactions requiring vitamin K in human metabolism are becoming clearer, the source of the vitamin K is still not completely understood. The presence of large numbers of bacteria in the human colon capable of synthesizing K2 would perhaps suggest that absorption of this bacterial byproduct might fulfill the human requirement. In fact, MK-6 is made by *Eubacterium lentum,* MK-7 by *Veillonella,* MK-8 by *Enterobacteria,* and MK-10 and MK-11 are made by *Bacteroides* [19, 20]. The bacterial contribution to vitamin K pools in humans is supported by studies done with gnotobiotic rats fed vitamin K-free diets. The rapidly developing hemorrhagic conditions in these rats could be reversed by supplying bacteria from conventionally raised rats, suggesting that absorption from the bowel provided sufficient quantities of K2 [21]. Concordant with this is the observation that taking broad-spectrum antibiotics can reduce vitamin K production by more than 70% [22]. However, K2 is embedded within the bacterial inner membrane, and as such would appear to be inaccessible to passive absorption. MKs have been shown to be secreted by some organisms [23], and it is also possible that water-soluble precursors of MK biosynthesis might be more readily available [7]. However, this scenario is further complicated by the fact that there is very little evidence that the large intestine is capable of absorption of MKs. Uptake has been shown to be poor in rats [24] and infants [25]. Even the finding that antibiotic treatment lowered vitamin K production does not conclusively identify bacteria as a major source of human vitamin K2 pools, as some antibiotics have been shown to inhibit the human enzymes necessary for recycling vitamin K2 [26]. The role of the microbiome in the production of K2 is therefore questionable and would suggest that perhaps vitamin K stores in humans might be the result of dietary intake.

Low concentrations of vitamin K2 can be found in dairy, meat, and fermented foods like natto [27], but makes up only 10% of total dietary vitamin K intake. While K1, found in a variety of green leafy plants and vegetable oils, is present in much higher amounts, it is not readily absorbed in the intestines as it is strongly bound to vegetable fiber [28]. Vitamin K is not transported by specific plasma carrier proteins like other fat-soluble vitamins, but is instead shuttled by lipoproteins. The small fraction of K1 that is absorbed is almost exclusively incorporated into the triacylglycerol-rich lipoprotein (TGRLP) fraction, while dietary K2 is associated with low-density lipoprotein (LDL) fraction [29]. These divergent pathways would deliver large amounts of K1 to the liver, but efficient delivery to extrahepatic tissues would only occur for K2. Measurements of the concentrations of vitamins K in various tissues mostly back this up, showing that K1 levels are low in the brain, kidneys, and lungs but high in the liver, heart, and pancreas; K2 (in the form of MK-4) was found to be in high concentration in the brain, kidneys, and pancreas but in low concentration in the liver, heart, and lungs. As for longer chain K2s, MK6-11 were found in the liver and trace amounts of MK6-9 were found in the heart and pancreas [30]. MK10 and MK11 may be major contributors to the hepatic pool of K2 [26], and the presence of these long-chain MKs again raise the possibility that the commensal population of colonic bacteria may somehow contribute to overall vitamin K levels in the host, as analysis of tissue samples has only shown the ability to synthesize MK-4 from K1. However, the presence of potential homologs for other prenyl diphosphate synthases in the genome further suggests that humans may be capable of producing longer chain MKs as well. Overall the data clearly indicate that dietary K1 is a major contributor to vitamin K levels in the body, but a full accounting of its sources has yet to emerge.

4. Uses of vitamin K

While the side chains of K1 and the various MKs differ, the redox-active portion of the molecules (the napthoquinone) remains unchanged. The reactivity of these various species should therefore be very similar, a fact underlined by the nearly identical mid-point redox potentials as determined by voltammetry [31, 32] (**Figure 3**). The degree of lipophilicity in the tails most likely dictates mobility of the quinones in the membrane, with the partially saturated isoprenyl tail of MK allowing for greater freedom of movement compared to the mostly unsaturated chain of K1. Additionally, longer chain MKs are likely stiffer and more viscous in the membrane due to the greater surface areas available for van der Waals interactions. For these reasons, the preferential incorporation of one MK over another into a redox-active enzyme is most likely due to availability within the membrane as well as the ability of the enzyme to accommodate different length side chains. In microsomal fractions, MK2 and MK3 were shown to have much higher activities than K1 [33], while a partially purified enzymatic system showed similar activities for MK2-6 compared to K1. MKs with seven or more isoprenoid units were not as active [34].

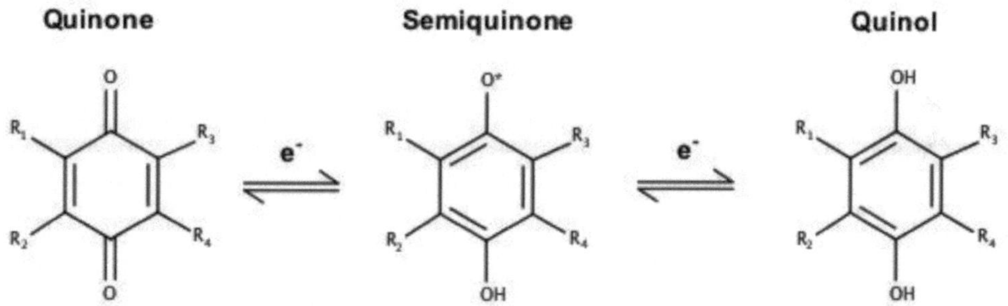

Figure 3. Resonance structures of napthoquinone species. Two electrons (e^-) can be accepted or donated in step-wise transfers from partner proteins.

Vitamin K2 has been found to play a role in protection against oxidative stress and inflammation in mammals [35], and improved locomotion defects in mutant fruit flies [36], suggesting that it might benefit human patients suffering mitochondrial pathologies. Mounting evidence suggests that MK-4 is an important component of sphingolipid biosynthesis and can inhibit the proliferation of several cancer cell lines [37]. The exact role of vitamin K2 in these processes is unknown however—its most thoroughly understood use is in protein modification.

Numerous proteins in vertebrates are modified post-translationally as a means of regulating and enhancing their activity. One such modification is the carboxylation of glutamate residues within Gla domains, which is mediated by the enzyme gamma-glutamyl carboxylase (GGCX). This modification allows for the high-affinity binding of calcium ions, which in turn mediates a conformational change necessary for proper folding of the protein. Gla-containing proteins play important roles in the venom of snakes and the toxins of cone snails [38], and they have numerous functions in humans including bone development, calcification, and sphingolipid metabolism [35, 39]. The cell-signaling activities of the vitamin K-dependent proteins Gas6 and protein S may also be crucial to cognitive processes [35]. Among the Gla-containing proteins, however, those involved in blood coagulation have received the most attention. Carboxylation of several of these factors activates them and thereby sets off a cascade leading to clotting. The

GGCX glycosylation reaction is coupled to the oxidation of vitamin K hydroquinone to vitamin K 2,3-epoxide, and it is this step that shows sensitivity to anticoagulants like warfarin. When this vitamin K cycle is disrupted or insufficient quantities of vitamin K are present in the diet, excessive bleeding can and does occur, as was the case in the initial discovery of vitamin K's role in nutrition.

While the flexibility of K2s is crucial to all of these redox-driven processes, short circuits occur wherein reduced menaquinones donate their electrons to "inappropriate" acceptors like oxygen. Such reactions result in the production of reactive oxygen species and can lead to massive damage to proteins and DNA [40, 41], underlining the importance of properly regulating the expression and distribution of MKs.

5. VKORs

Clearly, MKs play a critical role in mediating the activity of numerous proteins in mammals, yet the levels of this important cofactor in tissues is relatively low. After passing electrons onto the appropriate acceptors, MK is oxidized to its inactive, oxidized form. In bacteria, MK is quickly reduced again by the flow of electrons from the electron transport chain or to a lesser extent by the delivery of electrons from the disulfide bond pathway. To recharge and replenish their redox-active pool of MKs, mammals have evolved enzymes capable of reducing of vitamin K 2,3-epoxide (KO) to vitamin K and vitamin K hydroquinone (KH_2). These two steps occur via a warfarin-sensitive pathway as well as a warfarin-insensitive pathway, suggesting that two or more enzymes may be required to efficiently complete the reaction. While the enzymatic activity of vitamin K epoxide reductase (VKOR) had first been assayed in 1974 and VKOR had long been known to be the target of the anticoagulant warfarin, identification of the enzyme responsible for the regeneration of vitamin K did not come until 2004 [42, 43]. While this discovery set the stage for in-depth analysis of the kinetics of blood coagulation, one of the most surprising early findings was that VKOR homologs could be found not only in a large family of vertebrates, but also in insects, plants, bacteria, and archaea [44]. What role could VKOR possibly play in organisms that do not contain blood? The discovery of vitamin K-dependent proteins in sea squirts [45] suggests that this modification arose much earlier than the blood coagulation cascade and that vertebrates simply repurposed Gla-modified proteins.

To fully understand the function of a membrane-bound protein, it is important to determine the topology of the enzyme within the membrane. This allows for greater insights into the catalytic site as well as to possible interactions with partner proteins. The topology of VKOR in the endoplasmic reticulum (ER) membrane, however, has been fraught with controversy. Initial reports suggested an enzyme with 4 transmembrane domains (TM) [44], though there is also mounting evidence that VKOR may adopt a 3-TM structure (**Figure 4**). Of particular, importance to this debate is the potential positioning of critical cysteine residues. VKOR contains a total of four conserved cysteines, two of which are present in a C-X-X-C motif characteristic of redox-active thioredoxins. These two cysteines (C132 and C135) have been shown to be essential for the reduction of vitamin KO to vitamin K and vitamin KH_2 using purified VKOR [46]. The second set of conserved cysteines (C43 and C51) lie within a loop region between TMs. The 3-TM model for VKOR places the N-terminus of the protein and the active site cysteines on the ER side of the membrane, with the loop cysteines and C-terminus in the cytoplasm. On the other hand, the 4-TM model places both termini in the cytoplasm, while the active site and loop cysteines both face the ER lumen. This 4-TM topology would immediately suggest an enzymatic mechanism wherein the loop cysteines receive electrons from interactions with redox partners in the lumen, and then pass them on to the active site C-X-X-C. Because the 3 -TM model predicts that the two sets of cysteines are on opposite sides of the ER membrane, it is difficult to imagine how they might interact, and it suggests a distinctly different mechanism for reduction. The fact that several mutations encoding

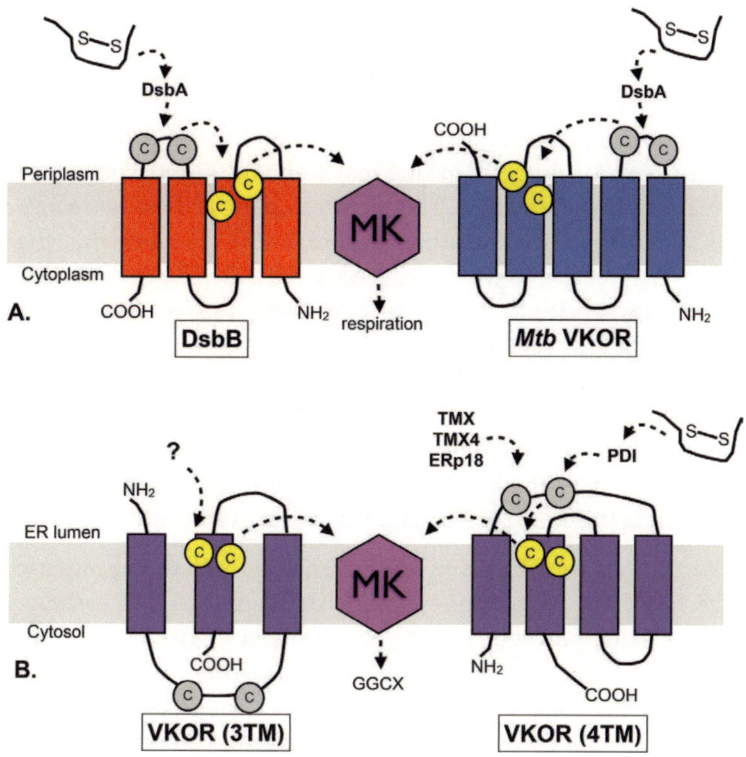

Figure 4. Representative topologies of membrane redox proteins. (A) Topologies of DsbB and *Mtb* VKOR in the bacterial plasma membrane. Critical cysteine resides required for transfer of electrons to menaquinone are represented by yellow circles, while loop cysteines required for accepting electrons from the periplasmic protein DsbA are represented as gray circles. Note that the order of these residues is reversed in the two enzymes with respect to the amino acid sequences. (B) Proposed topologies of the mammalian VKOR in the membrane of the endoplasmic reticulum. In the 3-TM structure, the loop cysteines and those required for transfer of electrons to menaquinone (vitamin K2) are located in different cellular compartments, while in the 4-TM model, all four are found near the interface of the membrane with the ER lumen. While several luminal proteins have been found to be capable of transferring electrons to the catalytic site of VKOR in the 4-TM model, it is not yet clear which proteins would perform this role in the 3-TM model. MK = menaquinone. GGCX = γ-glutamate carboxylase.

resistance to warfarin map to the loop region containing Cys43 and Cys51 further suggests that these loop cysteines may play a key role in VKOR activity.

The loop cysteines were not required for the enzymatic of activity of VKOR with the purified enzyme, though C51 was found to be important along with C132 and C135 for activity in cell extracts [47]. Expression of Cys 43 and Cys 51 mutants in reporter cells in which endogenous VKOR and VKORL1 were knocked out show that these mutant alleles retain ~90% activity [48]. However, challenges to such results have emerged. Results with purified VKOR showing the non-essentiality of the loop cysteines were obtained using dithiothreitol (DTT), a non-physiological reductant. Because DTT is membrane permeable, it is possible that Cys43 and Cys51 are important for shuttling electrons to the active site cysteines under physiological conditions, but DTT bypasses this necessity. To this end, experiments utilizing the membrane impermeable system of NADPH, thioredoxin, and thioredoxin to drive reduction gave results showing that the loop cysteines were actually required for VKOR activity [49].

The membrane topology of VKOR has been directly tested through a number of biochemical approaches. The Stafford lab fused green fluorescent protein (gfp) to either the N- or C-terminus and tested protease susceptibility. These studies showed that only the C-terminus was proteolytically cleaved, which suggested that while the N-terminus faced the ER lumen, the C-terminus must face the cytoplasm. This architecture placed the two confirmed active site cysteines (Cys-132 and Cys-135) on the luminal side of the membrane, while the two conserved loop cysteines (Cys-43 and Cys-51) were on the opposite side [50]. An important caveat to this work is that the protease sensitivity assay was performed after permeabilization with digitonin, a process not thought to affect the topology of membrane proteins. However, a very similar approach using live (i.e., non-permeabilized) cells and a redox-active gfp clearly demonstrated that both the N- and C-termini are located within the cytoplasm [51]. Further experiments showed that the loop region containing Cys43 and Cys51 could be glycosylated by machinery within the ER lumen and that the loop cysteines could form mixed disulfides with luminal proteins, thereby placing this loop region firmly in the ER in accordance with a 4-TM topology.

The overall architecture of VKOR becomes most germane when attempting to identify its redox partners. If VKOR had three membrane-spanning domains, the loop cysteines would not be accessible to soluble redox partners, yet the active site cysteines would need to be directly accessible to a partner. To achieve this, the partner must also be membrane bound, as has been suggested for GGCX [52], or must have a hydrophobic domain capable of inserting into the membrane during electron transfer. A potential membrane complex of VKOR and GGCX would explain how a transfer reaction between these two enzymes could be facilitated during blood coagulation, but it does not offer any insights into how electrons might be supplied to VKOR in the first place. Despite the fact that molecular dynamic simulations indicate that the 3-TM model of human VKOR has a structural advantage in terms of protein stability over a VKOR with 4 TM [53], questions regarding this model still remain.

In a 4 TM structure of VKOR, the active site cysteines face the ER lumen. It has therefore been hypothesized that VKOR's redox partner must be a luminal protein that most likely bears at least some homology to thioredoxin-like proteins, which encode cysteines in a C-X-X-C motif. The proposed reaction scheme posits that Cys-43 of VKOR forms a mixed disulfide with its redox partner, which subsequently attacks Cys-51 to form an intramolecular disulfide bond in VKOR and releases the redox partner. By mutating the resolving Cys-51, the mixed disulfide can be trapped, thus allowing identification of the redox partner. Such an approach identified several intriguing candidates including soluble proteins and the membrane-bound TMX, TMX4, and ERp18 as forming mixed disulfides with VKOR, although it is not clear what the downstream effects of such interactions might be [54].

Early studies with microsomal fractions indicated that protein disulfide isomerase (PDI) might be an important source of electrons for VKOR [55], which is consistent with PDI's localization to the ER lumen. PDI can act as an electron acceptor by interacting with proteins containing multiple cysteine residues. By accepting electrons from such proteins, disulfide bonds form between these cysteines, which can serve to stabilize or activate these substrates. Following this reaction, the reduced form of PDI is free to donate its electrons to other partner proteins. Studies confirmed that PDI could stimulate VKOR's reductive activity and went on to suggest that VKOR and PDI may even form a complex in the ER membrane [56]. In other words, the reduction of vitamin KO may be driven by the formation of disulfide bonds in the ER lumen. Such a mechanism appears to contribute to the overall redox homeostasis within the ER [57] and has also been suggested to operate in plants as well [58, 59].

While VKORs can be found in numerous classes of organisms, paralogs of VKOR are also quite prevalent. Known as "VKORLs" ("VKOR-like"), the exact role of these enzymes is still unclear. The human VKOR and VKORL1 share 42% identity and 60% similarity. Like VKOR itself, VKORL1 can reduce KO to vitamin K, which may explain why patients treated with anticoagulants do not exhibit significant side effects that would be expected from the inability to turn over vitamin K, like arterial calcifications [60]. Indeed, rat VKORL1 was shown to be up to 50-fold more resistant to warfarin as compared to VKORC1 in one study [61], although such a finding has been contested [62]. However, the rate at which VKORL1 reduces KO may be significantly slower than VKOR [60], and mice missing VKOR (but expressing VKORL1) bled to death shortly after birth [63], suggesting a different function for VKORL1. It has also been suggested that VKORL1 may play a role in the vitamin K cycle by reducing vitamin K to KH_2 [64], although such a suggestion may be premature, as comparisons of VKOR and VKORL1 activity can be problematic [62]. To gain further insight into the differential functions of VKORL1 versus VKOR, these authors looked at expression levels of the two genes in different tissues. They found clear evidence that VKOR and VKORL1 are differentially regulated in rats and mice, with VKOR showing higher expression in rat liver, lung, and kidney, VKORL1 showing higher expression in the brain, and similar expression profiles in the testis. Overall, the levels of VKORL1 were relatively constant across organs, while VKOR showed extremely high levels in the liver but much lower levels in the remaining tissues [61]. Such findings have led some to hypothesize that VKOR may have evolved to provide cofactor to the vitamin K-dependent proteins required for maintaining the high-flux environment of the circulatory system and the homeostasis of a calcified skeleton, while the ability of VKORL1 to reduce vitamin K may be ancillary to its role in antioxidant functions and disulfide bond formation [39]. The differential activity of VKORL1 compared to VKOR is supported by studies conclusively showing that the loop cysteines of VKORL1 are required for activity, in potential contrast to VKOR [62].

The quest to define a role for bacterial VKORs began with an observation that arose from studies of a well-defined, quinone-dependent pathway in bacteria responsible for catalyzing the formation of disulfide bonds in some periplasmic proteins. As covalent bonds between cysteine residues, disulfide bonds can stabilize otherwise energetically unfavorable conformations of certain proteins, thus promoting functionality, similar to binding of calcium ions in the Gla-dependent proteins of eukaryotes. While most disulfide bonds in the bacterial cytoplasm exist transiently as part of an enzyme catalytic cycle, disulfide bonds in the periplasm are much more stable. This is due to the activity of *d*isulfide-*b*onding protein *A* (DsbA), which utilizes two cysteine residues to accept electrons from substrate proteins [65, 66]. Such an electron-transfer reaction leads to the oxidation of the substrate protein (disulfide-bonded), while leaving the cysteine residues of DsbA in the reduced form (-SH). To regenerate active DsbA, electrons need to be transferred to another protein, DsbB [67, 68]. DsbB is localized to the plasma membrane via four transmembrane-spanning domains and utilizes two pairs of active-site cysteines to accept electrons from DsbA and pass them onto ubiquinone or menaquinone [69–71]. These quinone carriers deposit the electrons onto a final electron acceptor like oxygen or nitrate through the process of respiration, thus completing the disulfide bond generation cycle. While homologs of DsbA and DsbB have been identified in many bacteria, some do not encode these enzymes but contain disulfide-bonded proteins [72]. To identify the enzymes responsible for generating disulfide bonds in these organisms lacking DsbB, bioinformatics analysis was performed. The results of this analysis demonstrated that some bacteria encoded a DsbA-like protein fused to a homolog of eukaryotic VKOR [44, 72]. Because VKOR and DsbB both utilize catalytic cysteine residues in redox-dependent transfer reactions, have multiple membrane-spanning domains, and are known to reduce quinones, it suggested that the bacterial VKOR might function in a manner analogous to DsbB. Concordant with this, VKOR homologs were found only in bacteria- and archaea-lacking DsbB [72]. In fact, the VKOR encoded by *Mycobacterium tuberculosis* (*Mtb*) can restore disulfide bond activity to an *E. coli* strain missing *dsbB* [72], and a VKOR homologue has been shown to catalyze disulfide bond formation in cyanobacteria [73]. Therefore, despite the fact that DsbB and VKOR show no significant homology at the amino acid level, they perform analogous reactions. Perhaps even more striking, the VKOR-dependent disulfide bond activity in this *E. coli* strain can be inhibited by high concentrations of the anti-coagulant warfarin, providing another link between the bacterial and eukaryotic enzymes [74]. Such results are of course strikingly analogous to the potential role of VKORL1 in the formation of disulfide bonds in the ER lumen. The crystal structure of the fused DsbA-VKOR from *Synechococcus* suggests that the reactions performed by bacterial and eukaryotic VKORs may proceed via similar mechanisms and may provide insights into the inhibition by warfarin and other anti-coagulants [75, 76]. The *Synechococcus* VKOR (synVKOR) is, however, slightly different than the mammalian enzyme in that it utilizes ubiquinone as a cofactor rather than MK. The epoxide form of ubiquinone has not been found in *Synechococcus*, as with other bacteria, and synVKOR cannot reduce vitamin KO to the hydroquinone form [75]. While DsbB has been shown to encode four transmembrane domains, studies using fusion proteins show that the bacterial VKOR spans the plasma membrane five times, with the fifth-TM segment usually allowing for the fusion to the periplasmic DsbA. Despite this difference in topology, all four active site cysteines face the periplasm [77]. Like DsbB, the cysteines in bacterial VKOR are essential for the formation of a mixed disulfide with DsbA, but the pairs of cysteines are reversed in regards to their order within the amino acid sequence [77]. Phylogenetic analysis of VKOR and DsbB suggests that these two enzymes are in fact related evolutionarily, having diverged from a single lipidic quinone–disulfide oxidoreductase superfamily [78]. The diversion would appear to have led to differential function as well, for while eukaryotic VKORs reduce the epoxide form of K2 to the quinone and the hydroquinone form in order to provide substrate for the gamma-glutamate carboxylase reaction, there is no evidence for gamma-glutamate carboxylases in bacteria.

Despite the fact that there is little overall homology in the amino acid sequences of human VKOR and *Mtb* VKOR, the enzymes appear to catalyze very similar reactions, at least from a redox perspective. It is therefore interesting to test whether they are interchangeable. Expression of *Mtb* VKOR in mammalian cell lines confirmed that the bacterial VKOR is capable of reducing both vitamin K and vitamin KO to KH_2 [79]. While the loop cysteines of *Mtb* VKOR (Cys-57 and Cys-65) are essential for disulfide bond activity when expressed in *E. coli*, they may not be required for reduction of vitamin K epoxide in mammalian cells [79].

Studies of the functionality of mammalian VKORs in bacteria have been more problematic. Expression of hVKOR in *E. coli* results in the formation of inclusion bodies, thus preventing any attempts to assess function *in vivo*. Additionally, reconstitution of recombinant hVKOR activity from insoluble fractions was strictly dependent on the nature of the membrane composition [80]. Attempts to restore disulfide bond formation to a Δ*dsbB* strain of *E. coli* by expressing either the rat or human VKOR have so far been unsuccessful at least in part due to the lack of stable expression [81]. However, selection for a rat VKOR functional in substituting for DsbB in this system yielded a collection of mutants in the rat *vkor* gene that encoded amino acid changes in the protein. Notably, many of these gain of function mutations resulted in changes in the charge of what is predicted to be the first periplasmic loop of rat VKOR when it is expressed in *E. coli*. The charge distribution of amino acids is known to play an important role in establishing the proper topology of membrane proteins [82, 83]. Further, even higher levels of protein are detected when the mutant proteins are expressed in *E. coli* strains carrying mutations that alter the YidC insertase, a protein necessary for membrane localization of some proteins or mutations eliminating the cytoplasmic protease HslV [81]. Mutant strains harboring both *yidC* and *hslV* mutations showed significantly more VKOR activity and protein levels. These results suggest that while the bacterial VKOR can be properly inserted into the ER membrane for the reduction of vitamin KO, the mammalian VKOR may not be able to be properly inserted in the bacterial plasma membrane without initial changes in the protein itself. Nevertheless, the mutant versions of rat VKOR that are expressed in *E. coli* are sensitive to anticoagulants, suggesting that the functional expression even of the mutant enzyme may provide a powerful tool for its study.

6. Vitamin K2 as a target for inhibition

MKs are clearly critical components of many aspects of the growth and proliferation of bacterial and human cells, but most of the enzymes necessary for their biosynthesis are only bacterially encoded and are missing from humans. MK biosynthesis would appear to be an ideal target for the development of small molecule inhibitors as potent antibiotics. Among pathogenic bacteria, *Mtb* poses one of the most significant threats, as it accounts for nearly two million deaths annually. While combinatorial antibiotic therapies have been developed against *Mtb,* serious complications have arisen that compromise the efficacy of these treatments. In addition to excessive length of treatment, the side effects of these drugs can be debilitating, and antibiotic resistance has arisen at a startling rate. In addition, *Mtb* can remain for long periods of time in a dormant state in which traditional antibiotics are not effective. However, even in this quiescent state, *Mtb* requires an active electron transport chain to maintain adequate levels of ATP, and MKs therefore play a key role [84]. To this end, researchers have developed screens specifically targeting the MK biosynthetic pathway of *Mtb*. Early results show that compounds-targeting MenE show some promise [85], and the prenylating enzyme MenA is also being developed as a target [86, 87]. Most strikingly, one MenA inhibitor (allylaminomethanone-A) was shown to be up to 320-times more effective in killing non-replicating *Mtb* than first line drugs currently prescribed for infection [84], and MenA inhibitors have been shown to inhibit growth of *Mtb* resistant to commonly used antitubercular drugs [86]. Caution must be exercised in advancing such therapies, however, as the full scope of vitamin K metabolism in the body has not been elucidated. If gut bacteria do contribute significantly to vitamin K stores in the body, then inhibitors targeting MK biosynthesis may have significant effects on blood coagulation and bone calcification, for example. MenA inhibitors are particularly noteworthy, since off-target effects on the human homolog UBIAD1 could potentially disrupt a number of cellular processes that are only beginning to be understood.

As an inhibitor of vitamin K-dependent reactions, warfarin has long been used as an anticoagulant that at least in part targets human VKOR. While the mycobacterial VKOR has been shown to be sensitive to warfarin, the amount necessary to inhibit the bacterial enzyme is orders of magnitudes higher than the amount needed to prevent blood coagulation [74]. This would suggest that while the human and bacterial VKORs can perform similar functions and do so by similar mechanisms, the divergence in the amino acid sequence of the two is significant enough that treatment of mycobacterial infection with anticoagulants would not be an effective therapeutic strategy. However, ferulenol, an anticoagulant, shown to be approximately 20-fold more potent against human VKOR than warfarin, showed similar potency against the VKOR from *Synechococcus* [75]. It is therefore possible that drug discovery efforts to identify novel anticoagulants may impact the search for inhibitors of bacterial VKOR and vice versa.

Disulfide bond formation appears to be dispensable for *in vitro* aerobic growth of *E. coli* and other bacteria, although many virulence factors absolutely require disulfide bonds for proper assembly and function. Bacteria disrupted in the DSB pathway are rendered less virulent, and *E. coli* cannot grow anaerobically, suggesting that small molecule inhibitors of DsbB- or VKOR-dependent pathways may be potent anti-virulents and may prevent anaerobic growth of some pathogens. *Mtb* is especially vulnerable to such compounds, as VKOR is essential for growth of this organism, even in aerobic environments [88]. Bacterial DsbBs and VKORs therefore make attractive targets for antibiotic therapies.

The fact that DsbB and *Mtb*VKOR perform complementary functions but lack amino acid homology allowed our laboratory to develop a screen to identify potential small molecule inhibitors that specifically target DsbB or *Mtb*VKOR [89]. β-galactosidase (LacZ) is a cytoplasmic enzyme capable of cleaving the disaccharide lactose to yield galactose and glucose. The activity of this enzyme can be readily monitored in *E. coli* by using the lactose analog Isopropyl 5-bromo-4-chloro-3-indolyl-β-D-galactopyranoside (X-gal)—successful cleavage of X-gal yields an insoluble blue dye that can readily be distinguished by eye. When *lacZ* is fused to the gene encoding the membrane protein MalF, however, the enzyme is exported into the periplasm and is inactivated by the formation of inappropriate disulfide bonds. Strains expressing this fusion construct appear white on X-gal, as the substrate cannot be cleaved. However, strains lacking *dsbB* appear blue on X-gal when expressing this construct, as these cells lack the ability to catalyze the formation of the inappropriate disulfide bonds in LacZ. In such a case, LacZ is active and capable of cleaving X-gal. When the Δ*dsbB* strain is complemented with a construct expressing lowered-levels of *dsbB* or with *Mtb vkor*, the strains appear white again, as disulfide bond formation is restored. When libraries of small molecules are applied to *E. coli* strains differentially expressing *E. coli dsbB* (or the *dsbB* from another Gram-negative bacterium) or *Mtb vkor* along with the MalF-LacZ fusion in a high-throughput format, specific inhibitors of DsbB or VKOR can easily be identified by the appearance of a blue color. The differences in the primary structure of DsbB and VKOR would suggest that any compound that inhibits one should not inhibit the other. For this reason, each strain acts as a strong counter screen for the other. We have successfully employed this screen to identify several strong, specific inhibitors of the DsbB from *E. coli* as well as several other important pathogens, and we continue to use it to screen for potential inhibitors of the *Mtb vkor*. Further efforts to express functional mammalian VKOR and VKORL1s in the *E. coli* screening strain would not only provide a means by which to test potential side effects of compounds targeting the bacterial enzymes, but may offer a high-throughput approach to identifying new compounds capable of inhibiting VKOR-dependent processes in mammals. Additionally, because the screening system provides an easily monitored readout for VKOR activity, it might be used to study hVKOR variants shown to be resistant to anticoagulant therapies. Such studies could lead to more precisely targeted and potent blood thinners.

7. Conclusions

Since their incorporation into the electron transfer pathways of ancient microbes, menaquinones have become a cornerstone of redox-dependent reactions in almost every domain of life. Their ability to interact with a large variety of proteins, to readily accept and donate electrons, and to easily move within biological membranes have combined to make MKs flexible and efficient molecular wires. As such, organisms have evolved to integrate MKs into many metabolic processes, thus plugging into previously untapped sources of power. While researchers have seemed to only scratch the surface of the myriad uses for MKs to this point, further investigation will yield not only fascinating insights into the biochemical pathways critical to life, but may be a crucial starting point for the development of therapies designed to protect and enhance those pathways.

Acknowledgements

This work was supported by US National Institute of General Medical Sciences grant GMO41883 (J.B.) and Ruth L. Kirschstein National Research Service Award 1F32GM108443-01 (B.M.M.)

Author details

Brian M. Meehan and Jonathan Beckwith*

*Address all correspondence to: jon_beckwith@hms.harvard.edu

Harvard Medical School, Boston, Massachusetts, USA

References

[1] Wilson TH, Lin EC. Evolution of membrane bioenergetics. J Supramol Struct. 1980; 13:421–46.

[2] Falkowski PG. Evolution. Tracing oxygen's imprint on earth's metabolic evolution. Science. 2006;311(5768):1724–5.

[3] Imlay JA. Pathways of oxidative damage. Annu Rev Microbiol. 2003;57:395–418.

[4] Sessions AL, Doughty DM, Welander PV, Summons RE, Newman DK. The continuing puzzle of the great oxidation event. Curr Biol. 2009;19(14):R567–74.

[5] Raymond J, Segrè D. The effect of oxygen on biochemical networks and the evolution of complex life. Science. 2006;311(5768):1764–7.

[6] Hohmann-Marriott MF, Blankenship RE. Evolution of photosynthesis. Annu Rev Plant Biol. 2011;62:515–48.

[7] Bentley R, Meganathan R. Biosynthesis of vitamin K (menaquinone) in bacteria. Microbiol Rev. 1982;46(3):241–80.

[8] Han X, Chen C-C, Kuo C-J, Huang C-H, Zheng Y, Ko T-P, et al. Crystal structures of ligand-bound octaprenyl pyrophosphate synthase from *Escherichia coli* reveal the catalytic and chain-length determining mechanisms. Proteins. 2015;83(1):37–45.

[9] Nowicka B, Kruk J. Occurrence, biosynthesis and function of isoprenoid quinones. Biochim Biophys Acta. 2010;1797(9):1587–605.

[10] Johnson TW, Shen G, Zybailov B, Kolling D, Reategui R, Beauparlant S, et al. Recruitment of a foreign quinone into the A(1) site of photosystem I. I. Genetic and physiological characterization of phylloquinone biosynthetic pathway mutants in *Synechocystis* sp. pcc 6803. J Biol Chem. 2000;275(12):8523–30.

[11] Hiratsuka T, Furihata K, Ishikawa J, Yamashita H, Itoh N, Seto H, et al. An alternative menaquinone biosynthetic pathway operating in microorganisms. Science. 2008; 321(5896):1670–3.

[12] Gross J, Cho WK, Lezhneva L, Falk J, Krupinska K, Shinozaki K, et al. A plant locus essential for phylloquinone (vitamin K1) biosynthesis originated from a fusion of four eubacterial genes. J Biol Chem. 2006;281(25):17189–96.

[13] Reumann S. Biosynthesis of vitamin K1 (phylloquinone) by plant peroxisomes and its integration into signaling molecule synthesis pathways. Subcell Biochem. 2013;69:213–29.

[14] Nakagawa K, Hirota Y, Sawada N, Yuge N, Watanabe M, Uchino Y, et al. Identification of UBIAD1 as a novel human menaquinone-4 biosynthetic enzyme. Nature. 2010;468(7320):117–21.

[15] Okano T, Shimomura Y, Yamane M, Suhara Y, Kamao M, Sugiura M, et al. Conversion of phylloquinone (Vitamin K1) into menaquinone-4 (Vitamin K2) in mice: two possible routes for menaquinone-4 accumulation in cerebra of mice. J Biol Chem. 2008;283(17): 11270–9.

[16] Hirota Y, Tsugawa N, Nakagawa K, Suhara Y, Tanaka K, Uchino Y, et al. Menadione (vitamin K3) is a catabolic product of oral phylloquinone (vitamin K1) in the intestine and a circulating precursor of tissue menaquinone-4 (vitamin K2) in rats. J Biol Chem. 2013;288(46):33071–80.

[17] Davidson RT, Foley AL, Engelke JA, Suttie JW. Conversion of dietary phylloquinone to tissue menaquinone-4 in rats is not dependent on gut bacteria. J Nutr. 1998;128(2):220–3.

[18] Nakagawa K, Sawada N, Hirota Y, Uchino Y, Suhara Y, Hasegawa T, et al. Vitamin K2 biosynthetic enzyme, UBIAD1 is essential for embryonic development of mice. Plos One. 2014;9(8):e104078.

[19] Fernandez F, Collins MD. Vitamin K composition of anaerobic gut bacteria. FEMS Microbiol Lett. 1987;41:175–80.

[20] Shearer MJ, Newman P. Recent trends in the metabolism and cell biology of vitamin K with special reference to vitamin K cycling and MK-4 biosynthesis. J Lipid Res. 2014;55(3):345–62.

[21] Coates ME. Gnotobiotic animals in nutrition research. Proc Nutr Soc. 1973;32(2):53–8.

[22] Conly J, Stein K. Reduction of vitamin K2 concentrations in human liver associated with the use of broad spectrum antimicrobials. Clin Investig Med. 1994;6:531–9.

[23] Newman DK, Kolter R. A role for excreted quinones in extracellular electron transfer. Nature. 2000;405(6782):94–7.

[24] Groenen-van Dooren MM, Ronden JE, Soute BA, Vermeer C. Bioavailability of phylloquinone and menaquinones after oral and colorectal administration in vitamin K-deficient rats. Biochem Pharmacol. 1995;50(6):797–801.

[25] Beulens JWJ, Booth SL, van den Heuvel EGHM, Stoecklin E, Baka A, Vermeer C. The role of menaquinones (vitamin K2) in human health. Br J Nutr. 2013;110(8):1357–68.

[26] Suttie JW. The importance of menaquinones in human nutrition. Annu Rev Nutr. 1995;15:399–417.

[27] Walther B, Karl JP, Booth SL, Boyaval P. Menaquinones, bacteria, and the food supply: the relevance of dairy and fermented food products to vitamin K requirements. Adv Nutr. 2013;4(4):463–73.

[28] Gijsbers BL, Jie KS, Vermeer C. Effect of food composition on vitamin K absorption in human volunteers. Br J Nutr. 1996;76(2):223–9.

[29] Schurgers LJ, Vermeer C. Differential lipoprotein transport pathways of K-vitamins in healthy subjects. Biochim Biophys Acta. 2002;1570(1):27–32.

[30] Kurosu M, Begari E. Vitamin K2 in electron transport system: are enzymes involved in vitamin K2 biosynthesis promising drug targets? Molecules. 2010;15(3):1531–53.

[31] Schoepp-Cothenet B, Lieutaud C, Baymann F, Verméglio A, Friedrich T, Kramer DM, et al. Menaquinone as pool quinone in a purple bacterium. Proc Natl Acad Sci USA. 2009;106(21):8549–54.

[32] Wagner GC, Kassner RJ, Kamen MD. Redox potentials of certain vitamins K: implications for a role in sulfite reduction by obligately anaerobic bacteria. PNAS. 1974;71(2): 253–6.

[33] Friedman PA, Shia M. Some characteristics of a vitamin K-dependent carboxylating system from rat liver microsomes. Biochem Biophys Res Commun. 1976;70(2):647–54.

[34] Buitenhuis HC, Soute BA, Vermeer C. Comparison of the vitamins K1, K2 and K3 as cofactors for the hepatic vitamin K-dependent carboxylase. Biochim Biophys Acta. 1990;1034(2):170–5.

[35] Ferland G. Vitamin K, an emerging nutrient in brain function. Biofactors. 2012;38(2): 151–7.

[36] Vos M, Esposito G, Edirisinghe JN, Vilain S, Haddad DM, Slabbaert JR, et al. Vitamin K2 is a mitochondrial electron carrier that rescues pink1 deficiency. Science. 2012;336(6086):1306–10.

[37] Shearer MJ, Newman P. Metabolism and cell biology of vitamin K. Thromb Haemost. 2008;100(4):530–47.

[38] Czerwiec E, Begley GS, Bronstein M, Stenflo J, Taylor K, Furie BC, et al. Expression and characterization of recombinant vitamin K-dependent gamma-glutamyl carboxylase from an invertebrate, Conus textile. Eur J Biochem. 2002;269(24):6162–72.

[39] Oldenburg J, Watzka M, Bevans CG. VKORC1 and VKORC1L1: why do vertebrates have two vitamin K 2,3-epoxide reductases? Nutrients. 2015;7(8):6250–80.

[40] Wakeman CA, Hammer ND, Stauff DL, Attia AS, Anzaldi LL, Dikalov SI, et al. Menaquinone biosynthesis potentiates haem toxicity in *Staphylococcus aureus*. Mol Microbiol. 2012;86(6):1376–92.

[41] Korshunov S, Imlay JA. Detection and quantification of superoxide formed within the periplasm of *Escherichia coli*. J Bacteriol. 2006;188(17):6326–34.

[42] Li T, Chang C-Y, Jin D-Y, Lin P-J, Khvorova A, Stafford DW. Identification of the gene for vitamin K epoxide reductase. Nature. 2004;427(6974):541–4.

[43] Rost S, Fregin A, Ivaskevicius V, Conzelmann E, Hörtnagel K, Pelz H-J, et al. Mutations in VKORC1 cause warfarin resistance and multiple coagulation factor deficiency type 2. Nature. 2004;427(6974):537–41.

[44] Goodstadt L, Ponting CP. Vitamin K epoxide reductase: homology, active site and catalytic mechanism. Trends Biochem Sci. 2004;29(6):289–92.

[45] Kulman JD, Harris JE, Nakazawa N, Ogasawara M, Satake M, Davie EW. Vitamin K-dependent proteins in *Ciona intestinalis*, a basal chordate lacking a blood coagulation cascade. Proc Natl Acad Sci USA. 2006;103(43):15794–9.

[46] Jin D-Y, Tie J-K, Stafford DW. The conversion of vitamin K epoxide to vitamin K quinone and vitamin K quinone to vitamin K hydroquinone uses the same active site cysteines. Biochemistry. 2007;46(24):7279–83.

[47] Rost S, Fregin A, Hünerberg M, Bevans CG, Müller CR, Oldenburg J. Site-directed mutagenesis of coumarin-type anticoagulant-sensitive VKORC1: evidence that highly conserved amino acids define structural requirements for enzymatic activity and inhibition by warfarin. Thromb Haemost. 2005;94(4):780–6.

[48] Tie J-K, Jin D-Y, Tie K, Stafford DW. Evaluation of warfarin resistance using transcription activator-like effector nucleases-mediated vitamin K epoxide reductase knockout HEK293 cells. J Thromb Haemost. 2013;11(8):1556–64.

[49] Rishavy MA, Usubalieva A, Hallgren KW, Berkner KL. Novel insight into the mechanism of the vitamin K oxidoreductase (VKOR): electron relay through Cys43 and Cys51

reduces VKOR to allow vitamin K reduction and facilitation of vitamin K-dependent protein carboxylation. J Biol Chem. 2011;286(9):7267–78.

[50] Tie J-K, Jin D-Y, Stafford DW. Human vitamin K epoxide reductase and its bacterial homologue have different membrane topologies and reaction mechanisms. J Biol Chem. 2012;287(41):33945–55.

[51] Cao Z, van Lith M, Mitchell LJ, Pringle MA, Inaba K, Bulleid NJ. The membrane topology of vitamin K epoxide reductase is conserved between Human isoforms and the bacterial enzyme. Biochem J. 2016;:BJ20151223.

[52] Wu S, Liu S, Davis CH, Stafford DW, Kulman JD, Pedersen LG. A hetero-dimer model for concerted action of vitamin K carboxylase and vitamin K reductase in vitamin K cycle. J Theor Biol. 2011;279(1):143–9.

[53] Wu S, Tie J-K, Stafford DW, Pedersen LG. Membrane topology for human vitamin K epoxide reductase. J Thromb Haemost. 2014;12(1):112–4.

[54] Schulman S, Wang B, Li W, Rapoport TA. Vitamin K epoxide reductase prefers ER membrane-anchored thioredoxin-like redox partners. Proc Natl Acad Sci USA. 2010;107(34):15027–32.

[55] Soute BA, Groenen-van Dooren MM, Holmgren A, Lundström J, Vermeer C. Stimulation of the dithiol-dependent reductases in the vitamin K cycle by the thioredoxin system. Strong synergistic effects with protein disulphide-isomerase. Biochem J. 1992;281 (Pt 1):255–9.

[56] Wajih N, Hutson SM, Wallin R. Disulfide-dependent protein folding is linked to operation of the vitamin K cycle in the endoplasmic reticulum. A protein disulfide isomerase-VKORC1 redox enzyme complex appears to be responsible for vitamin K1 2,3-epoxide reduction. J Biol Chem. 2007;282(4):2626–35.

[57] Rutkevich LA, Williams DB. Vitamin K epoxide reductase contributes to protein disulfide formation and redox homeostasis within the endoplasmic reticulum. Mol Biol Cell. 2012;23(11):2017–27.

[58] Feng W-K, Wang L, Lu Y, Wang X-Y. A protein oxidase catalysing disulfide bond formation is localized to the chloroplast thylakoids. FEBS J. 2011;278(18):3419–30.

[59] Wan C-M, Yang X-J, Du J-J, Lu Y, Yu Z-B, Feng Y-G, et al. Identification and characterization of SlVKOR, a disulfide bond formation protein from *Solanum lycopersicum*, and bioinformatic analysis of plant VKORs. Biochem Mosc. 2014;79(5):440–9.

[60] Westhofen P, Watzka M, Marinova M, Hass M, Kirfel G, Müller J, et al. Human vitamin K 2,3-epoxide reductase complex subunit 1-like 1 (VKORC1L1) mediates vitamin K-dependent intracellular antioxidant function. J Biol Chem. 2011;286(17):15085–94.

[61] Hammed A, Matagrin B, Spohn G, Prouillac C, Benoit E, Lattard V. VKORC1L1, an enzyme rescuing the vitamin K 2,3-epoxide reductase activity in some extrahepatic tissues during anticoagulation therapy. J Biol Chem. 2013;288(40):28733–42.

[62] Tie J-K, Jin D-Y, Stafford DW. Conserved loop cysteines of vitamin K epoxide reductase complex subunit 1-like 1 (VKORC1L1) are involved in its active site regeneration. J Biol Chem. 2014;289(13):9396–407.

[63] Spohn G, Kleinridders A, Wunderlich FT, Watzka M, Zaucke F, Blumbach K, et al. VKORC1 deficiency in mice causes early postnatal lethality due to severe bleeding. Thromb Haemost. 2009;101(6):1044–50.

[64] Van Horn WD. Structural and functional insights into human vitamin K epoxide reductase and vitamin K epoxide reductase-like1. Crit Rev Biochem Mol Biol. 2013;48(4):357–72.

[65] Bardwell JC, McGovern K, Beckwith J. Identification of a protein required for disulfide bond formation in vivo. Cell. 1991;67(3):581–9.

[66] Kadokura H, Katzen F, Beckwith J. Protein disulfide bond formation in prokaryotes. Annu Rev Biochem. 2003;72:111–35.

[67] Bardwell JC, Lee JO, Jander G, Martin N, Belin D, Beckwith J. A pathway for disulfide bond formation in vivo. Proc Natl Acad Sci USA. 1993;90(3):1038–42.

[68] Missiakas D, Georgopoulos C, Raina S. Identification and characterization of the Escherichia coli gene dsbB, whose product is involved in the formation of disulfide bonds in vivo. Proc Natl Acad Sci USA. 1993;90(15):7084–8.

[69] Kadokura H, Beckwith J. Four cysteines of the membrane protein DsbB act in concert to oxidize its substrate DsbA. EMBO J. 2002;21(10):2354–63.

[70] Bader MW, Xie T, Yu CA, Bardwell JC. Disulfide bonds are generated by quinone reduction. J Biol Chem. 2000;275(34):26082–8.

[71] Takahashi Y-H, Inaba K, Ito K. Characterization of the menaquinone-dependent disulfide bond formation pathway of *Escherichia coli*. J Biol Chem. 2004;279(45):47057–65.

[72] Dutton RJ, Boyd D, Berkmen M, Beckwith J. Bacterial species exhibit diversity in their mechanisms and capacity for protein disulfide bond formation. Proc Natl Acad Sci USA. 2008;105(33):11933–8.

[73] Singh AK, Bhattacharyya-Pakrasi M, Pakrasi HB. Identification of an atypical membrane protein involved in the formation of protein disulfide bonds in oxygenic photosynthetic organisms. J Biol Chem. 2008;283(23):15762–70.

[74] Dutton RJ, Wayman A, Wei J-R, Rubin EJ, Beckwith J, Boyd D. Inhibition of bacterial disulfide bond formation by the anticoagulant warfarin. Proc Natl Acad Sci USA. 2010;107(1):297–301.

[75] Li W, Schulman S, Dutton RJ, Boyd D, Beckwith J, Rapoport TA. Structure of a bacterial homologue of vitamin K epoxide reductase. Nature. 2010;463(7280):507–12.

[76] Liu S, Cheng W, Fowle Grider R, Shen G, Li W. Structures of an intramembrane vitamin K epoxide reductase homolog reveal control mechanisms for electron transfer. Nat Commun. 2014;5:3110.

[77] Wang X, Dutton RJ, Beckwith J, Boyd D. Membrane topology and mutational analysis of *Mycobacterium tuberculosis* VKOR, a protein involved in disulfide bond formation and a homologue of human vitamin K epoxide reductase. Antioxid Redox Signal. 2011;14(8):1413–20.

[78] Bevans CG, Krettler C, Reinhart C, Watzka M, Oldenburg J. Phylogeny of the vitamin K 2,3-epoxide reductase (VKOR) family and evolutionary relationship to the disulfide bond formation protein B (DsbB) family. Nutrients. 2015;7(8):6224–49.

[79] Tie J-K, Jin D-Y, Stafford DW. *Mycobacterium tuberculosis* vitamin K epoxide reductase homologue supports vitamin K-dependent carboxylation in mammalian cells. Antioxid Redox Signal. 2012;16(4):329–38.

[80] Jaenecke F, Friedrich-Epler B, Parthier C, Stubbs MT. Membrane composition influences the activity of in vitro refolded human vitamin K epoxide reductase. Biochemistry. 2015;54(42):6454–61.

[81] Hatahet F, Blazyk JL, Martineau E, Mandela E, Zhao Y, Campbell RE, et al. Altered *Escherichia coli* membrane protein assembly machinery allows proper membrane assembly of eukaryotic protein vitamin K epoxide reductase. Proc Natl Acad Sci USA. 2015;112(49):15184–9.

[82] Heijne G. The distribution of positively charged residues in bacterial inner membrane proteins correlates with the trans-membrane topology. EMBO J. 1986;5(11):3021–7.

[83] Boyd D, Beckwith J. Positively charged amino acid residues can act as topogenic determinants in membrane proteins. PNAS. 1989;86(23):9446–50.

[84] Dhiman RK, Mahapatra S, Slayden RA, Boyne ME, Lenaerts A, Hinshaw JC, et al. Menaquinone synthesis is critical for maintaining mycobacterial viability during exponential growth and recovery from non-replicating persistence. Mol Microbiol. 2009;72(1):85–97.

[85] Lu X, Zhang H, Tonge PJ, Tan DS. Mechanism-based inhibitors of MenE, an acyl-CoA synthetase involved in bacterial menaquinone biosynthesis. Bioorg Med Chem Lett. 2008;18(22):5963–6.

[86] Kurosu M, Narayanasamy P, Biswas K, Dhiman R, Crick DC. Discovery of 1,4-dihydroxy-2-naphthoate [corrected] prenyltransferase inhibitors: new drug leads for multidrug-resistant gram-positive pathogens. J Med Chem. 2007;50(17):3973–5.

[87] Debnath J, Siricilla S, Wan B, Crick DC, Lenaerts AJ, Franzblau SG, et al. Discovery of selective menaquinone biosynthesis inhibitors against *Mycobacterium tuberculosis*. J Med Chem. 2012;55(8):3739–55.

[88] Sassetti CM, Boyd DH, Rubin EJ. Genes required for mycobacterial growth defined by high density mutagenesis. Mol Microbiol. 2003;48(1):77–84.

[89] Landeta C, Blazyk JL, Hatahet F, Meehan BM, Eser M, Myrick A, et al. Compounds targeting disulfide bond forming enzyme DsbB of Gram-negative bacteria. Nat Chem Biol. 2015;11(4):292.

Chapter 15

Vitamin K_2 Biosynthesis: Drug Targets for New Antibacterials

Michio Kurosu

Additional information is available at the end of the chapter

http://dx.doi.org/10.5772/65487

Abstract

In prokaryotes, vitamin K_2 (menaquinone) transfers two electrons in a process of aerobic or anaerobic respiration. Respiration occurs in the cell membrane of prokaryotic cells. Electron donors transfer two electrons to menaquinone (MK). Menaquinone in turn transfers these electrons to an electron acceptor. Menaquinones are vital for the electron transport chain. In the spectrum of Gram-positive bacteria and *Mycobacterium* spp., vitamin K_2 serves as the only quinone molecule in their electron shuffling systems. Hence, the bacterial enzymes associated with biosynthesis of the menaquinone(s) serve as potential target molecules for the development of new antibacterial drugs. This chapter summarizes the effects of vitamin K_2 in bacteria and describes in more detail the aspects of menaquinone in bacterial electron transport in general, while also featuring the discoveries of menaquinone biosynthesis inhibitors.

Keywords: vitamin K_1, vitamin K_2, menaquinone, menaquinone biosynthesis, electron transport systems, ATP biosynthesis, Gram-positive bacteria, *Mycobacterium tuberculosis*, antibacterial agents

1. Introduction

Vitamin K is a lipid-soluble vitamin that facilitates the process of blood clot formation. Henrick Dam and Edward Doisy were co-awarded the Nobel Prize in 1943 for the discovery of vitamin Ks and the discovery of their structure. Naturally existing forms of vitamin K, vitamin K_1 (i.e., phylloquinone), and vitamin K_2 (i.e., menaquinone) can be found in the human liver (some 10% of phylloquinone of the total vitamin K contents), as well as in other tissues in rather low amounts; vitamin K_1 accumulates in the liver, while vitamin K_2 is effectively distributed to other body organs [1]. Phylloquinone stems from the dietary intake, while menaquinones are synthesized

INTECH
open science | open minds

© 2017 The Author(s). Licensee InTech. This chapter is distributed under the terms of the Creative Commons Attribution License (http://creativecommons.org/licenses/by/3.0), which permits unrestricted use, distribution, and reproduction in any medium, provided the original work is properly cited. (cc) BY

by bacteria of the intestine. However, no direct evidence can be established for the biological effects of menaquinones in humans; however, it is surmised that menaquinones mainly are utilized in the synthesis of blood-clotting factors upon the depletion of phylloquinone [2]. Additionally, menaquinones have proven to be more efficient than phylloquinone as a bioactive molecule in processes such as osteroclast differentiation, lowering of blood cholesterol levels, as well as the slowing down of atherosclerotic progression.

The menaquinones have proven to be important in several biological reactions, for example electron transport, active transport, oxidative phosphorylation, as well as endospore formation in bacteria. Furthermore, variations in the inherent molecular structures of the menaquinones and their skewed distributions among bacterial strains are considered as an important marker in bacterial taxonomy [3]. The biosynthesis of menaquinone has been subject to considerable attention in the quest for drug targets displaying multidrug resistance toward Gram-positive pathogenic microorganisms, including *Mycobacterium tuberculosis.*

2. Vitamin K: general molecular structures

There are two naturally occurring forms of vitamin K. Plants and some cyanobacteria synthesize phylloquinone, which is also known as vitamin K_1. Bacteria synthesize a range of vitamin K, but not vitamin K_1, using the different lipophilic side chains derived from isoprene (5-carbon) units. Bacterial vitamin K(s) are designated menaquinone-n (MK-n), where "n" stands for the number of isoprenoid (5-carbon) units. Vitamin K_1 and K_2 share the common molecular structure, belonging to the 2-methyl-1,4-naphthoquinone system, and they appear to be differ structurally in their number of isoprene units within their side chain, as well as their degree of unsaturation [4] (**Figure 1**). Menaquinones, displaying side chains containing up to 15 isoprene units have been identified. For instance, MK-8 is predominantly seen in *E. coli,* while *M. tuberculosis* mainly utilizes MK-9 as the preferred electron carrier. The menaquinones, which possess from 2 to 13 isoprene units, have been detected in both human and animal tissues. A plethora of synthetic vitamin K(s) are now commercially available, and biochemically fabricated vitamin K(s), in the form of vitamin K_3, K_4, and K_5, are applied in several areas, which include pet food industry (e.g., VK_3), as well as human supplements (e.g., VK_5).

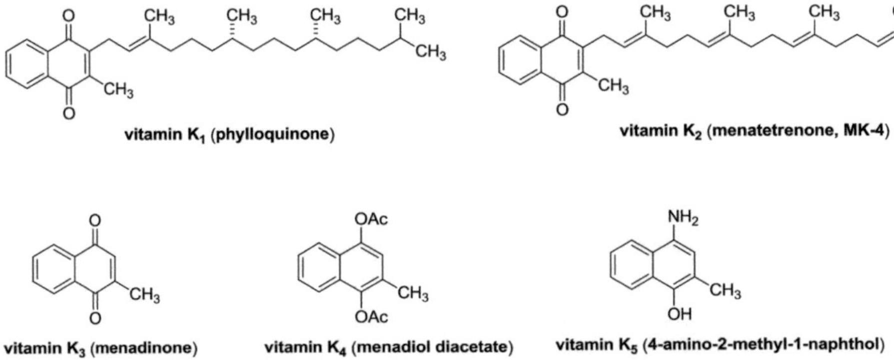

Figure 1. Structures of vitamin K_1 and K_2 and synthetic vitamin Ks (vitamin K_3, K_4, and K_5).

http://dx.doi.org/10.5772/65487

3. Vitamin K in humans

The role of vitamin K as a cofactor in blood coagulation stems from the post-translational modification of a number of plasma proteins such as factors II, VII, IX, X, proteins C, S, Gla (γ-carboxyglutamic acid) proteins has been well-documented. In addition to the essential role of vitamin K in the blood-clotting cascade, the potential role in the increase of bone mass [5, 6], antioxidant mechanisms [7], the biosynthesis of cholesterol and steroid hormones [8], and anticancer effects have been reported [9, 10].

Vitamin K is taken up by organs such as the liver and bones, but abundantly distributed in other organs such as the brain, the kidneys, and gonadal tissues [11]. However, the exact role of vitamin K in the present tissues is not well described. The distribution of vitamin K moieties is varying, which depends on the molecular structure of the side chain. As for humans, vitamin K_1 is distributed to all tissues and organs, with relatively large amounts to the liver, the heart, and the pancreas (some 10.6, 9.3, 28.4 pmol/g wet tissue weight, respectively). However, low levels (<2 pmol/g) were measured in the brain, kidney, and lung tissue specimens. Menaquinone-4 (MK-4) also seems to be distributed to a plethora of other tissues, that is, its levels exceed the levels of vitamin K_1 detected in the brain and kidneys (2.8 ng/g), which equals to that found in the pancreas. However, some organs, such as the liver, heart, and lung remain low in terms of MK-4 contents. The less known MK-6 ~ 11 menaquinone species are also found in the liver, while only small amounts of MK-6 ~ 9 are detected in organs such as the heart and pancreas. Finally, the total amount of vitamin K reported in human plasma was in the range of 0.47 ~ 1.19 nmol/L [1, 12].

Vitamin K can be absorbed well from diet; however, total vitamin K levels are depend greatly on the gut or digestive health. The intestinal bacteria (i.e., microbiota) influence human nutrition and metabolism in diverse ways. The gut microbiota produces menaquinone; thus, it is considered that vitamin K deficiency is quite rare for healthy humans. In addition, the vitamin K_1 is absolutely abundant in leafy and salad vegetables, and herbs. It is easy to understand that newborn babies are born with a vitamin K deficiency, giving newborns a vitamin K injection upon birth or oral vitamin K drops is established to prevent bleeding or a hemorrhagic disease development. In general, vitamin K deficiency results from extremely inadequate intake of fat (malabsorption) or use of coumarin anticoagulants.

4. The role of vitamin K_2 in electron transport system

Function of ubiquinone (Q) (coenzyme Q_{10}) as a component of the mitochondrial respiratory chain in human is well established ("the chemiosmotic theory," Mitchell, 1978). In prokaryotes, especially in Gram-positive bacteria, vitamin K_2 (menaquinone) will transfer two electrons in a process of aerobic or anaerobic respiration. Respiration occurs in the cell membrane of prokaryotic cells. Electron donors transfer two electrons to menaquinone. Menaquinone in turn transfer these electrons to an electron acceptor. Schematic electron flow mediated by menaquinone in *M. tuberculosis* is illustrated in **Figure 2A**. The exact organization of enzymes in respiratory chains will vary among different bacteria. Nicotinamide adenine dinucleotide phosphate (NADH) is the most important electron donor in eukaryotes (**Figure 2B**); however, bacteria can use a number of different electron donors, dehydrogenases, oxidases and reductases, and electron acceptors. Electrons are transported along the membrane through menaquinone and a series of protein carriers.

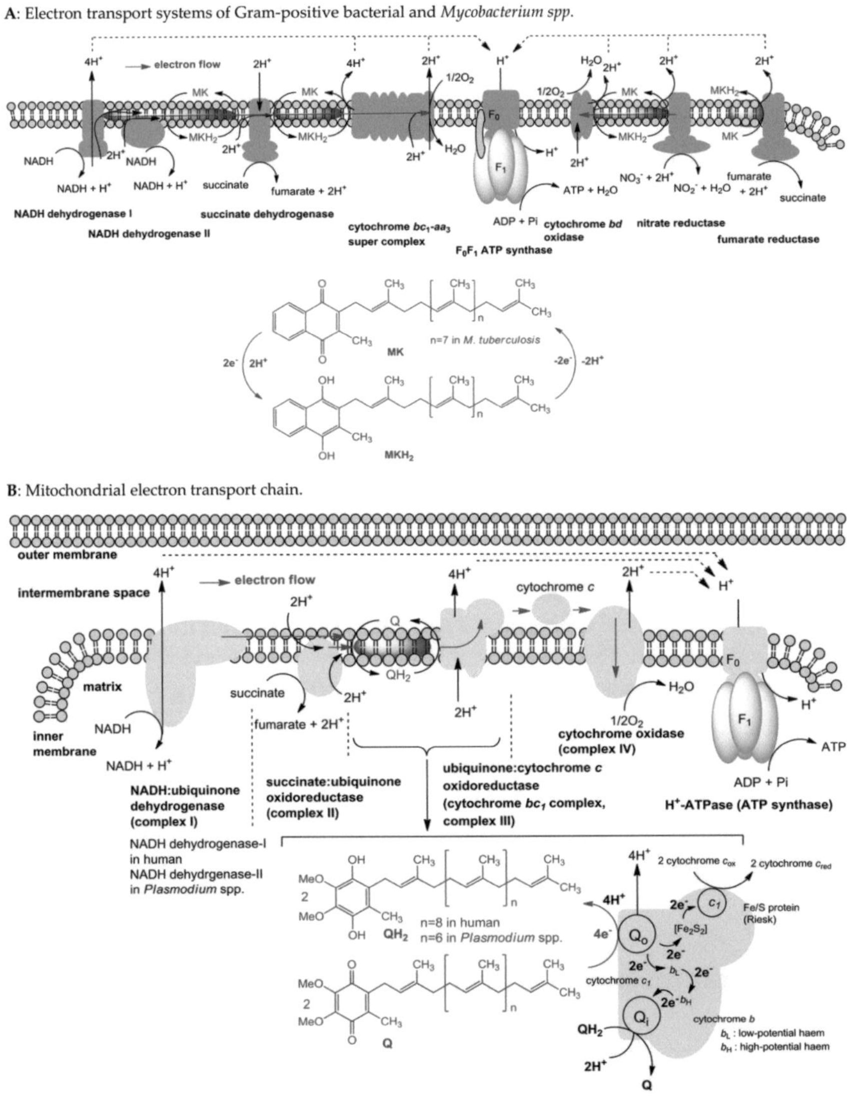

Figure 2. Schematic electron flow systems. (A) Electron transport systems of Gram-positive bacterial and *Mycobacterium* spp. (B) Mitochondrial electron transport chain.

Protons are translocated across the cell membrane (from the cytoplasm to the periplasmic space) concomitantly. Synthesis of Adenosine triphosphate (ATP) from ADP and phosphate is coupled with protons move through these complexes (**Figure 2**). Therefore, CoQ_{10} and menaquinone occupy a central and essential role in Adenosine triphosphate (ATP) synthesis [13, 14]. From the taxonomic studies, it is evident that the majority of Gram-positive bacteria including *Mycobacterium* spp. utilize only menaquinone in their electron transport systems [15], and menaquinone biosynthesis is essential for survival of Gram-positive bacteria [16, 17]. Several *in vitro* studies indicated that exogenous menaquinone did not rescue the bacteria treated with selective menaquinone biosynthesis inhibitors [18]. Menaquinone biosynthesis has been extensively studied in *E. coli.* A plethora of Gram-negative organisms use CoQ in their electron transport systems when aerobic conditions prevail, but menaquinone under anaerobic conditions. However, the reaction chain facilitating electron transport humans does not use menaquinone. Unquestionably, the chain of reactions funneling the transport of electrons serves as a central component in the synthesis of Adenosine triphosphate (ATP) and the subsequent multiplication of bacteria (**Figure 2**). Hence, inhibitors of the biosynthesis of menaquinone or inhibitors specifically targeting the enzymes linked to electron transport systems display the potential for the development of novel and selective drugs against multidrug-resistant (MDR) Gram-positive bacteria. Although the functions of vitamin K_1 in humans and vitamin K_2 in bacteria are entirely different, drug discovery targeting vitamin K_2 or its biosynthesis requires careful consideration of vitamin K distribution in tissue and selectivity against the target protein because essential vitamin K-dependent protein(s) may be interfered by vitamin K biosynthesis inhibitors. Nonetheless, many evidences support that menaquinone biosynthesis inhibitors can be developed into selective antibacterial agents for infections caused by Gram-positive bacteria and *Mycobacterium* spp.

5. Biosynthesis of menaquinone

Menaquinones play an important role in electron transport, and oxidative phosphorylation. In addition, they are responsible for active transport, and endospore formation in some *Bacillus* species [19]. The biosynthetic steps leading to menaquinone have been studied in *E. coli* (*vide supra*) [19, 20]. The synthesis of menaquinone is accomplished by MenA-MenG as illustrated in **Figure 3**. These enzymes are encoded by two clusters of genes. The men gene cluster consists of the *MenB, C, D, E, and F* and a separate cluster containing *MenA* and *MenG* [21, 22]. The biosynthesis of menaquinone is initiated from chorismate and proceeds through a series of menaquinone-specific reactions. MenF is isomerizing to chorismate in order to form isochorismate. MenD (a thiamine diphosphate-dependent enzyme) catalyzes a Stetter-like conjugate addition (a 1,4-addition of an carbonyl molecule to α β-unsaturated compound) of α-ketoglutarate with isochorismate, forming 2-succinyl-5-enolpyruvyl-6-hydroxy-3-cyclohexadiene-1-carboxylate. Its pyruvate moiety is eliminated by MenH to yield 2-succinyl-6-hydroxy-2,4-cyclohexadiene-1-carboxylate. MenC catalyzes aromatization of 2-succinyl-6-hydroxy-2,4-cyclohexadiene-1-carboxylate, forming *o*-succinylbenzoate. MenE is an *o*-succinylbenzoate-CoA ligase, which converts *o*-succinylbenzoate to *o*-succinylbenzoate-CoA. Thereafter, MenB catalyzes a formal Dieckmann type of condensation of *o*-succinylbenzoate-CoA to yield 1,4-dihydroxy-2-naphthoyl-CoA, which, in turn, is being hydrolyzed to 1,4-

dihydroxy-2-naphthoate (DHNA) thioeter-splitting enzyme encoded by *yfbB*. In contrast, the prenyl diphosphate with appropriate size (i.e., n = 7 in *E. coli*) is being biosynthesized by the iterative reaction of allyl diphosphate in the presence of isopentenyl diphosphate. Then, DHNA is prenylated and methylated by MenA and MenG, respectively, yielding menaquinones as the end product. The side chains of menaquinones vary in different species and even within the same organisms. The more common of menaquinones display 7, 8, and 9 isoprene (C-5) units; MK-7 serves as the major menaquinone entity in several Gram-positive spore-forming bacteria. MK-8 can be found in *E. coli*, while MK-9 in common in *M. tuberculosis*. However, menaquinones containing 4, 5, 6, 10, 11, 12, and 13 isoprene units have been reported in bacteria.

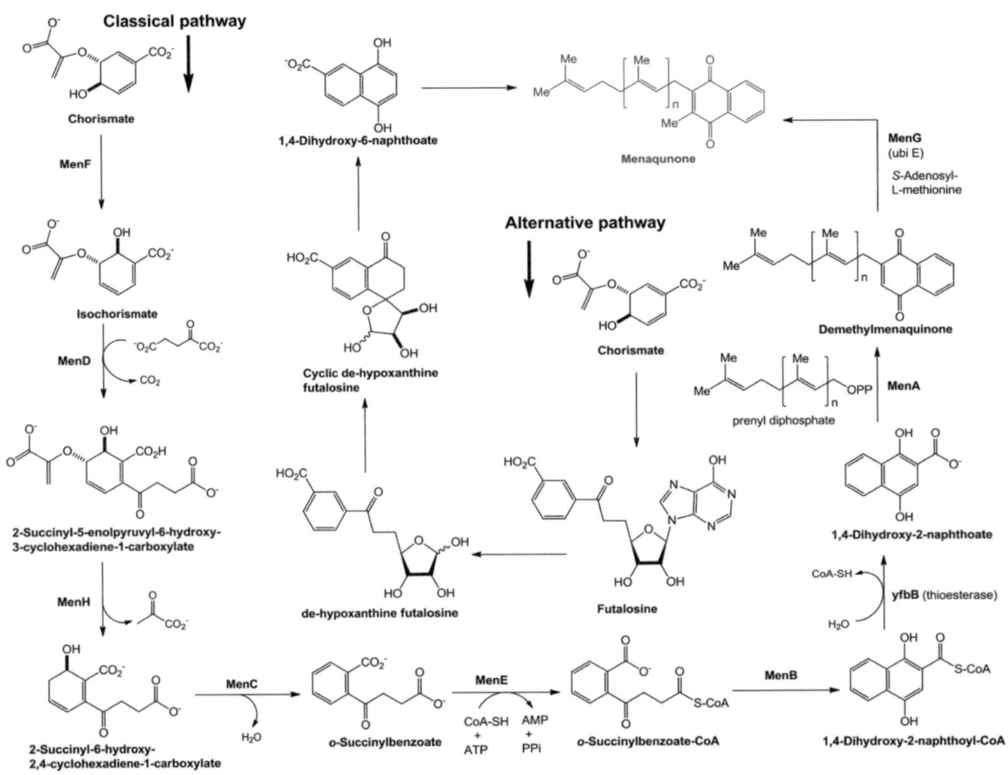

Figure 3. Biosynthesis of menaquinone.

Menaquinones are the predominant isoprenoid lipoquinones of Gram-positive bacteria, whereas Gram-negative bacteria and enterobacteria use menaquinone (MK), demethylmenaquinone (DMK), and ubiquinone (Q) in their electron transport chains (**Figure 2**). Recent studies have shown that several γ-proteobacteria appear to share the similar electron transport system to that observed in *E. coli* [23–26]. Several studies indicated that the regulation of menaquinone biosynthesis of aerobically growing bacteria is different from those of bacteria under anaerobic respiratory conditions being controlled by FNR as the general regulator.

http://dx.doi.org/10.5772/65487

Additionally, the MK/DMK ratio is not dependent on the fur locus, and substantial amounts of naphthoquinones (MK and DMK) are retrieved only during anaerobic conditions in *E. coli*. The menaquinones were detected almost no exception within the bacterial membrane. The total amount of naphthoquinones was found to be some 0.60 ~ 1.09 μmol/g cell. It has become evident that several bacterial species do not have methylase (or MenG) and thus produce DMK as their sole quinone [19]. Conversion from MK to DMK is the last step in the biosynthesis. The activity of the DMK methylase is likely to be regulated by the presence or absence of the electron carriers or by the supply of *S*-adenosylmethionine [20]. A bioinformatic analysis of whole-genome sequences suggested that some microorganisms, including *Helicobacter pylori* and *Campylobacter jejuni*, and lactobacilli do not have orthologs of the men genes, although they synthesize menaquinone. These bacteria synthesize menaquinones in an alternative pathway via futalosine via alternative pathway (**Figure 3**) [23].

6. Antibacterial drug discovery by targeting menaquinone biosynthesis

Menaquinone is the sole quinone in the electron transport chain in the majority of Gram-positive bacteria including *Mycobacterium* spp. The biosynthetic pathway leading to menaquinone is absent in humans; therefore, the bacterial enzymes responsible for menaquinone biosynthesis are potential drug targets for development of novel antibacterial agents. It is speculated that dormant (non-replicating) *M. tuberculosis* displays a less active metabolism and also diminished energy reserves; however, Adenosine triphosphate (ATP) synthesis during oxidative phosphorylation is active during the dormant state. Therefore, inhibition of the menaquinone biosynthesis might exert serious effects on the maintenance of dormancy in *M. tuberculosis*. This concept emphasized the reports that phenothiazines block the type II Nicotinamide adenine dinucleotide phosphate (NADH): menaquinone oxidoreductase (**Figure 4**) in the bacterial respiratory chain and also were effective in killing non-replicating *M. tuberculosis* [27]. Interestingly, it was demonstrated that inhibition of MenA (1,4-dihydroxy-2-naphthoate prenyltransferase) (**Figure 3**) showed significant growth inhibitory activities against drug-resistant *Mycobacterium* spp. and Gram-positive bacteria [28]. MenA inhibitors

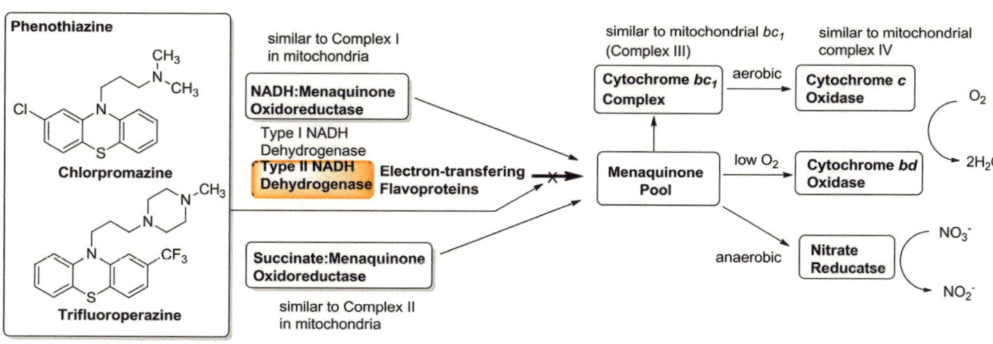

Figure 4. Electron flow in *M. tuberculosis* and type II Nicotinamide adenine dinucleotide phosphate (NADH) dehydrogenase inhibitors.

effectively killed non-replicating *M. tuberculosis in vitro*. Several other promising biological date were generated on MenA as a new antibacterial drug target; (1) *M. tuberculosis* growth *in vitro* could not be rescued by exogenous vitamin K_2 supplementation, (2) all Gram-positive bacteria tested (e.g., *Staphylococcus aureus, Enterococcus faecalis,* and *Clostridium difficile*) were susceptible to MenA inhibitors, whereas Gram-negative bacteria (*E. coli, Klebsiella pneumoniae, Pseudomonas aeruginosa,* and *Acinetobacter baumannii*) were not susceptible under aerobic conditions, and (3) MenA inhibitors are effective in killing *E. coli* under anaerobic conditions [18]. To date, several menaquinone biosynthesis enzymes have been studies for the development of novel antibacterial agents.

7. Menaquinone biosynthesis inhibitors

7.1. MenA inhibitors

Among the menaquinone biosynthesis enzymes, MenA is a membrane-associated protein that catalyzes prenylation of demethylmenaquinone (DMK), forming 1,4-dihydroxy-2-naphtoate (DHNA). Analyses of the amino acid sequence of MenA were revealed that MenA displays five transmembrane segments, and that there exists highly conserved aspartate (D), which would be localized to the inner-plasma membrane, which was being predicted by the aid of a prediction program (Sosui) [14]. The activity is totally dependent on the presence of divalent cations, such as Mg^{2+}. It is therefore likely that these divalent cations produce ion pairs with Asp residues contained within the catalytic site within MenA. A library of DMMK mimics possessing the amino group(s) were generated and evaluated in an enzymatic assay *in vitro* (IC_{50}) against *Mtb* MenA. Identified MenA inhibitors were evaluated in bacterial growth inhibitory assays (MIC). The *tertiary* or *secondary* amine-containing benzophenone derivatives **1** and **2** are the first-generation MenA inhibitors that exhibited bactericidal activities against *M. tuberculosis* (MIC 1–1.5 μg/mL for **1**) and Methicillin-resistant Staphylococcus aureus (MRSA) (MIC 4.0 μg/mL for **2**), respectively. These molecules are inhibited growth of drug-resistant *Mycobacterium* spp. and drug-resistant Gram-positive bacteria (vancomycin-resistant *S. aureus,* vancomycin-resistant *E. faecalis,* and linezolid-resistant Methicillin-resistant Staphylococcus aureus (MRSA)) at low concentrations [29, 30]. Significantly, the MenA

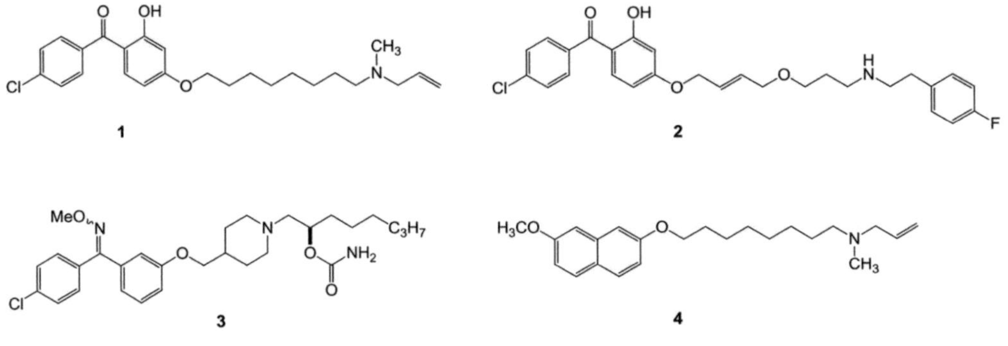

Figure 5. MenA inhibitors.

inhibitor **1** killed non-replicating *M. tuberculosis* much faster than first-line Tuberculosis (TB) drug, rifampicin at lower concentrations. Later, the same group developed selective antimycobacterial MenA inhibitor **3**. The optically pure **3** is the most active molecule in killing non-replicating Mtb. *In vitro* data with selective MenA inhibitors suggested that menaquinone biosynthesis is important in maintaining mycobacterial viability under conditions of restricted oxygen [18]. MenA inhibitors are likely to block the electron flow, consequently inhibiting the bacterial growth. The other group identified 7-methoxy-2-naphthol-based MenA inhibitors (e.g., **4**) that killed *M. tuberculosis* and Gram-positive bacteria with the MIC level of 3–25 μg/mL (**Figure 5**) [31].

7.2. MenB inhibitors

One of at least seven enzymes in menaquinone biosynthesis, MenB (1,4-dihydroxy-2-naphthoyl-CoA synthase) forms the bicyclic ring system by catalyzing the Dieckmann type reaction of *o*-succinylbenzoyl-coenzyme A to 1,4-dihydroxy-2-naphthoic acid. A high-resolution co-crystal structure of *E. coli* MenB with a stabilized analog of *o*-succinylbenzoylCoA was successfully diffracted. The MenB X-ray structure provides important insight into the catalytic mechanism. Similarly, the X-ray structure of MenB from *M. tuberculosis* was characterized. A high-throughput screen (HTS) against *M. tuberculosis* MenB led to the discovery of 2-amino-4-oxo-4-phenylbutanoic acid (**5**) that inhibits MenB at low nanomolar concentrations [32]. Later, methyl 4-(4-chlorophenyl)-4-oxobut-2-enoate (**6**) was reported to inhibit MenB by forming the CoA adduct, **7**. The adduct **7** binds to the *S. aureus* MenB with a K_d value of 2 μM, and also killed drug-sensitive and drug-resistant *S. aureus* strains at 0.35–0.75 μg/mL concentrations [33]. Bacterial growth inhibitory assays of **6** against a battery of bacteria concluded that **6** is effective only against bacteria that utilize menaquinone for respiration. *In vivo* efficacy of **1** using the mouse models of Methicillin-resistant Staphylococcus aureus (MRSA) infection revealed that **6** increased survival in a systemic infection model and resulted in a dose-dependent decrease in bacterial load [34]. These *in vitro* and *in vivo* studies came to the conclusion that MenB is a valid target for the development of new anti-Methicillin-resistant Staphylococcus aureus drug and infections caused by Gram-positive bacterial infections (**Figure 6**).

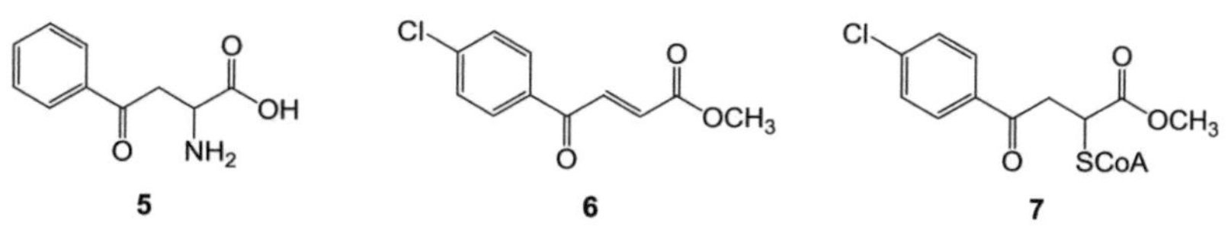

Figure 6. MenB inhibitors.

7.3. MenD inhibitors

MenD (2-succinyl-5-enolpyruvyl-6-hydroxy-3-cyclohexadiene-1-carboxylate synthase) catalyzes a thiamin diphosphate-dependent decarboxylative carboligation of α-ketoglutarate and isochorismate via a Stetter-like conjugate addition. MenD is also essential for menaquinone biosynthesis in some bacteria and has been recognized as an antibacterial drug target. A succinylphosphonate ester, 4-(methoxyoxidophosphoryl)-4-oxobutanoate (**8**) was reported to be a competitive inhibitor of MenD (Ki values ~700 nM) [35]. An analog of the cofactor, thiamine diphosphate, oxythiamine **9** was reported to exhibit MenD enzyme and *S. aureus* growth inhibitory activities (**Figure 7**) [36].

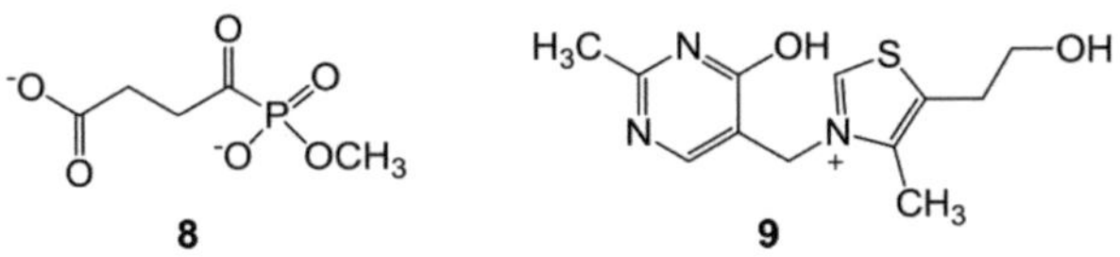

Figure 7. MenD inhibitors.

7.4. MenE inhibitors

MenE (*o*-succinylbenzoate-CoA synthetase) is an essential adenylate-forming enzyme that is also a promising target for development of novel antibiotics in the menaquinone biosynthesis. Adenosylsulfonamide, adenosylsulfamate, and adenosylsulfamide analogs **10**, **11**, and **12** were developed as inhibitors of MenE enzymes based on the structure of *o*-succinylbenzoate-CoA. The vinyl sulfonamide **10** was found to be the most potent MenE inhibitor (IC_{50} ~ 5.7 μM against *M. tuberculosis* MenE) [37]. The vinyl sulfonamide **10** is a competitive inhibitor of *M. tuberculosis* MenE with respect to Adenosine triphosphate (ATP) (Ki = 5.4 ± 0.1 nM) (**Figure 8**).

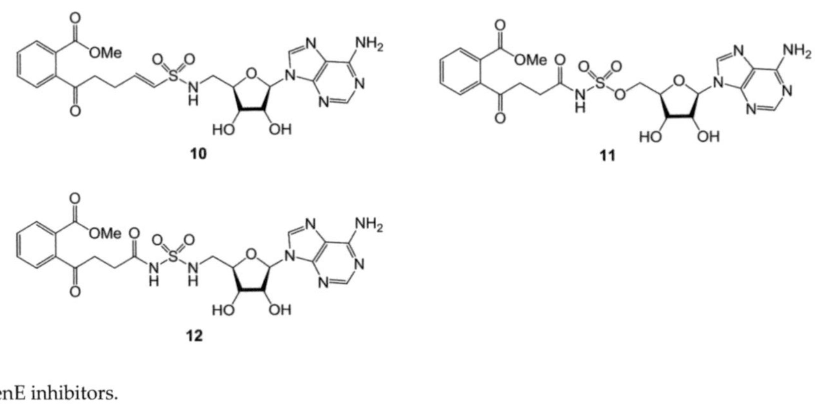

Figure 8. MenE inhibitors.

8. Conclusion

Bacterial Adenosine triphosphate (ATP) synthase, F_1F_0-ATPase, is a viable target for treatment of *M. tuberculosis* infections. Diarylquinolone, an inhibitor of *M. tuberculosis* Adenosine triphosphate (ATP) synthase exhibited a remarkable activity against *Mycobacterium* spp. Electron transport systems associated with Adenosine triphosphate (ATP) synthases are established drug targets for antibacterial and antiprotozoal infections. For example, Nicotinamide adenine dinucleotide phosphate (NADH) hydrogenase is a sustainable drug target for the malaria parasite and is also a promising target for *M. tuberculosis* infections. On the other hand, menaquinone biosynthesis has recently been received attention for the development of novel antibacterial agents. Menaquinone is a key component of the electron transport systems in the majority of Gram-positive bacteria including *M. tuberculosis*. As summarized in Chapter 6, inhibitors of menaquinone biosynthesis have been identified and several compounds are also effective inhibitors of bacterial growth. In development of new drugs for *M. tuberculosis* infections, it is the ultimate goal to discover a Tuberculosis (TB) drug that is effective against human latent tuberculosis infection. Among the menaquinone biosynthesis enzymes, MenA has been extensively studies as drug target for *M. tuberculosis* and Gram-positive bacteria. Selective MenA inhibitors inhibit the growth of non-replicating *M. tuberculosis*, suggesting that menaquinone is essential in maintaining the bacterial viability during conditions of restricted levels of oxygen. A large set of data suggest that the DosR/DosS/DosT signaling pathway is mandatory for *M. tuberculosis'* genetic response to hypoxic conditions and nitric oxide, in the adaptation of *M. tuberculosis* to conditions triggering a reversible bacteriostasis. In this way, the DosR/DosS/DosT signaling pathway may contribute to the latency seen *in vivo* [38]. MenA inhibitors display the ability to block or hamper the flow of electrons without inducing a dormancy response in *M. tuberculosis*. Consequently, menaquinone biosynthesis inhibitors have the potential to kill *M. tuberculosis* at any states by inhibiting Adenosine triphosphate (ATP) synthesis non-directory.

Recently, a narrow-spectrum antibiotic, siamycin I was reported to kill *Helicobacter* and *Campylobacter* by inhibiting the futalosine (alternative) pathway (**Figure 3**). Branched fatty acids (12- or 13-methyltetradecanoic acids) also inhibit the biosynthesis of menaquinone biosynthesis of *H. pylori*. Because of advances of biological assays of menaquinone biosynthesis, new inhibitor molecules that possess drug-like characteristics will be identified. It is import to prove the efficacy of menaquinone biosynthesis inhibitor using an appropriate infected animal model. To date, *in vivo* efficacy of a MenB inhibitor was demonstrated using the mouse model of Methicillin-resistant Staphylococcus aureus (MRSA) infection. A Tuberculosis (TB) drug, SQ109, a strong inhibitor of trehalose monomycolate (TMM) transporter, was reported to exhibit inhibitory activities of MenA and MenG enzymes from *M. tuberculosis*. Discovery of a pharmacologically acceptable menaquinone biosynthesis inhibitor, which possesses a significant antibacterial activity against replicating and non-replicating *M. tuberculosis*, has been highlighted in Tuberculosis (TB) drug development. It is worthwhile mentioning that menaquinone biosynthesis inhibitors are also promising agents to kill Gram-negative bacteria growing under oxygen-depleted or anaerobic conditions in which menaquinone is utilized for their respiration.

Acknowledgements

I thank CORNET award (University of Tennessee Health Science Center) for generous financial support.

Author details

Michio Kurosu*

Address all correspondence to: mkurosu@uthsc.edu

Department of Pharmaceutical Sciences, College of Pharmacy, University of Tennessee Health Science Center, Memphis, TN, USA

References

[1] Usui Y, Tanimura H, Nishimura N, Kobayashi N. Vitamin K concentrations in the plasma and liver of surgical patients. Am. J. Clin. Nutr. 1990; 51: 846–852. doi:10.3945/an.111.001800

[2] Kindberg CG, Suttie JW. Effect of various intakes of phylloquinone on signs of vitamin K deficiency and serum and liver phylloquinone concentrations in the rat. J. Nutr. 1989; 119: 175–180. doi:10.2527/jas.2009-2759

[3] Collins MD, Jones D. Distribution of isoprenoid quinone structural types in bacterial and their taxonomic implications. Microbiol. Rev. 1981; 45: 316–354. doi:10.1007/978-3-642-13612-2_2

[4] The nomenclature of quinones with isoprenoid side-chains. Pure Appl. Chem. 1974; 38: 439–447. doi:10.1111/j.1432-1033

[5] Vermeer C, Gijsbers BL, Craciun AM, Groenen-van Dooren MM, Knapen MH. Effects of vitamin K on bone mass and bone metabolism. J. Nutr. 1996; 126: 1187S–1191S.

[6] Iwamoto J, Yeh JK, Takeda T, Ichimura S, Sato Y. Comparative effects of vitamin K and vitamin D supplementation on prevention of osteopenia in calcium-deficient young rats. Bone 2003; 33: 557–566. doi:10.1016/S8756-3282(03)00249-7

[7] Visser CM. Some speculations on the mechanisms of the vitamins E and K starting from origin of life considerations and the antioxidant theory. Bioorg. Chem. 1980; 9: 411–422. doi:10.1016/0045-2068(80)90001-2

[8] Shirakawa H, Ohsaki Y, Minegishi Y, Takumi N, Ohinata K, Furukawa Y, Mizutani T, Komai M. Vitamin K deficiency reduces testosterone production in the testis through down-regulation of the Cyp11a a cholesterol side chain cleavage enzyme in rats. Biochim. Biophys. Acta 2006; 1760: 1482–1488. doi:10.1016/j.bbagen.2006.05.008

[9] Munter G, Hershko C. Increased warfarin sensitivity as an early manifestation of occult prostate cancer with chronic disseminated intravascular coagulation. Acta Haematol. 2001; 105: 97–99. doi:10.1159/000046542

[10] Nakamura M, Nagano H, Noda T, Wada H, Ota H, Damdinsuren B, Marubashi S, Miyamoto A, Takeda Y, Doki Y, Umeshita K, Dono K, Sakon M, Monden M. Vitamin K_2 has growth inhibition effect against hepatocellular carcinoma cell lines but does not enhance anti-tumor effect of combination treatment of interferon- and fluorouracil *in vitro*. Hepatol. Res. 2006; 35: 289–295. doi:10.1016/j.hepres.2006.04.014

[11] Thijssen HHW, Drittij-Reijnders MJ. Vitamin K status in human tissues: tissue-specific accumulation of phylloquinone and menaquinone-4. British J. Nutr. 1996; 75: 121–127. doi:10.1079/BJN19960115

[12] Toshio O, Kimie N, Maya K. Vitamin K and bone update. *In vivo* metabolism of vitamin K: in relation to the conversion of vitamin K_1 to MK-4. Clin. Calcium. 2009; 19: 1779–1787. doi:CliCa091217791787

[13] Haddock BA, Colin JW. Bacterial respiration. Bacteriol. Rev. 1977; 41: 47–99.

[14] Yasuhiro A. Bacterial electron transport chains. Annu. Rev. Biochem. 1988; 57: 101–312. doi:10.1146/annurev.bi.57.070188.000533

[15] Suvana KD, Stevenson R, Meganathan R, Hudspeth MES. Menaquinone (vitamin K_2) biosynthesis: localization and characterization of the menA gene from *Escherichia coli*. J. Bacteriol. 1998; 180: 2782–2787. doi:10.1016/0378-1119(95)00721-0

[16] Truglio JJ, Theis K, Feng Y, Gajda R, Machutta C, Tong PJ, Kisher C. Crystal structure of *Mycobacterium tuberculosis* MenB, a key enzyme in vitamin K_2 biosynthesis. J. Biol. Chem. 2003; 24: 42352–42360. doi:10.1021/bi200877x

[17] Bishop DHL, King HK. Ubiquinone and vitamin K in bacteria. 2. Intracellular distribution in *Escherichia coli* and *Micrococcus lysodeikticus*. Biochem. J. 1962; 83: 606–614.

[18] Debnath J, Siricilla S, Wan B, Crick DC, Lenaerts AJ, Franzblau SG, Kurosu M. Discovery of selective menaquinone biosynthesis inhibitors against *Mycobacterium tuberculosis*. J. Med. Chem. 2012; 55: 3739–3755. doi:10.1021/jm201608g

[19] Hollfinder R. The dependence on quinone specificity of terminal electron transport of bacteria. Curr. Microbiol. 1981; 6: 155–159. doi:10.1007/BF01642390

[20] Kröger A, Unden G. The function of menaquinone in bacterial electron transport. In Coenzyme Q; Editor, Lenz G. John Wiley & Sons Ltd. 1985. p. 285–300.

[21] Shineberg B, Young IG. Biosynthesis of bacterial menaquinones: the membrane-associated 1,4-dihydroxy-2-naphthoate octaprenyltransferase of *Escherichia coli*. Biochemistry 1976; 15: 2754–2758. doi:10.1021/bi00658a007

[22] Schoepp-Cotheneta B, Lieutauda C, Baymanna F, Vermégliob A, Friedrichc T, Kramerd DM, Nitschke W. Menaquinone as pool quinone in a purple bacterium. PNAS 2009; 106: 8549–8554. doi:10.1073/pnas.0813173106

[23] Hiratsuka T, Furihata K, Ishikawa J, Yamashita H, Itoh N, Seto H, Dair T. An alternative menaquinone biosynthetic pathway operating in microorganisms. Science 2008; 321: 1670–1673. doi: 10.1073/pnas.0813173106

[24] Shaw DJ, Rice DW, Guest JR. Homology between CAP and Fnr, a regulator of anaerobic respiration in *Escherichia coli*. J. Mol. Biol. 1983; 166: 241–247. doi:10.1016/S0022-2836(83)80011-4

[25] Kaiser M, Sawers G. Overlapping promoters modulate Fnr- and ArcA-dependent anaerobic transcriptional activation of the focApfl operon in *Escherichia coli*. Microbiology 1997; 143: 775–783. doi:10.1099/00221287-143-3-775

[26] Zigha A, Rosenfeld E, Schmitt P, Duport C. The redox regulator Fnr is required for fermentative growth and enterotoxin synthesis in *bacillus cereus* F4430/73. J. Bacteriol. 2007; 189: 2813–2824. doi:10.1128/JB.01701-06

[27] Weistein EA, Yano T, Li LS, Avarbock D, Avarbock A, Helm D, McColm AA, Duncan K, Lonsdale JT, Rubin H. Inhibitors of type II Nicotinamide adenine dinucleotide phosphate (NADH): menaquinone oxidoreductase represent a class of antitubercular drugs. PNAS 2005; 102: 4548–4553. doi:10.1073/pnas.0500469102

[28] Kurosu M, Narayanasamy P, Biswas K, Dhiman R, Crick DC. Discovery of 1,4-dihydroxy-2-naphthoate prenyltransferase inhibitors: new drug leads for multidrug-resistant Gram-positive pathogens. J. Med. Chem. 2007; 50: 3973–3975. doi:10.1021/jm070638m

[29] Kurosu M, Crick DC. MenA is a promising drug target for developing novel lead molecules to combat *Mycobacterium tuberculosis*. Med. Chem. 2009; 5: 197–207. doi: 10.2174/157340609787582882

[30] Dhiman RK, Mahapatra S, Slayden RA, Boyne ME, Lenaerts A, Hinshaw JC, Angala SK, Chatterjee D, Biswas K, Narayanasamy P, Kurosu M, Crick DC. Menaquinone synthesis is critical for maintaining mycobacterial viability during exponential growth and recovery from non-replicating persistence. Molecul. Microbiol. 2009; 72: 85–97. doi: 10.1111/j.1365-2958.2009.06625.x.

[31] Choi S-R, Larson MA, Hinrichs SH, Bartling AM, Frandsen J, Narayanasamy P. Discovery of bicyclic inhibitors against menaquinone biosynthesis. Future Med. Chem. 2016; 8: 11–16. doi:10.4155/fmc.15.168

[32] Lu X, Zhou R, Sharma I, Li X, Kumar G, Swaminathan S, Tonge PJ, Tan DS. Stable analogues of OSB-AMP: potent inhibitors of MenE, the *o*-succinylbenzoate-CoA synthetase from bacterial menaquinone biosynthesis. Chembiochem. 2012; 13: 129–136. doi:10.1002/cbic.201100585

[33] Li X, Zhang H, Tonge PJ. Inhibition of 1,4-dihydroxynaphthoyl-CoA synthase (MenB), an enzyme drug target bacterial menaquinone biosynthesis pathway. 236th ACS National Meeting, Philadelphia, PA, USA, 2008; August 17–21.

[34] Matarlo JS, Lu Y, Daryaee F, Daryaee T, Ruzsicska B, Walker SG, Tonge PJ. A methyl 4-oxo-4-phenylbut-2-enoate with *in vivo* activity against Methicillin-resistant Staphylococcus aureus (MRSA) that inhibits MenB in the bacterial menaquinone biosynthesis pathway. ACS Infect. Dis. 2016; 2: 329–340. doi:10.1021/acsinfecdis.6b00023

[35] Fang M, Toogood DR, Macova A, Ho K, Franzblau SG, McNeil MR, Sanders DAR, Palmer DRJ. Succinylphosphonate esters are competitive inhibitors of MenD that show active-site discrimination between homologous α-ketoglutarate-decarboxylating enzymes. Biochemistry 2010; 49: 2672–2679. doi:10.1021/bi901432d

[36] Xu H, Graham M, Karelis J, Walker SG, Tonge PJ. Mechanistic studies of MenD, 2-succinyl-5-enoylpyruvyl-6-hydroxy-3-cyclohexene-1-carboxylic acid synthase from *Staphylococcus aureus*. 237th ACS National Meeting, Salt Lake City, UT, USA, 2009; March 22–26.

[37] Lu X, Zhang H, Tonge PJ, Tan DS. Mechanism-based inhibitors of MenE, an acyl-CoA synthetase involved in bacterial menaquinone biosynthesis. Bioorg. Med. Chem. Lett. 2008; 18: 5963–5966. doi:10.1016/j.bmcl.2008.07.130

[38] Honaker RW, Leistikow RL, Bartek IL, Voskuil MI. Unique roles of DosT and DosS in DosR regulon induction and *Mycobacterium tuberculosis* dormancy. Infect. Immun. 2009; 77: 3258–3263. doi:10.1128/IAI.01449-08

Chapter 16

Toxicological and Pharmacological Effects of VKOR Inhibitors

Yohei Miyamoto

Additional information is available at the end of the chapter

http://dx.doi.org/10.5772/63512

Abstract

Vitamin K_1 2,3-epoxide reductase (VKOR) inhibition is the main pharmacological effect of warfarin, an anticoagulant that is typically used in the prevention of thrombosis and thromboembolism. The repeated oral administration of sodium dehydroacetate (DHA-S), which has been used as a food additive, preservative, and antimicrobial agent, induced severe hemorrhage in multiple organs and prolonged blood coagulation factors with VKOR inhibition in rats. On the other hand, VKOR and the vitamin K–dependent growth arrest–specific gene 6 (Gas6)/Axl pathway play a key role in mesangial cell proliferation in glomerulonephritis (GN). We herein indicated the potential of the VKOR inhibitor, 3-acetyl-5-methyltetronic acid (AMT), to prevent the proliferation of glomerular mesangial cells and suppress the progression of GN. DHA-S-induced hemorrhage was caused by the depletion of blood VK, associated with any factors including VKOR inhibition. The novel VKOR inhibitor, AMT, reduced renal mesangial cell proliferation and may be a supportive treatment for GN.

Keywords: 3-acetyl-5-methyltetronic acid (AMT), hemorrhage, glomerulonephritis, sodium dehydroacetate (DHA-S), vitamin K_1 2,3-epoxide reductase (VKOR)

1. Introduction

The anticoagulant warfarin, which has a coumarin structure, causes hemorrhage by inhibiting the vitamin K (VK)–dependent synthesis of blood coagulation factors [1]. Blood coagulation factors undergo γ-carboxylation and have the ability to bind to calcium ions on the platelet surface, under the presence of VK and molecular oxygen. Vitamin K_1 2,3-epoxide (VKO), which is produced while this γ-carboxylation, is converted to vitamin K_1 (VK_1) by vitamin K_1 2,3-

INTECH
open science | open minds

© 2017 The Author(s). Licensee InTech. This chapter is distributed under the terms of the Creative Commons Attribution License (http://creativecommons.org/licenses/by/3.0), which permits unrestricted use, distribution, and reproduction in any medium, provided the original work is properly cited. (cc) BY

epoxide reductase (VKOR) and is then recycled. Warfarin arrests the regeneration of VK by inhibiting VKOR and induces hemorrhage due to a VK deficiency.

Sodium dehydroacetate (DHA-S) has a wide spectrum of antimicrobial activity that accounts for its use in cosmetic products, as a preservative, and as an antimicrobial agent [2]. In Japan, it has also been used as a food preservative for cheese, butter, and margarine at a concentration of 0.5 g/kg or less as dehydroacetic acid [3, 4]. This concentration is the highest dose allowed under the Food Sanitation Law; however, the risk assessment of DHA-S has not been sufficiently performed for a food additive, and its acceptable daily intake has not yet been established. On the other hand, the toxicity of DHA-S has been widely studied. In an acute toxicity study of DHA-S using dogs, clinical signs such as salivation, vomiting, convulsions, and ataxia were reported [5]. In a subchronic toxicity study of DHA-S using dogs, body weight loss, gastric hemorrhage, and an increase in blood urea nitrogen were observed [6]. These findings were considered to be primarily due to a lack of appetite and subsequent weight loss [6]. However, the main structure of DHA-S, 2H-pyran-2-on, is included in a coumarin structure. It is reported that some derivatives of 4-hydroxy-2-pyrone, which includes similar structure of DHA-S, exhibit anti-blood coagulant activities in rats [7]. These findings suggest that DHA-S also shows the anticoagulant activity. Therefore, we investigated the effects of repeated administration of DHA-S in rats, and the amelioration by VK against DHA-S-induced hemorrhage [8].

In Thy-1 glomerulonephritis (GN) rats, the expression of growth arrest-specific gene 6 (Gas6) and Axl was found to be markedly increased in glomeruli and paralleled the proliferation of mesangial cells [9, 10]. VKOR is also related with mesangial cell proliferation through the vitamin K–dependent activation of Gas6 [10]. Gas6 is as an autocrine growth factor for mesangial cells through binding to its cell surface receptor Axl. These findings suggest that the Gas6/Axl pathway plays an important role in mesangial cell proliferation in GN [11]. VKOR inhibitors including warfarin is effective at blocking mesangial cell proliferation and improving renal function [11]; however, it inhibits the production of blood coagulation factors in the liver and has been able to cause bleeding [12]. Therefore, low-dose treatments of warfarin are used in renal failure patients, and the control of its blood concentration or the conformation of blood coagulation activities have to been required [13]. We are discovering new compounds that inhibit mesangial cell proliferation, but not hepatic production of blood coagulation factors using Thy-1 GN rats **Table 2**. Thy-1 GN rats have been used as an acute model of GN accompanied with mesangial cell proliferation, matrix expansion, and moderate decline of renal function [14]. We indicated the potential of a new VKOR inhibitor, 3-acetyl-5-methyltetronic acid (AMT), as an inhibitor of mesangial cell proliferation and suppressor of renal disease [15].

2. Main body

In a 14-day repeated toxicity study using rats, animals that received 200 or 400 mg/kg/day of DHA-S died from severe hemorrhage in various organs, such as the stomach, intestines, testes, epididymides, subcutaneous tissue, and subdural area. The profile of hemorrhage caused by

http://dx.doi.org/10.5772/63512

DHA-S, which occurred in multiple organs and showed differences in sensitivity among individuals, was similar to that caused by warfarin. The prothrombin time (PT) and activated partial thromboplastin time (APTT) were significantly prolonged in each surviving rat in the 100 and 200 mg/kg/day groups (**Figure 1**). These results indicated that the repeated dosing of DHA-S-induced hemorrhage in rats [8].

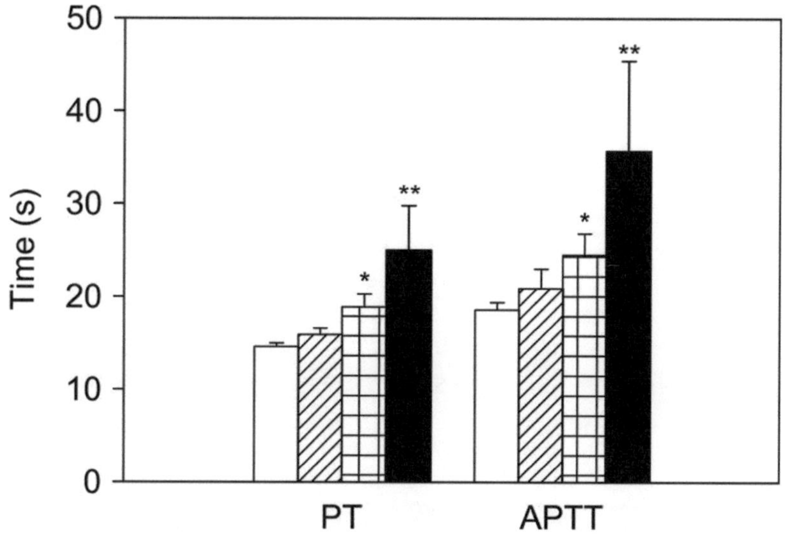

Figure 1. Effects on PT and APTT in male rats received the repeated oral administration of DHA-S at doses of 50, 100, and 200 mg/kg/day. (□) Vehicle; (▨) 50 mg/kg/day; (⊞) 100 mg/kg/day; (■) 200 mg/kg/day. Each value is presented as the mean ± SD. * $P < 0.05$, ** $P < 0.01$: Significantly different from controls at 5% and 1%.

On the other hand, DHA-S-treated rats showed no significant decrease in PLT. In the histopathological examination of the liver, marked changes, such as fibrin thrombi in the vascular system and single cell necrosis, were not observed. Therefore, hemorrhage was assumed to be caused by a deficiency in VK. In order to demonstrate this hypothesis, the effects of VK on hemorrhage were investigated using rats that received a single subcutaneous injection of VK_2 following the repeated oral administration of DHA-S at 200 mg/kg/day for 5 days. In rats, dietary VK_1 was converted to menadione in the gut and menadione was absorbed and converted to VK_2 [16]. Consequently, VK_2 (menaquinone-4) reached a maximum concentration in the liver in rats faster than after the intravenous or intraperitoneal administration of VK_1 [17]. VK_2 was also reported to exert stronger preventive effects against hemorrhage than other VK products, such as VK_1 and VK_3. As a result, DHA-S prolonged blood coagulation parameters (PT, APTT, the thrombo test (TTO), and hepaplastin test (HPT), in rats that received 200 mg/kg/day for 5 days (**Figure 2**). However, the prolongation of these parameters was suppressed in animals injected subcutaneously with VK_2 after DHA-S dosing. Since TTO and HPT are sensitive indexes of a deficiency in VK-dependent blood coagulation factors [18, 19], it was concluded that DHA-S induced hemorrhage via a deficiency in VK.

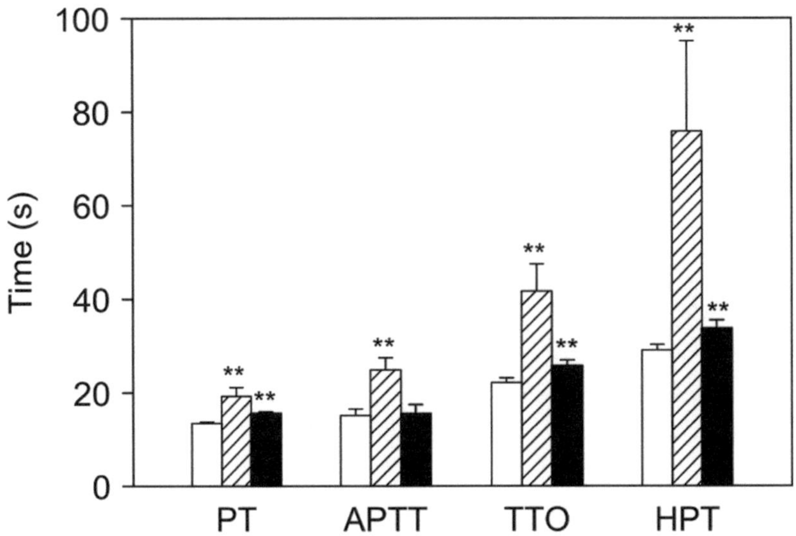

Figure 2. Effects on blood coagulation parameters (PT, APTT, TTO, and HPT) in male rats received DHA-S for 5 days following a single subcutaneous injection of vitamin K_2 (VK_2). (□) Vehicle; (▨) DHA-S at 200 mg/kg/day for 5 days; (■) DHA-S at 200 mg/kg/day for 5 days and VK_2 at 1 mg/kg after the final DHA-S dosing on day 5. Each value is presented as the mean ± SD. ** $P < 0.01$: Significantly different from controls at 1%.

A VK deficiency is caused by anticoagulants, hepatobiliary disease, the inhibition of VK production by the intestinal flora, and the poor absorption of VK from the gut [20]. Prolongation of PT and APTT were also observed in the 100 mg/kg group, which did not significantly decrease in body weight. In the macroscopic observation of surviving rats, no remarkable changes were detected in the gastro-intestinal tracts. While DHA-S exhibited weak antimicrobial activity against obligatory anaerobes, the major intestinal flora [2], the effects of a high dose of DHA-S on the intestinal flora currently remain unknown. VK is a cofactor for enzymatic conversion (γ-carboxylation) of glutamic acid residues in the inactive precursors of VK-dependent proteins (blood coagulation factors) in order to convert them to their active forms. In VK cycle, accompanying with this reaction, VK_1H_2 (hydroquinone form), is oxidized to VKO, and subsequently reduced the VK_1 (quinone form). The anticoagulant warfarin is known to be a VK antagonist and is used as a rodenticide to induce hemorrhage in rats [21]. A warfarin-induced VK deficiency occurs through the inhibition of VKOR that then converts VKO to the quinone form of VK_1 in the VK cycle [1, 22, 23]. The coumarin structure of warfarin involves the main structure of DHA-S, and some derivatives of 4-hydroxy-2-pyrone have been reported to exhibit anticoagulant activities in rats [7]. The hemorrhage induced by DHA-S, which occurs in multiple organs and shows individual differences in sensitivity, is similar to that by warfarin.

The effects of DHA-S on VKOR activity were investigated using male rat liver microsomes. VKOR activity, expressed as the quantitative amount of VK_1 converted from VKO, was induced *in vitro* by a VKOR reaction and the inhibition by each compound was estimated from the percentage of VKOR activity measured in an identical incubation containing no compound. DHA-S and warfarin both showed dose-dependent suppression of VKOR activities, and DHA-

S inhibited clearly VKOR activity as well as warfarin [8]. The IC_{50} values of DHA-S and warfarin were 3149 and 2.15 μmol/L, respectively. The inhibitory effect of DHA-S on VKOR activity was expected to be approximately 1400-fold as compared with warfarin. The treatment with DHA-S at dose of 100 mg/kg (480 μmol/kg) for 5 days prolonged PT and APTT in rats by approximately 1.3-fold against that of the control. On the other hand, the repeated oral administration of warfarin at 0.2 mg/kg (0.6 μmol/kg) for 5 days elongated PT in rats [24]. Furthermore, TTO activity was reduced by approximately 50% in rats under a controlled rate (0.15 μmol/kg/day) of warfarin administration by osmotic pumps implanted subcutaneously [25]. The dose which DHA-S inhibits blood coagulation factors in rats is 800–3200-fold higher than that of warfarin. This fact explains the difference between the VKOR inhibitory activities of DHA-S and warfarin. In humans, the daily intake level of DHA is estimated to be 0.0303 mg [26] which is approximately 16,500-fold less than that inducing hemorrhage in rats. Therefore, the potential of DHA-S to induce hemorrhage in humans is low [8]. In conclusion, DHA-S induces severe hemorrhage in rats through the depletion of blood VK associated with any factors including the inhibition of VKOR [8].

On the other hand, we demonstrated that the novel VKOR inhibitor, AMT prevented renal mesangial cell proliferation, and then suppressed glomerular injury in Thy-1 GN rats [15]. The low molecular weight compound, AMT exhibits inhibitory activity against rat liver and kidney VKOR *in vitro* at IC_{50} values of 3.18 and 3.20 nmol/mL, respectively [15]. VKOR is a protein that converts the epoxide of VK back to VK, a co-factor that is essential for the post-translational γ-carboxylation of several blood coagulation factors, and is also involved in mesangial cell proliferation via the VK-dependent activation of the Gas6/Axl pathway [11, 27]. The gene of VKOR was identified and VKOR was expressed ubiquitously in the liver, kidney, lung, heart, and skeletal muscle [27, 28]. However, no information exists on the specific isoforms of VKOR in any tissue. Therefore, AMT showed no selectivity in its inhibitory activity toward rat kidney and liver VKOR. It is known that there are no selective compounds including warfarin. These findings suggest that kidney and liver VKOR have similar substrate recognition or that there was no specific isoform of VKOR, at least in the rat liver and kidney.

In Thy-1 GN rats, kidney weights decreased significantly in a dose-dependent manner after the AMT treatment (**Table 1**). No significant changes were observed in light microscopy, while the slight enlargement and rough surface of kidneys were noted in the control [15]. Renal enlargement followed by mesangial cell proliferation is frequently observed in Thy-1 GN rats and is mostly due to edema associated with inflammation [29, 30]. Thickening of the glomerular basement membrane and fibrocellular crescent of the glomerulus were histopathologically suppressed by the AMT treatment (**Figure 3**). These pathological findings associated with the progression of renal injury [14, 29] may be caused by the inhibitory effects of AMT on mesangial cell proliferation and glomerular inflammation [15]. Additionally, tubular degeneration such as hyaline casts, hyaline droplets, tubular dilatation, and basophilic tubules followed by glomerular damage were found to be dose-dependently suppressed by the AMT treatment in Thy-1 GN rats [15].

		Relative kidney weight (g/100 g body weight)					
Group	Dose	Right		Left		Total	
Normal[a]	–	0.377 ± 0.033	(6)	0.383 ± 0.025	(6)	0.760 ± 0.057	(6)
Thy-1-administered	0 mg/kg (control)	0.478 ± 0.016	(6)	0.469 ± 0.017	(6)	0.947 ± 0.033	(6)
	10 mg/kg	0.417 ± 0.016*	(6)	0.409 ± 0.017*	(6)	0.826 ± 0.033*	(6)
	30 mg/kg	0.395 ± 0.009**	(6)	0.383 ± 0.010**	(6)	0.779 ± 0.018**	(6)

Values are means ± S.E. for the numbers of rats indicated in parentheses.
*,** Significantly different from control, $P < 0.05$ (*), $P < 0.01$ (**).
[a]Normal rats without anti-Thy-1 injection.

Table 1. Relative organ weights of kidney following after 12-day repeated intravenous administrations of AMT (0, 10 and 30 mg/kg/day) in Thy-1 glomerulonephritis rats.

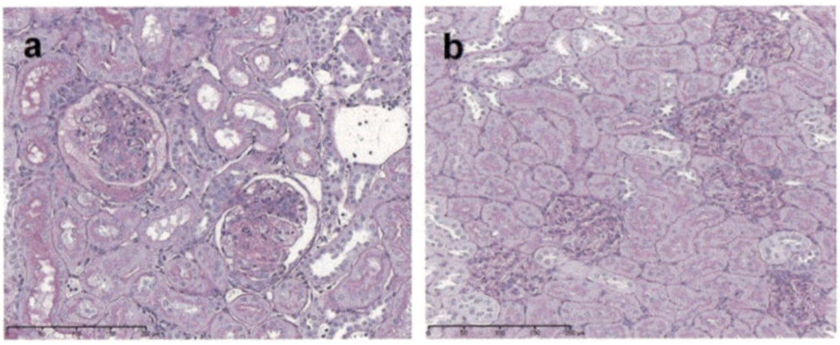

Figure 3. Light microscope of kidneys after the 12-day repeated intravenous administration of AMT to Thy-1 GN rats. Periodic acid-Schiff (PAS)-stained sections show thickening of the glomerular basement membrane and fibrocellular crescent of the glomerulus in the 0 mg/kg group (a), and no significant change in the glomerulus or tubules in the 30 mg/kg group (b).

The Thy-1 GN rat is one of the reversible models of kidney injury. An anti-Thy-1 injection induces the destruction of glomerular mesangial cells and leads to the thickening of the glomerular basement membrane in rats [14, 31]. A decline in renal function has been observed 1 week after its injection; however, renal function recovers completely 5–6 weeks later [14, 31]. In addition, marked mesangial cell proliferation has been reported in patients with chronic GN [32, 33]. Therefore, the Thy-1 GN rat is considered to be an adequate animal model of GN and is suitable for assessing the effectiveness of chemical compounds as suppressors toward the disease [34–36].

The typical VKOR inhibitor, warfarin, inhibits mesangial cell proliferation and prevents the progression of kidney diseases by inhibition of the activation of the Gas6/Axl pathway [11]. Both levels of Gas6 and its receptor Axl increased in various kidney diseases [11, 37]. Gas6 is activated by vitamin K–dependent γ-carboxylase, which converts the glutamic acid of the N-terminal domain of Gas6 to the γ-carboxyl glutamate. VKOR regulates the concentration of VK in the VK cycle, and then controls γ-carboxylase activity [9, 10]. Therefore, VKOR controls the activation of the Gas6/Axl pathway, and its inhibition has potential as a therapeutic treatment for kidney diseases [11]. Warfarin has been used as a supportive medication for human GN in clinical. However, it inhibits not only renal VKOR, but also hepatic VKOR, which is involved in the production of blood coagulation factors, thereby increasing the bleeding risk [12].

In our experiment, following the 14-day repeated oral administration of AMT up to 1500 mg/kg to rats, no significant toxicities including hemorrhage and blood coagulation were noted in our study (data not shown). AMT, which exerts potent inhibitory effects on hepatic VKOR, did not affect blood coagulation in Thy-1 GN rats. Nevertheless, anti-inflammatory effects in kidney were observed in the AMT-treatment group, which may be caused by the inhibition of renal VKOR. We measured AMT concentrations in the kidney and liver after the intravenous administration of AMT to rats. Kidney and liver concentrations of AMT 5 min after the single intravenous administration of 30 mg/kg were 4.26 nmol/g tissue and 0.26 nmol/g tissue, respectively [15], and revealed that the concentration in the kidney reached the IC_{50} value (3.2 nmol/L) against VKOR, whereas that in liver did not (3.18 nmol/L). Renal anti-inflammatory effects and the lack of an effect on anticoagulation were both explained by VKOR inhibitory activity and tissue concentrations of AMT in Thy-1 GN rats. AMT had no selectivity against kidney and liver VKOR; however, the tissue distribution of AMT was unique, and its concentration was higher in the kidney than in the liver. At the maximum concentration after its oral administration, AMT concentrations were higher in the kidney than in the liver (data not shown). AMT is very small and highly polar, and, as such, is likely to be distributed to the kidneys and excreted. Furthermore, it has no radical scavenging potential itself (data not shown). Therefore, the suppression of mesangial cell proliferation and glomerular inflammation in Thy-1 GN rats was considered to be based on the VKOR inhibitory effects of AMT. In Thy-1 GN rat following the AMT treatment, creatinine clearance (CCr) significantly increased, and the urinary albumin-to-creatinine ratio (ACR) significantly decreased [15].

	0 mg/kg		10 mg/kg		30 mg/kg	
Total protein (mg/day)	106 ± 24	(6)	129 ± 51	(6)	91 ± 119	(6)
Albumin (mg/day)	74.8 ± 9.1	(6)	75.9 ± 22.9	(6)	59.8 ± 19.1	(6)
Creatinine (mg/day)	8.10 ± 0.19	(6)	10.5 ± 0.9*	(6)	12.8 ± 0.4**	(6)
ACR[a] (mg/mg)	9.26 ± 1.18	(6)	6.98 ± 1.69	(6)	4.77 ± 1.61	(6)
CCr[b] (ml/min/kg)	1.91 ± 0.08	(6)	2.76 ± 0.22**	(6)	2.87 ± 0.20**	(6)

Values are means ± S.E. for the numbers of rats indicated in parentheses.
[a]ACR is albumin-to-creatinine ratio.
[b]CCr is calculated from the urine creatinine concentration, urine volume and the plasma creatinine concentration.
*,** Significantly different from control, $P < 0.05$ (*), $P < 0.01$ (**).

Table 2. Urinalysis after 12-day repeated intravenous administrations of AMT (0, 10 and 30 mg/kg/day) in Thy-1 glomerulonephritis rats.

In Thy-1 rats, a marked decline in CCr and increase in ACR involved with the glomerular disorder have been reported [9, 17, 37]. In our study, serum creatinine levels were only suppressed by the 10 mg/kg AMT treatment, whereas urine creatinine levels were suppressed and CCr was subsequently increased by the 10 and 30 mg/kg treatments in Thy-1 rats. In addition, ACR was also slightly decreased in a dose-dependent manner. Previous studies demonstrated that ACR is the most important index in GN and accurately assesses the progression of kidney injury [18, 24]. Therefore, we consider that the CCr and ACR to be more sensitive than plasma and/or serum creatinine. On the other hand, no significant change was observed in blood coagulation tests such as PT, APTT, TTO, and HPT. Based on results, AMT suppressed glomerular mesangial cell proliferation and the progression of glomerular disease in Thy-1 GN rats at a non-toxic dose. Recently, warfarin-related nephropathy (WRN) has been reported in patients with and without chronic kidney disease (CKD) [38, 39]. Clinical studies showed that mortality rates in 1 year were higher in WRN patients than in other patients. This result coincides with previous findings of increased mortality rates in warfarin-treated chronic hemodialysis patients [40, 41]. However, the increased mortality rate associated with WRN is related to the complications of diabetes, hypertension, cardiovascular disease, etc., and the mechanisms and risks of WRN currently remain unclear. Warfarin is a widely used anticoagulant for thrombotic complications and increases mortality rates in CKD patients. Further studies are necessary to assess how these complications are related to the mechanisms and risks [41].

GN is a major process of end-stage renal disease (ESRD) along with diabetes and hypertension [42]. However, there are no appropriate therapeutic treatments for GN. In therapy for GN, steroids are widely used as anti-inflammation drugs for patients, particularly those with IgA nephropathy, which is typical mesangial proliferative GN [43, 44]. As steroids reduce proteinuria, they prevent the progression of kidney failure. However, extensive and severe side effects concerning the immune, circulating, and metabolic systems have been reported in patients with steroid therapy [45, 46]. Both angiotensin-converting enzyme (ACE) inhibitors and angiotensin receptor blockers (ARBs) are widely treated for kidney disease in patients with diabetes and hypertension [47–49]. ACE inhibitors and ARBs control blood pressure by inhibiting the production of angiotensin II or preventing it from binding to angiotensin II receptors in the renin-angiotensin system, and, thus, reduce hypertension. In addition, nonclinical and clinical studies have demonstrated that ACE inhibitors and ARBs decrease proteinuria and prevent kidney failure [47, 50, 51]. In Thy-1 GN rats, treatments with ACE inhibitors or ARBs were found to inhibit increases in blood pressure and proteinuria, and subsequently the decline in renal function [52]. Persistent hypertension and glomerular hypertension may stretch the glomerular capillary wall, and result in endothelial damage and glomerular sclerosis, followed by a rise in protein glomerular filtration [53]. Schmieder [54] and Mochizuki et al. [55] reported that these compounds inhibited the kidney-specific renin-angiotensin aldosterone system and prevented end-organ damage in the kidney beyond these antihypertensive effects. On the other hand, the VKOR inhibitor, AMT, prevents mesangial cell proliferation in the kidneys and suppresses the progression of GN. Unfortunately, AMT is not as effective by oral administration due to its low bioavailability. If improvements are archived in the biological stability of AMT, which has unique pharmacokinetic properties and is distributed to the kidney at a higher concentration than to the liver, VKOR inhibitors may be useful for the treatment for GN, particularly in combination with existing medications such as ACE inhibitors or ARBs.

http://dx.doi.org/10.5772/63512

In conclusion, the novel VKOR inhibitor, AMT, reduced renal mesangial cell proliferation and may be a supportive treatment for GN.

Acknowledgements

Reprinted with permission from the Japanese Society of Veterinary Science (The Journal of Veterinary Medical Science, 2012; 74(7): 863–869) and John Wiley & Sons Inc. (Journal of Applied Toxicology, 2008; 28(4): 524–529).

Author details

Yohei Miyamoto*

Address all correspondence to: Youhei_Miyamoto@nts.toray.co.jp

Pharmaceutical Clinical Research Department, Toray Industries, Inc., Tokyo, Japan

References

[1] Sutcliffe FA, MacNicoll AD, Gibson GG. Aspects of anticoagulant action: a review of the pharmacology, metabolism and toxicology of warfarin and congeners. Rev Drug Metab Drug Interact. 1987;5(4):225–272. doi:10.1515/DMDI.1987.5.4.225.

[2] Cosmetic Ingredient Review. Final report on the safety assessment of sodium dehydroacetate and dehydroacetic acid. Int J Toxicol. 1985;4(3):123–159. doi:10.3109/10915818509078671.

[3] Kanai K, Asada-Kubota M, Kanamura S. Morphological studies on the liver of mice administered with dehydroacetic acid. Kankyo Kagaku Sogo Kenkyujo Nenpo. 1981;8:27–31.

[4] Uchida O, Naito K, Yasuhara K, Ohba S, Sato C, Shimonuri S, Kobayashi K, Tobe M. Studies on the acute oral toxicity of dehydroacetic acid, sorbic acid and their combination compound in rats. Eisei Shikenjo Hokoku. 1985;103:166–171.

[5] Seevers MH, Shideman FE, Woods LA, Weeks JR, Kruse WT. Dehydroacetic acid (DHA). II. General pharmacology and mechanism of action. J Pharmacol Exp Ther. 1950;99(1):69–83.

[6] Spencer HC, Rowe VK, McCollister DD. Dehydroacetic acid (DHA). I. Acute and chronic toxicity. J Pharmacol Exp Ther.1950;99(1):57–68.

[7] Rehse K, Schinkel W, Siemann U. Anticoagulant 4-hydroxy-2-pyrones. Arch Pharm (Weinheim). 1980;313(4):344–351. doi:10.1002/ardp.19803130411.

[8] Sakaguchi Y, Suga S, Oshida K, Miyamoto-Kuramitsu K, Ueda K, Miyamoto Y. Anticoagulant effect of sodium dehydroacetate (DHA-S) in rats. J Appl Toxicol. 2008;28(4):524–529. doi:10.1002/jat.1304.

[9] Yanagita M, Arai H, Ishii K, Nakano T, Ohashi K, Mizuno K, Varnum B, Fukatsu A, Doi T, Kita T. Gas6 regulates mesangial cell proliferation through Axl in experimental glomerulonephritis. Am J Pathol. 2001;158(4):1423–1432. doi:10.1016/S0002-9440(10) 64093-X.

[10] Yanagita M. The role of the vitamin K-dependent growth factor Gas6 in glomerular pathophysiology. Curr Opin Nephrol Hypertens. 2004;13(4):465–470.

[11] Yanagita M. Gas6, warfarin, and kidney diseases. Clin Exp Nephrol. 2004;8(4):304–309. doi:10.1007/s10157-004-0305-z.

[12] Limdi NA, Beasley TM, Baird MF, Goldstein JA, McGwin G, Arnett DK, Acton RT, Allon M. Kidney function influences warfarin responsiveness and hemorrhagic complications. J Am Soc Nephrol. 2009;20(4):912–921. doi:10.1681/ASN.2008070802.

[13] Woo KT. Current therapeutic strategies in glomerulonephritis. Ann Acad Med Singapore. 1999;28(2):272–278.

[14] Jefferson JA, Johonson RJ. Experimental mesangial proliferative glomerulonephritis (the anti-Thy-1.1 model). J Nephrol. 1999;12(5):297–307.

[15] Uchida M, Sakaguchi Y, Miyamoto Y. A novel vitamin K_1 2,3-epoxide reductase (VKOR) inhibitor, 3-acetyl-5-methyltetronic acid, reduces experimental glomerulonephritis. J Vet Med Sci. 20121;74(7):863–869. doi:10.1292/jvms.11-0530.

[16] Billeter M, Bollinger W, Martius C. Untersuchungen uber die umwandlung von verfutterten K-vitaminen durch austausch der seitenkette und die rolle der darmbakterien hierbei. Biochem Z. 1964;340:290–303.

[17] Sakamoto N, Kimura M, Hiraike H, Itokawa Y. Changes of phylloquinone and menaquinone-4 concentrations in rat liver after oral, intravenous and intraperitoneal administration. Int J Vitam Nutr Res. 1996;66(4):322–328.

[18] Motohara K, Endo F, Matsuda I. Effect of vitamin K administration on carboxy prothrombin (PIVKA-II) levels in newborns. Lancet. 1985; 3(8449):242–244. doi:10.1016/S0140-6736(85)90291-0.

[19] Roncaglioni MC, Reyers I, Cerletti C, Donati MB, De Gaetano G. Moderate anticoagulation by salicylate prevents thrombosis without bleeding complications. An experimental study in rats. Biochem Pharmacol. 1988;37(24):4743–4745. doi:10.1016/0006-2952(88)90346-2.

[20] Shearer MJ. Vitamin K and vitamin K-dependent proteins. Br J Haematol. 1990;75(2): 156–162. doi:10.1111/j.1365-2141.1990.00156.x

[21] Bentley EW. A review of anticoagulant rodenticides in current use. Bull World Health Organ. 1972;47(3):275–280.

[22] Park BK. Warfarin: metabolism and mode of action: a review. Biochem Pharmacol. 1988;37(1):19–27. doi:10.1016/0006-2952(88)90750-2.

[23] Cain D, Hutson SM, Wallin R. Warfarin resistance is associated with a protein component of the vitamin K 2,3-epoxide reductase enzyme complex in rat liver. Thromb Haemost. 1998;80(1):128–133.

[24] Chan K, Lo AC, Yeung JH, Woo KS. The effects of Danshen (Salvia miltiorrhiza) on warfarin pharmacodynamics and pharmacokinetics of warfarin enantiomers in rats. J Pharm Pharmacol. 1995;47(5):402–406. doi:10.1111/j.2042-7158.1995.tb05819.x.

[25] Mosterd JJ, Thijssen HH. The relationship between the vitamin K cycle inhibition and the plasma anticoagulant response at steady-state S-warfarin conditions in the rat. J Pharmacol Exp Ther. 1991;260(3):1081–1085.

[26] Ishiwata H, Nishijima M, Fukasawa Y. Estimation of preservative concentration in foods and their daily intake based on official inspection results in Japan in fiscal year 1998. Shokuhin Eiseigaku Zasshi 2001;42(6):404–412.

[27] Wang Y, Zhen Y, Shi Y, Chen J, Zhang C, Wang X, Yang X, Zheng Y, Liu Y, Hui R. Vitamin K epoxide reductase: a protein involved in angiogenesis. Mol. Cancer Res. 2005;3(6): 317–323. doi:10.1158/1541-7786.MCR-04-0221.

[28] Li T, Chang CY, Jin DY, Lin PJ, Khvorova A, Stafford DW. Identification of the gene for vitamin K epoxide reductase. Nature. 2004;427(6974):541–544. doi:10.1038/nature02254.

[29] Yamamoto T, Mundy CA, Wilson CB, Blantz, RC. Effect of mesangial cell lysis and proliferation on glomerular hemodynamics in the rat. Kidney Int. 1991;40(4):705–713. doi:10.1038/ki.1991.264.

[30] Zoran LB. Principles of genitourinary radiology. 2nd ed. Thieme Medical Publishers; 1994.

[31] Shimizu F, Kawachi H, Orikasa M. Role of mesangial cell damage in progressive renal disease. Kidney Blood Press Res. 1999;22(2):5–12. doi:10.1159/000025903.

[32] Ito S, Kuriyama H, Iino N, Iguchi S, Shimada H, Ueno M, Narita I, Nakano M, Nishi S, Karasawa R, Gejyo F. Patient with diffuse mesangial and endocapillary proliferative glomerulonephritis with hypocomplementemia and elevated anti-streptolysin O treated with prednisolone, angiotensin-converting enzyme inhibitor, and angiotensin II receptor antagonist. Clin Exp Nephrol. 2003;7(4):290–295.

[33] Noël LH, Gubler M. C. Histological classification of chronic glomerular diseases. Rev. Prat. 2003;53(18):2005–2012.

[34] Gohda T, Ra C, Hamada C, Tsuge T, Kawachi H, Tomino Y. Suppressive activity of pemirolast potassium, an antiallergic drug, on glomerulonephritis. Studies in glomerulonephritis model rats and in patients with chronic glomerulonephritis concurrently affected by allergic rhinitis. Arzneimittelforschung. 2008;58(1):18–23. doi:10.1055/s-0031-1296461.

[35] Mii A, Shimizu A, Masuda Y, Ishizaki M, Kawachi H, Iino Y, Katayama Y, Fukuda Y. Angiotensin II receptor blockade inhibits acute glomerular injuries with the alteration of receptor expression. Lab. Invest. 2009;89(2):164–177. doi:10.1038/labinvest.2008.128.

[36] Sadakane C, Kase Y, Koseki J, Hasegawa Y, Shindo S, Maruyama H, Takeda S, Takeda H, Hattori T. Effects of TJN-598, a new selective phosphodiesterase type IV inhibitor on anti-Thy1 nephritis in rats. Clin Exp Nephrol. 2011;15(1):14–24.

[37] Eng PC, Chua WC, Suk Peng Chew V, Wong PT, Yin JL, Hambly B, McLachlan CS. Chronic angiotensin-converting enzyme inhibition up-regulates mouse kidney growth arrest specific-6 protein and the AXL subfamily of receptor tyrosine kinases. J Renin Angiotensin Aldosterone Syst. 2008;9(4):238–241. doi:10.1177/1470320308098342.

[38] Makibayashi K, Tatematsu M, Hirata M, Fukushima N, Kusano K, Ohashi S, Abe H, Kuze K, Fukatsu A, Kita T, Doi T. A vitamin D analog ameliorates glomerular injury on rat glomerulonephritis. Am J Pathol. 2001;158(5):1733–1741. doi:10.1016/S0002-9440(10)64129-6.

[39] Pavkov ME, Knowler WC, Hanson RL, Bennett PH, Nelson RG. Predictive power of sequential measures of albuminuria for progression to ESRD or death in Pima Indians with type 2 diabetes. Am J Kidney Dis. 2008;51(5):759–766. doi:10.1053/j.ajkd.2008.01.011.

[40] Brodsky SV, Satoskar A, Chen J, Nadasdy G, Eagen JW, Hamirani M, Hebert L, Calomeni E, Nadasdy T. Acute kidney injury during warfarin therapy associated with obstructive tubular red blood cell casts: a report of 9 cases. Am J Kidney Dis. 2009;54(6): 1121–1126. doi:10.1053/j.ajkd.2009.04.024.

[41] Brodsky SV, Nadasdy T, Rovin BH, Satoskar AA, Nadasdy GM, Wu HM, Bhatt UY, Hebert LA. Warfarin-related nephropathy occurs in patients with and without chronic kidney disease and is associated with an increased mortality rate. Kidney Int. 2011;80(2):181–189. doi:10.1038/ki.2011.44.

[42] Constantini E. United States Renal Data System 2006 Annual Data Report: Atlas of Chronic Kidney Disease & End-Stage Renal Disease in the United States. USRDS 2006 ADR: Precis-Background on the US ESRD program. Am J Kidney Dis. 2007;49(Suppl 1):S17–S32. doi:10.1053/j.ajkd.2006.11.004.

[43] Nolin L, Courteau M. Management of IgA nephropathy: evidence-based recommendations. Kidney Int. 1999;55(Suppl 70):S56-S62. doi:10.1046/j.1523-1755.1999.07008.x.

[44] Zhou YH, Tang LG, Guo SL, Jin ZC, Wu MJ, Zang JJ, Xu JF, Wu CF, Qin YY, Cai Q, Gao QB, Zhang SS, Yu DH, He J. Steroids in the treatment of IgA nephropathy to the improvement of renal survival: a systematic review and meta-analysis. PLoS ONE 2011;6(4):e18788. doi:10.1371/journal.pone.0018788.

[45] Hodson EM, Knight JF, Willis NS, Craig JC. Corticosteroid therapy for nephrotic syndrome in children. Cochrane Database Syst Rev. 2001;3:CD001533. doi:10.1002/14651858.CD001533.pub5.

[46] Olney RC. Mechanisms of impaired growth: effect of steroids on bone and cartilage. Horm Res. 2009;72(Suppl 1):30–35. doi:10.1159/000229761.

[47] Brewster UC, Perazella MA. The renin-angiotensin-aldosterone system and the kidney: effects on kidney disease. Am J Med. 2004;116(4):263–272. doi:10.1016/j.amjmed.2003.09.034.

[48] Ruggenenti P, Perna A, Gherardi G, Benini R, Remuzzi G. Chronic proteinuric nephropathies: outcomes and response to treatment in a prospective cohort of 352 patients with different patterns of renal injury. Am J Kidney Dis. 200;35(6):1155–1165. doi:10.1016/S0272-6386(00)70054-0.

[49] THE GISEN GROUP. Randomised placebo-controlled trial of effect of ramipril on decline in glomerular filtration rate and risk of terminal renal failure in proteinuric, non-diabetic nephropathy. Lancet. 1997;349(9069?):1857–1863. doi:10.1016/S0140-6736(96)11445-8.

[50] Pugsley MK. The angiotensin-II (AT-II) receptor blocker olmesartan reduces renal damage in animal models of hypertension and diabetes. Proc West Pharmacol Soc. 2005;48:35–38.

[51] Tomita M, Sogabe H, Nakazato S, Nakatsuji S, Noto T, Hamada K, Kawachi H, Shimizu F, Matsuo M, Mutoh S. Monoclonal antibody 1-22-3-induced glomerulonephritis in uninephrectomized rats as a model of progressive renal failure. Nephrol Dial Transplant. 2005;20(11):2358–02367. doi:10.1093/ndt/gfi062.

[52] Villa L, Boor P, Konieczny A, Kunter U, van Roeyen CR, Denecke B, Gan L, Kupper MB, Hoffmann K, Eitner F, Ostendorf T, Floege J. Effects and mechanisms of angiotensin II receptor blockade with telmisartan in a normotensive model of mesangioproliferative nephritis. Nephrol Dial Transplant. 2011;26(10):3131–3143. doi:10.1093/ndt/gfr096.

[53] Qian Y1, Feldman E, Pennathur S, Kretzler M, Brosius FC 3rd. From Fibrosis to Sclerosis; Mechanisms of glomerulosclerosis in diabetic nephropathy. Diabetes. 2008;57(6):439–1445. doi:10.2337/db08-0061.

[54] Schmieder RE. Mechanisms for the clinical benefits of angiotensin II receptor blockers. Am J Hypertens. 2005;18(5):720–730. doi:10.1016/j.amjhyper.2004.11.032.

[55] Mochizuki S, Dahlöf B, Shimizu M, Ikewaki K, Yoshikawa M, Taniguchi I, Ohta M, Yamada T, Ogawa K, Kanae K, Kawai M, Seki S, Okazaki F, Taniguchi M, Yoshida S, Tajima N; Jikei Heart Study group. Valsartan in a Japanese population with hypertension and other cardiovascular disease (Jikei Heart Study): a randomised, openlabel, blinded endpoint morbidity-mortality study. Lancet. 2007;369(9571):1431–1439. doi: 10.1016/S0140-6736(07)60669-2.

Chapter 17

Enhanced Intracellular Delivery and Improved Antitumor Efficacy of Menaquinone-4

Kazuhisa Matsunaga, Munechika Enjoji,
Yoshiharu Karube and Jiro Takata

Additional information is available at the end of the chapter

http://dx.doi.org/10.5772/63343

Abstract

Hepatocellular carcinoma (HCC) is a major malignant tumor type that occurs globally. HCC incidence is increasing, especially in Asian countries. Despite many therapeutic approaches, the long-term prognosis of HCC remains poor because of frequent recurrence due to intrahepatic metastasis or multicentric carcinogenesis. Therefore, it is necessary to develop effective and safe chemopreventive agents to improve the prognosis of HCC. Menaquinone-4 (MK-4) has a suppressive effect on HCC, but cellular delivery is poor. We hypothesized that effective cellular delivery of menahydroquinone-4 (MKH), a fully reduced form of MK-4, would regulate HCC growth and metastasis. We developed a bioreductive activation-independent delivery system with the *N,N*-dimethylglycine ester of MKH (MKH-bis-DMG) to deliver MKH to HCC cells without any bioreductive processing of MK-4. MKH-bis-DMG inhibited the proliferation of both DCP-positive and DCP-negative HCC cell lines in a time- and dose-dependent manner via G1/S cell-cycle arrest. We assessed the effect of MKH-derivatives on HCC metastasis using a mouse model of spleen-liver metastasis. The mean tumor hepatic replacement area of MKH-bis-DMG treated mice was significantly less than that of untreated mice. In conclusion, MKH-bis-DMG may be beneficial as a chemopreventive agent for recurrent HCC.

Keywords: menahydroquinone-4, antitumor agent, prodrug, drug delivery, hepatocellular carcinoma

1. Introduction

Vitamin K has two types of molecular homologues: phylloquinone (vitamin K1, PK) and menaquinone (vitamin K2, MK-n). These homologues have the same aromatic naphthoquinone

INTECH
open science | open minds

© 2017 The Author(s). Licensee InTech. This chapter is distributed under the terms of the Creative Commons Attribution License (http://creativecommons.org/licenses/by/3.0), which permits unrestricted use, distribution, and reproduction in any medium, provided the original work is properly cited. (cc) BY

"head" but different hydrocarbon "tails": PK with a phytyl tail and MK-n with an unsaturated isoprenoid chain. MK-n can be classified into 14 types based on the length of the unsaturated isoprenoid chain, where "n" quantifies the repeating isoprenyl units. MK-4 is normally synthesized from PK in certain animal tissues by removal of the phytyl tail of PK to produce menadione (vitamin K3, MD) as an intermediate. The intermediate is then condensed with a geranylgeranyl tail. UbiA prenyltransferase domain-containing protein 1 (UBIAD1) converts MD to MK-4 with geranylgeranyl diphosphate [1]. For this reason, MK-4 appears to be the most important form of vitamin K.

Vitamin K (PK and MK-n) plays a major role in the clotting cascade by acting as a coenzyme for a vitamin K-dependent carboxylase. The carboxylase catalyzes the carboxylation of glutamic acid (Glu) residues to produce γ-carboxyglutamic acid (Gla). Vitamin K also appears to play a role in the regulation of bone metabolism through a similar mechanism that involves γ-carboxylation of pro-osteocalcin. Interestingly, MK-4 intake seems to be associated with greater effects of reduced bone resorption compared with PK consumption [2]. Clinically, high doses (45 mg daily) of MK-4 have been used as an approved treatment for osteoporosis in Japan since 1995. Therefore, the safety of long-term administration of MK-4 has been established in Japanese patients with osteoporosis.

Over the last decade, many reports have shown that MK-4 has antioncogenic effects within various cancer cell lines, including leukemia, lung cancer, ovarian cancer, prostate cancer, and hepatocellular carcinoma (HCC) [3–6]. Specifically, numerous articles describe the effects of MK-4 against HCC. This is because des-γ-carboxy prothrombin (DCP, PIVKA-II), an abnormal prothrombin that is not completely carboxylated, is a well-recognized HCC-specific tumor marker, and a predictor of vascular invasion, metastasis, and tumor recurrence [7]. It has been reported that apoptosis, cell-cycle arrest, and autophagy are involved in the antitumor activity of MK-4 [8–10]. Although the possible mechanisms of the antitumor effect of MK-4 have been investigated previously, they remain unclear.

2. Mechanisms of growth inhibition of HCC cells by MK-4

Otsuka et al. reported that MK-4 inhibited the growth and invasion of HCC via activation of protein kinase A (PKA) and the subsequent inhibition of Rho activation. It has been reported that PKA induces cell-cycle arrest at the G1 phase and the G2/M phase [11]. Hitomi et al. [12] reported that MK-4-induced G1 arrest of the cell cycle via significantly reduced protein expression of cyclin D1 and cyclin-dependent kinase 4 (Cdk4), but not the p16INK4a Cdk inhibitor in PLC/PRF/5 HCC cells *in vivo*. Cyclin D1 promotes the G1/S phase of the cell cycle and is frequently overexpressed in many human cancers. Matsumoto et al. [13] reported that MK-4-induced arrest of the G1 phase of the cell cycle and apoptosis via activated extracellular signal-regulated kinase ½ (ERK1/2) in a mitogen-activated ERK-regulating kinase-dependent manner in Hep3B HCC cells. Ozaki et al. [10] reported that MK-4 inhibited the growth of HCC cells via suppression of cyclin D1 expression through inhibited IκB kinase activity, and therefore suppressed IκB phosphorylation and NF-κB activation.

In addition, Xia et al. [14] reported that the inhibitory effect on NF-κB activity by MK-4 is mediated through the inhibition of protein kinase C α and ε kinase activities, as well as subsequent inhibition of protein kinase D1 activation. Kaneda et al. proposed that MK-4 suppressed HuH7 HCC tumor malignancy via induced Cx32 expression through the reduction of Cx43 expression. Consequently, gap junctional intercellular communication through Cx32 is activated. Normal hepatocytes communicate with neighboring cells via Cx32-containing gap junction communication, a process essential for suppressing tumorigenesis [15]. Yamamoto et al. suggested that regulation of the expression of the hepatoma-derived growth factor gene is one of the crucial mechanisms of MK-4-induced cell growth suppression in HCC. Hepatoma-derived growth factor stimulates the proliferation of HCC cells after its translocation to the nucleus by use of bipartite nuclear localization signals [16]. Azuma et al. [17] suggested that activation of steroid and xenobiotic receptors (SXR) by MK-4 contributes to the tumor suppressive effects on HCC cells. Li et al. suggested that MK-4 inhibited the growth of Smmc-7721 HCC cells by induction of apoptosis involving caspase-8 activation and p53. This apoptotic process was not mediated by the caspase-9 pathway [18]. Yao et al. suggested that the mechanism involved induced p53 and increased p21 levels that eventually lead to cell-cycle arrest in the G2 phase. In addition, they suggested that the antitumor effect of MK-4 may be improved by silencing BCL-2 expression in SMMC-7721 HCC cells [19].

3. Clinical trials of MK-4 to treat HCC

In clinical trials with cirrhotic women, Habu et al. demonstrated that daily doses of 45 mg MK-4 decreased the risk of HCC to about 20% compared with the control group. Twenty-one women were in the treatment group and 19 in the control group [20]. Mizuta et al. reported that a daily dose of 45 mg MK-4 suppressed the recurrence of HCC in HCC patients who had undergone curative resection or percutaneous local ablation therapy. Thirty-two patients were in the treatment group and 29 in the control group [21]. From the results of these small-scale clinical trials, it is expected that MK-4 acts as a chemopreventive agent for HCC. However, a recent larger scale study that enrolled 548 patients at 31 study sites, and included a placebo-controlled, double-blind trial, demonstrated that the efficacy of vitamin K2 in suppressing HCC recurrence could not be confirmed [22]. The poor anticancer activity of MK-4 observed in this Japanese trial may have been a consequence of the large study design and meant that MK-4 could not be developed as an anticancer drug. However, various attempts are being made to try to improve the anticancer effect of MK-4 in HCC.

4. Improvement of the antiproliferative effect of vitamin K2

4.1. A novel chemosynthetic vitamin K derivative

Carr et al. demonstrated that a new, chemically synthesized vitamin K analog, compound 5 (Cpd5), inhibited Cdc25A phosphatase activity and particularly reduced HCC cell growth through arrest of the G1/S phase of the cell cycle. Inhibition of Cdc25A by Cpd5 results in prolonged tyrosine phosphorylation and activation of ERK1/2, which could be triggered by upstream epidermal growth factor receptor signaling pathway molecules [23]. Suhara et al. synthesized vitamin K2 analogues with hydroxyl or phenyl groups at the ω-terminal of the side chain, and with dual side chains at the C-2 and C-3 positions. They found that modifying the side chain of vitamin K affects the SXR-mediated transcriptional activity [24, 25]. The novel biological activities of MK-4 include tumor suppressive effects related to gene transcription through the SXR. The new derivatives of MK-4 have shown some efficacy against HCC, but a lengthy development process is still necessary to yield safe and effective clinical products.

4.2. Synergistic drug combinations

Yoshiji et al. [26, 27] reported that combined treatment with MK-4 and angiotensin-converting enzyme inhibitor significantly suppressed experimental hepatocarcinogenesis. Further to this, they reported that the combined treatment with MK-4 and angiotensin-converting enzyme inhibitor may suppress the cumulative recurrence of HCC after the curative therapy, at least partly through suppression of the vascular endothelial growth-mediated neovascularization. Kanamori et al. [28] demonstrated that a combination of MK-4 and acyclic retinoid synergistically inhibited the growth of Huh7 HCC cells by increasing apoptosis. When combined with acyclic retinoid, MK-4 synergistically inhibits Ras activation and inhibits phosphorylation of retinoid X receptor α. Zhang H et al. [29] showed that MK-4 enhanced the inhibition of 5-fluorouracil-induced cell growth in HepG2, Huh7, HLE, and Hep3B HCC cells, and via G1 cell-cycle arrest through induced expression of p21 and p27 and inhibited expression of cyclin D1. Zhang et al. reported that a combination of MK-4 and sorafenib work synergistically to inhibit growth of HepG2, Hep3B, and HuH7 HCC cells. They also demonstrated that the levels of cyclin D1 expression are clearly reduced in HepG2 cells treated with a combination treatment of MK-4 and sorafenib [30]. A clinical trial to test the efficacy of the combination of MK-4 and sorafenib in HCC was attempted [31].

4.3. Delivery of menahydroquinone-4, the active form of MK-4

We have previously synthesized the ester derivatives of menahydroquinone-4 (MKH), the fully reduced form of MK-4, and revealed their effective antiproliferative activity against HCC cell lines [32–35]. The menaquinone (MK-4 to MK-10) concentrations were significantly lower in HCC tissues, from patients with or without increased plasma concentrations of DCP, than in the surrounding normal liver tissue. There was no significant difference between PK and PK epoxide concentration in HCC tissues without increased plasma concentrations of DCP and normal liver tissue [36]. Furthermore, the rate of uptake into MH7777 cells (vitamin K2-sensitive HCC) was lower than for normal hepatocytes. In addition, the rate of uptake into H4IIE cells (vitamin K2-resistant HCC) was negligible compared with that for MH7777 cells and normal hepatocytes. Further, hepatocytes from diethylnitrosamine-induced liver nodules exhibited a significantly lower rate of vitamin K2 uptake than that for normal hepatocytes [37].

MKH acts as a cofactor for γ-glutamyl carboxylase (GGCX), which catalyzes the carboxylation of specific Glu residues (γ-carboxylation) of substrate proteins such as prothrombin. Thus, decreased MKH availability in HCC cells is a possible causative mechanism of DCP production in HCC. MKH is mainly generated by a vitamin K 2, 3-epoxide reductase complex subunit 1-like-1 (VKORC1L1) in mammalian cells [38]. Key reasons for further development are that MK-4 delivery to HCC cells is poor and reductive activation of MK-4 is low in HCC cells. We hypothesized that effective delivery of MKH into HCC cells is essential in regulating HCC growth and metastasis. However, MKH cannot be used as a therapeutic agent, because it is easy to be oxidized and converted to MK-4. In addition, the production of MKH depends on VKORC1L1 activity in HCC cells. As such, MK-4 does not display sufficient antitumor activity even at high doses.

In light of these findings, we synthesized three types of *N,N*-dimethylglycine esters of MKH (MKH-1-DMG, MKH-4-DMG, and MKH-bis-DMG) and assessed their potential as water-soluble prodrugs for a bioreductive activation-independent delivery system of MKH [33–35] (**Figure 1**).

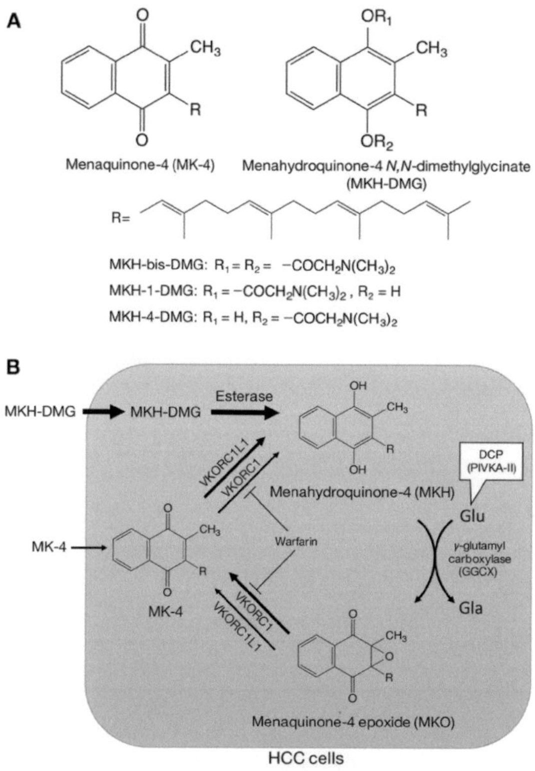

Figure 1. Structure of the MKH derivatives and schematic illustration of the concept of the MKH delivery system. (A) Chemical structures of menaquinone-4 (MK-4), menahydroquinone-4 (MKH) and MKH *N,N*-dimethylglycinate (MKH–DMG). (B) Schematic illustration of the vitamin K cycle and the concept of the MKH delivery system. Adapted from Ref. [32].

5. Water solubility and hydrolysis of the MKH esters

The hydrochloride salts of the esters showed much improved aqueous solubility. The solubilities of MKH-1-DMG and MKH-4-DMG in water were 24 and 5.7 mM respectively and that of MKH-bis-DMG was >50 mM. In contrast, the solubility of MK-4 in water was <2.3 mM. Improving the low water solubility of MK-4 enables a bolus dose necessary for cancer chemotherapy without any surfactant. *In vitro* studies have confirmed that MKH derivatives can be hydrolyzed with esterases located in the rat and human liver and that the resultant MKH acts as cofactor for GGCX without the reductive activation process. The first-order rate constants observed for the hydrolysis of the MKH derivatives in the liver homogenate supernatant are listed in **Table 1**, along with the degradation rate constants of the derivatives in phosphate buffer. The rate of hydrolysis of MKH-1-DMG was about 24-fold and 5.7-fold faster than that of MKH-4-DMG in rat and human liver homogenate supernatant, respectively.

Compound	Without physostigmine ($\times 10^{-2}$min^{-1})	With physostigmine ($\times 10^{-2}$min^{-1})	Regeneration half-life (min)
Rat liver homogenate			
MKH-1-DMG	27.2	0.261	2.55
MKH-4-DMG	1.14	0.315	60.6
MKH-bis-DMG	17.2[a]	0.328[a]	39.7[a]
Human liver homogenate			
MKH-1-DMG	2.70	0.0714	25.7
MKH-4-DMG	0.476	0.0431	146
MKH-bis-DMG	1.00[a]	0.117[a]	227[a]
Isotonic phosphate buffer of pH7.4			
MKH-1-DMG	0.0203		
MKH-4-DMG	0.0295		
MKH-bis-DMG	0.0500[a]		

[a]Disappearance of MKH-bis-DMG.
Adapted from Refs. [33, 34].

Table 1. Apparent first-order rate constants for the hydrolysis of the MKH derivatives, and regeneration half-lives of MK-4 in human and rat liver homogenate supernatant and phosphate buffer (pH 7.4).

The formation of MK-4 from MKH-bis-DMG should proceed through the intermediates MKH-1-DMG, MKH-4-DMG, and MKH. The pseudo-first-order rate constants for the interconversion of the species are assumed, as shown in **Figure 2**. Detection of MKH was unsuccessful because of its high susceptibility to oxidation.

http://dx.doi.org/10.5772/63343

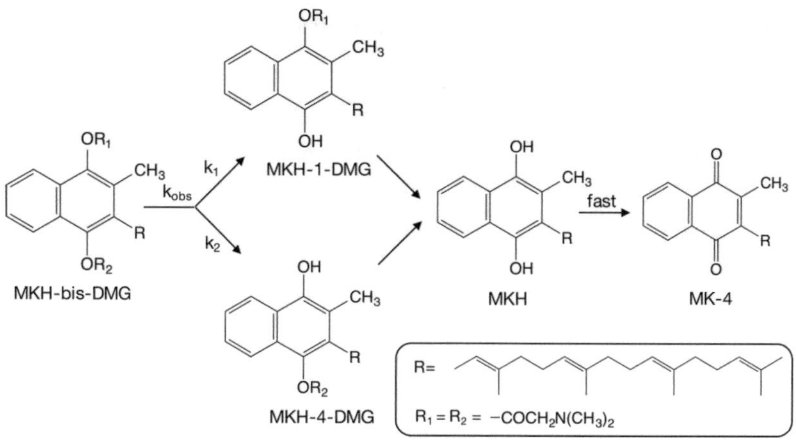

Figure 2. Hydrolytic pathway of MKH-bis-DMG. Adapted from Ref. [34].

As shown in **Table 2**, the rate of hydrolysis of MKH-bis-DMG at the 1-position (k_1) was about 2.6-fold and 5.1-fold faster than that at the 4-position (k_2) in rat and human liver homogenate supernatant, respectively. The rates of hydrolysis of MKH-1-DMG, MKH-4-DMG, and MKH-bis-DMG in the liver homogenate supernatant were significantly reduced in the presence of physostigmine, a liver carboxylesterase inhibitor. Consequently, MKH esters were hydrolyzed by rat and human liver esterases.

Medium	k_{obs}[a] ($\times 10^{-2}$min^{-1})	k_1 ($\times 10^{-2}$min^{-1})	k_2 ($\times 10^{-2}$min^{-1})
Rat liver homogenate	17.2 ± 1.0	11.9 ± 0.39	4.39 ± 0.18
Human liver homogenate	1.00	0.794 ± 0.019	0.156 ± 0.008

k_1 and k_2 accordance with **Figure 2**.
[a]Disappearance of MKH-bis-DMG.
Adapted from Refs. [33, 34].

Table 2. Rate constants for the hydrolysis of MKH–DMG in human and rat liver homogenate supernatant.

6. Vitamin K-dependent carboxylation *in vitro*

To provide evidence that confirmed the bioreductive activation-independent delivery system of MKH was working properly, carboxylation activity was measured with the incorporation of $^{14}CO_2$ in the synthetic tripeptide BOC-Glu-Glu-Leu-OMe. The accelerated carboxylation of MK-4 was only observed in the presence of dithiothreitol (DTT), an artificial reducing agent for MK-4, and not in its absence. Conversely, MKH esters stimulated carboxylase activity in the absence of DTT [33], clearly indicating that the MKH esters can stimulate carboxylation without the reductive activation process of MK-4.

7. Evaluation of MKH delivery in HCC cells with MKH esters and the antiproliferative effect

MKH-bis-DMG inhibited the proliferation of both DCP-positive (PLC/PRF/5, Hep3B) and DCP-negative (SK-Hep-1) HCC cell lines in a time- and dose-dependent manner, and exhibited lower IC_{50} values (range from 14–37 μmol/L), and a fourfold to 18-fold increase in growth-inhibitory activity compared with MK-4. MKH-bis-DMG showed a rapid and strong growth-inhibitory effect after only 48 h of treatment. In contrast, MK-4 had little inhibitory effect on cell proliferation, and its effects appeared after 72 h of treatment (**Figure 3**).

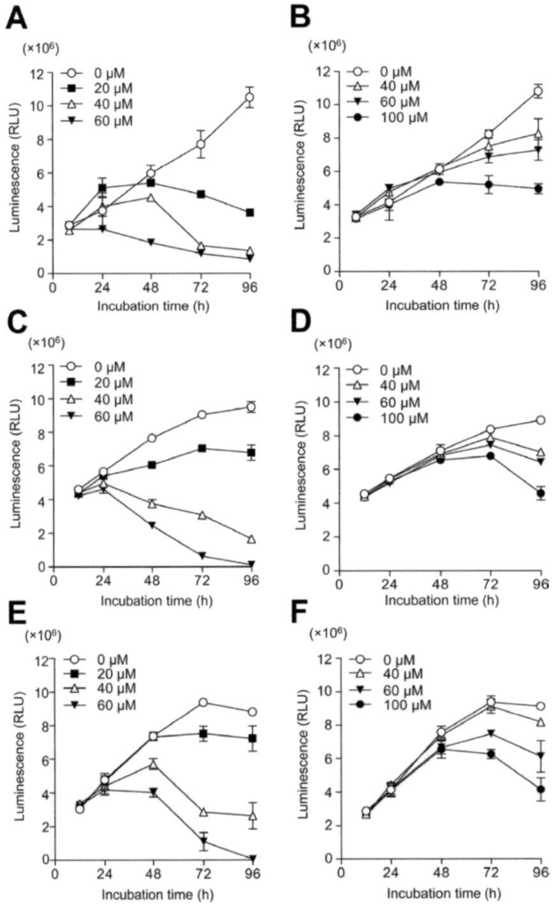

Figure 3. Inhibitory effects of MKH–DMG and MK-4 on DCP-positive and DCP-negative HCC cell proliferation. MKH–DMG treatment of PLC/PRF/5 (A), Hep3B (C), and SK-Hep-1 (E) cell lines, and MK-4 treatment of PLC/PRF/5 (B), Hep3B (D), and SK-Hep-1 (F) cells. Symbols: ○, 0 μM; ■, 20 μM; △, 40 μM; ▼, 60 μM after MKH–DMG treatment. Symbols: ○, 0 μM; △, 40 μM; ▼, 60 μM; ●, 100 μM after MK-4 treatment. Error bars indicate mean ± SD (n = 3). Adapted from Ref. [32].

MKH was susceptible to oxidation, so we measured the levels of menaquinone-4 epoxide (MKO) in HCC cells to assess the function of MKH-bis-DMG as a delivery system for MKH in HCC cells. Concomitant with vitamin K-dependent carboxylation of Glu to Gla by GGCX, MKH is stoichiometrically converted to MKO and that means MKO levels in HCC cells reflect the levels of MKH. The AUC_{MKH} values after MKH—DMG administration in three types of HCC cell lines were 3.5-fold to 15-fold higher than those after MK-4 administration (**Table 3**). Based on these results, it was clearly confirmed that MKH-bis-DMG works as an effective delivery prodrug of MKH into HCC cells. The resultant MKH may exhibit excellent antiproliferative activity against HCC cells despite their DCP-positive and DCP-negative status.

HCC cell line	Compound	AUC_{0-72h} for MKO (nmol·h·mg protein^{-1})	AUC_{0-72h} for MK-4 (nmol·h·mg protein^{-1})	AUC_{0-72h} for MKH (nmol·h·mg protein^{-1})
PLC/PRF/5	MK-4	22.2 ± 3.72	47.7 ± 7.25	22.2 ± 3.72[a]
	MKH–DMG	193 ± 25.1	143 ± 13.6	336 ± 37.2[b]
Hep3B	MK-4	113 ± 4.87	74.5 ± 17.0	113 ± 4.87[a]
	MKH–DMG	371 ± 31.0	25.9 ± 5.04	397 ± 34.2[b]
SK-Hep-1	MK-4	38.4 ± 4.44	122 ± 19.5	38.4 ± 4.44[a]
	MKH–DMG	136 ± 14.3	193 ± 21.5	329 ± 35.4[b]

Doses are 25 μM (at near IC_{50} value).
[a]MKH value after MK-4 administration: MKO.
[b]MKH value after MKH–DMG administration: sum of MKO and MK-4.
Adapted from Ref. [32].

Table 3. Area under the curve for intracellular concentration versus time (AUC) after treatment with MK-4 or MKH–DMG in HCC cell lines.

8. Mechanism of growth inhibition of HCC cells by MKH esters

One of the mechanisms of the antiproliferative effect of MK-4 was thought to involve G1/S cell-cycle arrest via reduced protein expression of cyclin D1 and Cdk4, and through suppression of NF-κB activation [10, 12]. We investigated whether the antiproliferative activity of MKH-bis-DMG was via cell-cycle arrest in HCC cells using flow cytometry and Western blotting [32]. MKH-bis-DMG-treated PLC/PRF/5 cells showed an increase in G1 phase cells and a decrease in S phase cells in flow cytometric analysis. Treatment of both DCP-positive and DCP-negative HCC cells with MKH-bis-DMG downregulated cyclin D1, cyclin D3, and Cdk4 expression after 24 h, and almost completely removed expression after 48 h. In comparison, the modest downregulation of cyclin D1, cyclin D3, and Cdk4 expression was observed after 48 h of MK-4 treatment in all tested HCC cell lines. NF-κB was downregulated after MKH-bis-DMG treatment in all tested HCC cell lines, but no effect was observed after MK-4 treatment in PLC/PRF/5 and SK-Hep-1 cell lines at this dose.

These findings strongly support our hypothesis that the rapid and strong growth-inhibitory effects on cells resulted from the rapid and effective delivery of MKH into HCC cells by an MKH prodrug. The mechanism of the MKH-bis-DMG antiproliferative effect is the same as that of MK-4 and involves cell-cycle arrest. Therefore, MKH-bis-DMG is expected to be a safe antitumor agent and chemopreventive agent.

9. Pharmacokinetics of MKH esters

Plasma MKO levels can reflect the levels of MKH not only *in vitro* but also *in vivo*. This also assists the function of GGCX at the active site. The relative bioavailability for MKH (F_{MKH}), after the parenteral administration of the MKH esters, relative to MK-4 solubilized with HCO-60, was calculated using AUC_{MKO} as in Eq. (1), and is shown in **Table 4**. In Eq. (1), $AUC_{MKO,\,MKH-DMG}$ and $AUC_{MKO,\,MK\text{-}4}$ are the AUC_{MKO} values after the administration of MKH–DMG and MK-4, respectively. $D_{MK\text{-}4}$ and $D_{MKH-DMG}$ are the doses of MK-4 and MKH–DMG, respectively. MKH-1-DMG and MKH-bis-DMG, but not MKH-4-DMG, showed an improvement in bioavailability compared with the MK-4 injection.

$$F_{MKH} = \frac{AUC_{MKO,MKH-DMG} \cdot D_{MK-4}}{AUC_{MKO,MK-4} \cdot D_{MKH-DMG}} \tag{1}$$

	MK-4	MKH-1-DMG	MKH-4-DMG	MKH-bis-DMG
for MK-4				
C_{max} (nmol mL^{-1})	107 ± 2.07	2.41 ± 0.826	20.8 ± 5.24	11.1 ± 3.03
t_{max} (h)	0.125	0.125	0.125	0.125
$AUC_{MK\text{-}4}$ (nmol h mL^{-1})	31.7 ± 0.526	2.52 ± 0.387	6.01 ± 1.26	10.1 ± 1.15
$MRT_{MK\text{-}4x}$ (h)	0.338 ± 0.018	2.44 ± 0.032	0.924 ± 0.004	1.32 ± 0.063
for MKO				
C_{max} (nmol mL^{-1})	1.92 ± 0.172	0.991 ± 0.131	0.518 ± 0.039	0.968 ± 0.180
t_{max} (h)	1	2	0.5	1
AUC_{MKO} (nmol h mL^{-1})	2.37 ± 0.115	4.45 ± 0.510	2.06 ± 0.322	3.20 ± 0.467
MRT_{MKO} (h)	2.14 ± 0.091	3.78 ± 0.404	3.93 ± 0.463	2.80 ± 0.066
F_{MKH}	100	188	87	135

[a]The values are the mean ± S.D. of 3 rats at a dose of 5 mg/kg equivalent for MK-4.
[b]Calculated from Eq. (1) using the mean values.
Adapted from Ref. [35].

Table 4. Pharmacokinetic parameters for MKO and MK-4 in plasma after the intravenous administration of the prodrugs, and for MK-4 in vitamin K cycle-inhibited rats.[a]

http://dx.doi.org/10.5772/63343

The distribution of MKO in the liver after the injection of MKH esters is the most important indicator for assessing the potential of MKH esters as the MKH delivery system for HCC. The values of AUC_{MKO} and MRT_{MKO} of MKH-1-DMG were larger than that for MK-4, which indicates that the MKH esters distribute successfully and provide prolonged deliver MKH to the liver. For the evaluation of MKH-1-DMG as a liver-specific delivery system for MKH, the selective advantage value was defined as in Eq. (2). In Eq. (2), $AUC^{Liver}_{MKO, MKH-DMG}$ and $AUC^{Liver}_{MKO, MK-4}$ are the AUC_{MKO} values in the liver after the administration of MKH–DMG and MK-4, respectively. $AUC^{Plasma}_{MKH-DMG, MKH-DMG}$ and $AUC^{Plasma}_{MK-4, MK-4}$ are the AUC values of plasma levels of MKH–DMG and MK-4 after the administration of MKH–DMG and MK-4, respectively. Remarkable site-specific delivery of MKH was observed after the intravenous injection of MKH-1-DMG. The selective advantage of MKH-1-DMG was 5.7 [35].

$$\text{Selective advantage} = \left\{ \frac{AUC^{Liver}_{MKO,MKH-DMG}}{AUC^{Liver}_{MKO,MK-4}} \right\} \Bigg/ \left\{ \frac{AUC^{Plasma}_{MKH-DMG,MKH-DMG}}{AUC^{Plasma}_{MK-4,MK-4}} \right\} \tag{2}$$

It is proposed that the intravenous injection of MKH esters is appropriate when a rapid and large quantity is necessary for cancer treatment, whereas oral administration of MKH esters is otherwise appropriate for cancer prevention. We performed a pharmacokinetic study of MKH-bis-DMG after oral administration, and found that MKH-bis-DMG was absorbed in the ester form, distributed to the liver and converted to MKH *in vivo* [32].

10. Antiproliferative effects of MKH-bis-DMG in a spleen–liver metastasis mouse model

To assess the pharmacological effects of the MKH delivery system *in vivo*, we assessed the effects of oral administration of MKH-bis-DMG on hepatic metastasis and proliferation of PLC/PRF/5 cells in a spleen–liver metastasis model [39]. MKH-bis-DMG treatment significantly suppressed the increase of liver weight caused by tumor growth (**Figure 4A**). The percentage surface area of the cancer compared with the total surface area of the liver was significantly lower in the spleen-liver metastasis model treated with the MKH-bis-DMG than with the vehicle (**Figure 4B**).

Plasma DCP production was completely suppressed after MKH–DMG administration, while liver metastasis of HCC was not completely prevented (**Figure 4C**). These results suggest that the level of influence of DCP suppression on the antiproliferative effects of MKH-bis-DMG was severely limited. No obvious side effects, such as body weight loss, were observed in the MKH-di-DMG treated animals.

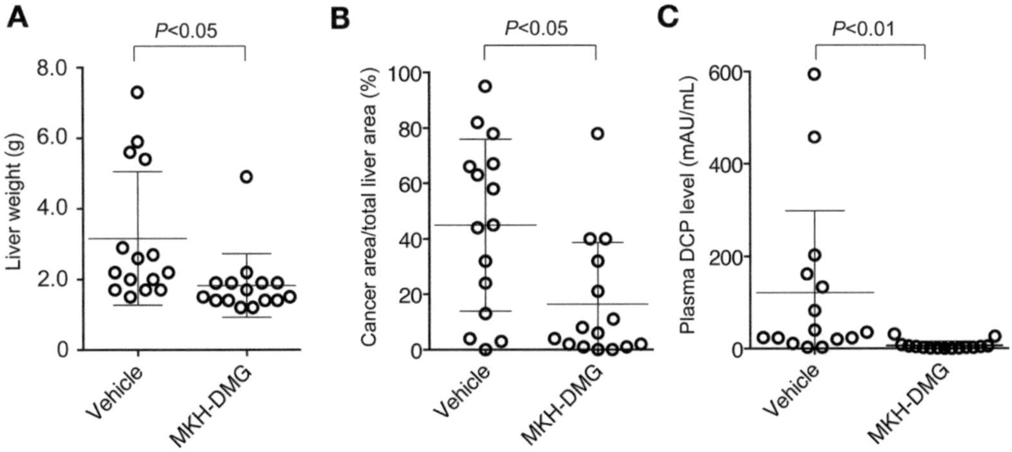

Figure 4. HCC growth inhibitory effects of MKH–DMG in a spleen-liver metastasis mouse model. (A) Total liver weight. (B) Percentage of cancer surface area/total liver surface area. (C) DCP levels in plasma. Central horizontal line, mean; error bar, SD. Vehicle group, n = 15; MKH–DMG group, n = 15. Doses were 0.2 μmol/head/day for 50 days. Adapted from Ref. [32].

11. Conclusion and perspectives

In previous studies, we have indicated that MKH-bis-DMG can be delivered to the liver through intravenous and oral administration and that thereafter it is transported and converted to MKH with enzymatic hydrolysis in HCC cells. Regarding the effective delivery of MKH into the HCC cells, MKH-bis-DMG enhanced the inhibition of HCC cell growth compared with that from MK-4 and suppressed metastasis of HCC. Given these results, we suggest that MKH-bis-DMG is a potential candidate for chemoprevention that can be safely administered over long periods, and can reduce or eliminate the recurrence and metastasis of HCC. This MKH prodrug approach was our original method to improve the antitumor effect of MK-4, and this approach may be applied to improve HCC treatment in the future. However, further studies are required to develop MKH–DMG for use in human HCC treatment.

Author details

Kazuhisa Matsunaga*, Munechika Enjoji, Yoshiharu Karube and Jiro Takata

*Address all correspondence to: k-matsu@fukuoka-u.ac.jp

Faculty of Pharmaceutical Sciences, Fukuoka University, Fukuoka, Japan

References

[1] Nakagawa K, Hirota Y, Sawada N, Yuge N, Watanabe M, Uchino Y, et al. Identification of UBIAD1 as a novel human menaquinone-4 biosynthetic enzyme. Nature. 2010;468(7320):117–21.

[2] Hara K, Akiyama Y, Nakamura T, Murota S, Morita I. The inhibitory effect of vitamin K2 (menatetrenone) on bone resorption may be related to its side chain. Bone. 1995;16(2):179–84.

[3] Jinghe X, Mizuta T, Ozaki I. Vitamin K and hepatocellular carcinoma: the basic and clinic. World Journal of Clinical Cases. 2015;3(9):757–64.

[4] Yokoyama T, Miyazawa K, Yoshida T, Ohyashiki K. Combination of vitamin K2 plus imatinib mesylate enhances induction of apoptosis in small cell lung cancer cell lines. International Journal of Oncology. 2005;26(1):33–40.

[5] Shibayama-Imazu T, Sakairi S, Watanabe A, Aiuchi T, Nakajo S, Nakaya K. Vitamin K(2) selectively induced apoptosis in ovarian TYK-nu and pancreatic MIA PaCa-2 cells out of eight solid tumor cell lines through a mechanism different from geranylgeraniol. Journal of Cancer Research and Clinical Oncology. 2003;129(1):1–11.

[6] Sakai I, Hashimoto S, Yoda M, Hida T, Ohsawa S, Nakajo S, et al. Novel role of vitamin K2: a potent inducer of differentiation of various human myeloid leukemia cell lines. Biochemical and Biophysical Research Communications. 1994;205(2):1305–10.

[7] Koike Y, Shiratori Y, Sato S, Obi S, Teratani T, Imamura M, et al. Des-gamma-carboxy prothrombin as a useful predisposing factor for the development of portal venous invasion in patients with hepatocellular carcinoma: a prospective analysis of 227 patients. Cancer. 2001;91(3):561–9.

[8] Enomoto M, Tsuchida A, Miyazawa K, Yokoyama T, Kawakita H, Tokita H, et al. Vitamin K2-induced cell growth inhibition via autophagy formation in cholangiocellular carcinoma cell lines. International Journal of Molecular Medicine. 2007;20(6):801–8.

[9] Kawakita H, Tsuchida A, Miyazawa K, Naito M, Shigoka M, Kyo B, et al. Growth inhibitory effects of vitamin K2 on colon cancer cell lines via different types of cell death including autophagy and apoptosis. International Journal of Molecular Medicine. 2009;23(6):709–16.

[10] Ozaki I, Zhang H, Mizuta T, Ide Y, Eguchi Y, Yasutake T, et al. Menatetrenone, a vitamin K2 analogue, inhibits hepatocellular carcinoma cell growth by suppressing cyclin D1 expression through inhibition of nuclear factor kappaB activation. Clinical Cancer Research. 2007;13(7):2236–45.

[11] Otsuka M, Kato N, Shao RX, Hoshida Y, Ijichi H, Koike Y, et al. Vitamin K2 inhibits the growth and invasiveness of hepatocellular carcinoma cells via protein kinase A activation. Hepatology (Baltimore, Md). 2004;40(1):243–51.

[12] Hitomi M, Yokoyama F, Kita Y, Nonomura T, Masaki T, Yoshiji H, et al. Antitumor effects of vitamins K1, K2 and K3 on hepatocellular carcinoma in vitro and in vivo. International Journal of Oncology. 2005;26(3):713–20.

[13] Matsumoto K, Okano J, Nagahara T, Murawaki Y. Apoptosis of liver cancer cells by vitamin K2 and enhancement by MEK inhibition. International Journal of Oncology. 2006;29(6):1501–8.

[14] Xia J, Matsuhashi S, Hamajima H, Iwane S, Takahashi H, Eguchi Y, et al. The role of PKC isoforms in the inhibition of NF-kappaB activation by vitamin K2 in human hepatocellular carcinoma cells. The Journal of Nutritional Biochemistry. 2012;23(12): 1668–75.

[15] Kaneda M, Zhang D, Bhattacharjee R, Nakahama K, Arii S, Morita I. Vitamin K2 suppresses malignancy of HuH7 hepatoma cells via inhibition of connexin 43. Cancer Letters. 2008;263(1):53–60.

[16] Yamamoto T, Nakamura H, Liu W, Cao K, Yoshikawa S, Enomoto H, et al. Involvement of hepatoma-derived growth factor in the growth inhibition of hepatocellular carcinoma cells by vitamin K(2). Journal of Gastroenterology. 2009;44(3):228–35.

[17] Azuma K, Urano T, Ouchi Y, Inoue S. Vitamin K2 suppresses proliferation and motility of hepatocellular carcinoma cells by activating steroid and xenobiotic receptor. Endocrine Journal. 2009;56(7):843–9.

[18] Li L, Qi Z, Qian J, Bi F, Lv J, Xu L, et al. Induction of apoptosis in hepatocellular carcinoma Smmc-7721 cells by vitamin K(2) is associated with p53 and independent of the intrinsic apoptotic pathway. Molecular and Cellular Biochemistry. 2010;342(1–2): 125–31.

[19] Yao Y, Li L, Zhang H, Jia R, Liu B, Zhao X, et al. Enhanced therapeutic efficacy of vitamin K2 by silencing BCL-2 expression in SMMC-7721 hepatocellular carcinoma cells. Oncology Letters. 2012;4(1):163–7.

[20] Habu D, Shiomi S, Tamori A, Takeda T, Tanaka T, Kubo S, et al. Role of vitamin K2 in the development of hepatocellular carcinoma in women with viral cirrhosis of the liver. JAMA. 2004;292(3):358–61.

[21] Mizuta T, Ozaki I, Eguchi Y, Yasutake T, Kawazoe S, Fujimoto K, et al. The effect of menatetrenone, a vitamin K2 analog, on disease recurrence and survival in patients with hepatocellular carcinoma after curative treatment: a pilot study. Cancer. 2006;106(4):867–72.

[22] Yoshida H, Shiratori Y, Kudo M, Shiina S, Mizuta T, Kojiro M, et al. Effect of vitamin K2 on the recurrence of hepatocellular carcinoma. Hepatology (Baltimore, Md). 2011;54(2):532–40.

[23] Carr BI, Wang Z, Kar S. K vitamins, PTP antagonism, and cell growth arrest. Journal of Cellular Physiology. 2002;193(3):263–74.

[24] Suhara Y, Watanabe M, Motoyoshi S, Nakagawa K, Wada A, Takeda K, et al. Synthesis of new vitamin K analogues as steroid and xenobiotic receptor (SXR) agonists: insights into the biological role of the side chain part of vitamin K. Journal of Medicinal Chemistry. 2011;54(13):4918–22.

[25] Suhara Y, Watanabe M, Nakagawa K, Wada A, Ito Y, Takeda K, et al. Synthesis of novel vitamin K2 analogues with modification at the omega-terminal position and their biological evaluation as potent steroid and xenobiotic receptor (SXR) agonists. Journal of Medicinal Chemistry. 2011;54(12):4269–73.

[26] Yoshiji H, Kuriyama S, Noguchi R, Yoshii J, Ikenaka Y, Yanase K, et al. Amelioration of carcinogenesis and tumor growth in the rat liver by combination of vitamin K2 and angiotensin-converting enzyme inhibitor via anti-angiogenic activities. Oncology Reports. 2006;15(1):155–9.

[27] Yoshiji H, Noguchi R, Yamazaki M, Ikenaka Y, Sawai M, Ishikawa M, et al. Combined treatment of vitamin K2 and angiotensin-converting enzyme inhibitor ameliorates hepatic dysplastic nodule in a patient with liver cirrhosis. World Journal of Gastroenterology. 2007;13(23):3259–61.

[28] Kanamori T, Shimizu M, Okuno M, Matsushima-Nishiwaki R, Tsurumi H, Kojima S, et al. Synergistic growth inhibition by acyclic retinoid and vitamin K2 in human hepatocellular carcinoma cells. Cancer Science. 2007;98(3):431–7.

[29] Zhang H, Ozaki I, Hamajima H, Iwane S, Takahashi H, Kawaguchi Y, et al. Vitamin K2 augments 5-fluorouracil-induced growth inhibition of human hepatocellular carcinoma cells by inhibiting NF-kappaB activation. Oncology Reports. 2011;25(1):159–66.

[30] Zhang Y, Zhang B, Zhang A, Zhao Y, Zhao J, Liu J, et al. Synergistic growth inhibition by sorafenib and vitamin K2 in human hepatocellular carcinoma cells. Clinics (Sao Paulo, Brazil). 2012;67(9):1093–9.

[31] Jung DH, Hwang S, Song GW, Ryoo BY, Kim N, Tak E, et al. An interim safety analysis of hepatocellular carcinoma patients administrating oral vitamin K with or without sorafenib. Korean Journal of Hepato-Biliary-Pancreatic Surgery. 2015;19(1):1–5.

[32] Setoguchi S, Watase D, Matsunaga K, Matsubara M, Kubo Y, Kusuda M, et al. Enhanced antitumor effects of novel intracellular delivery of an active form of menaquinone-4, menahydroquinone-4, into hepatocellular carcinoma. Cancer Prevention Research (Philadelphia, Pa). 2015;8(2):129–38.

[33] Takata J, Karube Y, Hanada M, Matsunaga K, Iwasaki H. Prodrug for bioreductive activation-independent delivery of menahydroquinone-4: human liver enzymatic activation and its action in warfarin-poisoned human liver. Biological & Pharmaceutical Bulletin. 1999;22(2):172–8.

[34] Takata J, Karube Y, Hanada M, Matsunaga K, Matsushima Y, Sendo T, et al. Vitamin K prodrugs: 1. Synthesis of amino acid esters of menahydroquinone-4 and enzymatic reconversion to an active form. Pharmaceutical Research. 1995;12(1):18–23.

[35] Takata J, Karube Y, Hanada M, Matsunaga K, Matsushima Y, Sendo T, et al. Vitamin K prodrugs: 2. water-soluble prodrugs of menahydroquinone-4 for systemic site-specific delivery. Pharmaceutical Research. 1995;12(12):1973–9.

[36] Huisse MG, Leclercq M, Belghiti J, Flejou JF, Suttie JW, Bezeaud A, et al. Mechanism of the abnormal vitamin K-dependent gamma-carboxylation process in human hepatocellular carcinomas. Cancer. 1994;74(5):1533–41.

[37] Li ZQ, He FY, Stehle CJ, Wang Z, Kar S, Finn FM, et al. Vitamin K uptake in hepatocytes and hepatoma cells. Life Sciences. 2002;70(18):2085–100.

[38] Westhofen P, Watzka M, Marinova M, Hass M, Kirfel G, Muller J, et al. Human vitamin K 2,3-epoxide reductase complex subunit 1-like 1 (VKORC1L1) mediates vitamin K-dependent intracellular antioxidant function. The Journal of Biological Chemistry. 2011;286(17):15085–94.

[39] Fidler IJ. Rationale and methods for the use of nude mice to study the biology and therapy of human cancer metastasis. Cancer Metastasis Reviews. 1986;5(1):29–49.